Transforming Computer Technology

Johns Hopkins Studies in the History of Technology

Merritt Roe Smith, Series Editor

Transforming Computer Technology

Information Processing for the Pentagon, 1962–1986

ARTHUR L. NORBERG

and

JUDY E. O'NEILL

with contributions by

KERRY J. FREEDMAN

The Johns Hopkins University Press

Baltimore and London

© 1996 The Johns Hopkins University Press
All rights reserved. Published 1996
Printed in the United States of America on acid-free paper
05 04 03 02 01 00 99 98 97 96   5 4 3 2 1

The Johns Hopkins University Press
2715 North Charles Street
Baltimore, Maryland 21218–4319
The Johns Hopkins Press Ltd., London

ISBN 0–8018–5152–1

Library of Congress Cataloging-in-Publication Data will be found
at the end of this book.

A catalog record for this book is available from the British Library.

ISBN 0-8018-6369-4 (pbk)
paperback edition, 1999

Contents

Preface and Acknowledgments

We need at the outset to distinguish this book from an earlier report we wrote on the history of IPTO, *A History of the Information Processing Techniques Office of the Defense Advanced Research Projects Agency* (Minneapolis, Minn.: Charles Babbage Institute, 1992), researched and written under a contract issued by the Department of Defense (NASA-Ames Research Grant NAG 2-532, subcontract USC/PO 473764). The original idea for a history of IPTO came from Saul Amarel when he was director of IPTO in 1985–87. He assembled a small steering group to help him define an approach for a historical study, and the deliberations of this group became the starting point for the Charles Babbage Institute proposal. Arthur Norberg and William Aspray cooperated on the project's concept and worked on its early definition. Before leaving the project after six months, Aspray participated in archival research and conducted a number of interviews that proved important to the study. Thereafter, preparation of this history involved the research and writing efforts of a number of people associated with the Charles Babbage Institute of the University of Minnesota.

In preparing the Department of Defense (DOD) report, Norberg concentrated on the overall administrative history, the history of the program, analysis of the influence of IPTO, the area of artificial intelligence, and the conclusions; he also directed the project. O'Neill was primarily responsible for the areas of time sharing and networking, and for the background material on computing in the 1950s. Freedman researched and wrote on graphics. We conducted forty-five formal tape-recorded interviews, many comprising more than one session. (For a list of the interviewees, with the names of the interviewers and the dates of the interviews, see the Appendix.) The edited interview transcripts are open to researchers in the CBI Collection in Minneapolis. We always carefully compared interviewees' accounts with contemporary documents. Where disagreements existed between these two kinds of sources, we cast our conclusions so as to be consistent with the written record. We then circulated the text of the report to the DOD to a relatively large group of reviewers, who carefully read the manuscript and offered suggestions for enhancement.

After submission of the report to DOD in October 1992, we set about in a second phase to integrate the various parts of the report into a coherent historical manuscript. Norberg rewrote the material—integrating its two parts into this uniform and revised work, placing the history of IPTO into a larger context of developments in computer science and Cold War activities, and enlarging and strengthening the conclusions. O'Neill and Freedman offered editorial commentary on every part of the reconstructed and rewritten manuscript.

The DOD report and revised manuscript largely broke new ground, for published comments on the history of IPTO have been brief and have usually been associated with descriptions of technical developments. See, for example, Lawrence G. Roberts, "The ARPANET and Computer Networks," in *A History of Personal Workstations,* edited by Adele Goldberg (New York: Association for Computing Machinery, 1988); and Joseph C. R. Licklider, "The Early Years: Founding IPTO," in *Expert Systems and Artificial Intelligence: Applications and Management,* edited by Thomas C. Bartee (Indianapolis, Ind.: Howard W. Sams, 1988). Some IPTO history can be found in studies of DARPA. However, even those studies have been brief and have focused on policy issues, not technical accomplishments. One exception is a 1975 study by Richard Barber Associates, *The Advanced Research Projects Agency, 1958–1974* (Washington, D.C.: Barber Associates, 1975), which supplied summaries of activities during the tenures of DARPA directors. Attention to IPTO in all of these studies is brief and focused entirely on Department of Defense activities. *Transforming Computer Technology* provides a necessary bridge from assertions about accomplishments to an understanding of how those accomplishments were achieved. Little additional literature on DARPA, IPTO, the DOD, and defense-related university activities has appeared. The few items extant receive attention in the notes to this volume.

We pursued our investigation of IPTO with two different but parallel analyses in view: (1) a history of the research programs, management of the programs, and the influence of IPTO; and (2) an in-depth investigation of the larger context of the development of four significant areas of computing that IPTO actively funded—time sharing, networking, graphics, and some areas of artificial intelligence. In researching both the report and the book, we uncovered a substantial body of useful records, much of it deposited at the National Archives and Records Agency site in Suitland, Maryland (Record Group 330).

But we encountered some problems with these and other sources. We also found useful, and were permitted to use, records from the 1980s at the IPTO office in Arlington, Virginia.

Not only are the IPTO records incomplete but, more important, they vary in content and quality over the twenty-five years of this study. Records for the 1960s and (for selected projects only) for the 1980s contain materials that describe the funding justifications of IPTO directors. Little information about justifications was available for the 1970s. As is the case for many policy decisions in government and business, arguments between levels in the hierarchy and between agencies about programmatic and budget decisions are missing from the record. Our informants told us that such debate took place in meetings in which decisions were made without notes, after oral discussions. The fast pace of the evaluation and funding process meant that the reasons for decisions were rarely written down, especially as the IPTO-supported community became larger and programs more carefully specified in advance. We did not have access to top-level DARPA files, so we found DARPA directors' evaluations for only a few IPTO projects. It would have been valuable to read program evaluations and assessments in cases in which IPTO programs competed with other DARPA office programs for resources. Again, our informants insisted on the oral nature of internal DARPA discussions of such matters. We therefore occasionally had no choice but to rely on the memory of informants who participated in these decisions and evaluations; the reader will note in discussions of the early years a greater reliance on written sources, and in discussions of the later years a greater reliance on oral sources. On the basis of documents from the 1970s and 1980s which we examined, it appears that later decisions were similar to early ones; and our conversations with IPTO personnel have convinced us that the oral discussions of the 1970s and 1980s were similar in kind to those of the 1960s, and that decisions taken in the later years are compatible to those indicated in written documents of the 1960s. Indeed, we have found a remarkable consistency in the pattern of evaluation and decision making throughout the entire history of IPTO. While a greater range of records might to some degree alter our description of later events, we believe that our conclusions would remain basically intact.

In our use of sources of all kinds, we looked for patterns of two sorts. First, where we had to lay much weight on a single document, we looked for agreement or convergence with the overall record sample. Second, we looked for agreement between written and oral sources. We thus constantly examined convergences and divergences between information stemming from different organizations and different informants, so as to make judgments that shaped our conclusions. A retrospective narrative can develop from any assortment of information that helps to tell a story; the narrative need not—cannot—be a complete catalog of events and projects. We have highlighted various activities in and about IPTO which were not selected at random. They substantiate our interpretation of significant trends in the program, of the program's management, and of the important influence of IPTO on computer technology in the mid- to late twentieth century.

Since the project to prepare a history of the Information Processing Techniques Office began in November 1988, we have had a wide range of advice from participants in IPTO's history and from our historian colleagues. Serving as an advisory committee were Saul Amarel, I. Bernard Cohen, Robert M. Fano, Robert E. Kahn, J.C.R. Licklider, Allen Newell, Ronald B. Ohlander, Merritt Roe Smith, and Keith Uncapher (chair). Some members participated in oral interviews on their part in IPTO's history; all offered technical and other kinds of advice along the way, and reviewed several early drafts of our history. At times, committee meetings were attended by Barry Boehm, Lou Kallis, and Stephen Squires of DARPA.

Historians, too, offered sharp, insightful comments about our strategy during the research and writing. We are extremely grateful for the advice and solicitude of William Aspray, I. Bernard Cohen, Edwin T. Layton Jr., Michael Mahoney, and Robert W. Seidel. We were especially gratified with the efforts of Alex Roland and Merritt Roe Smith. We wish to thank these colleagues for their patience in examining several drafts of the manuscript and for their helpful analyses of the text. However, we take full responsibility for whatever errors and misjudgments still remain in the text.

We also acknowledge material assistance from Duane A. Adams, William Bandy, Paul Baran, Martin Campbell-Kelly, David Carlstrom, Vinton G. Cerf, Fernando J. Corbato, Larry E. Druffel, Robert Engelmore, Floyd H. Hollister, Richard Y. Kain, Leonard Kleinrock, Joshua Lederberg, Paul Losleben, Alexander A. Mackenzie, Alan J. McLaughlin, Donald Mengel, Douglas T. Ross, Ivan E. Sutherland, Keith Uncapher, Richard Van Atta, David C. Walden, and Joel Yudken. During both phases, archivists in several institutions gave freely of their time and knowledge to help the project. At DARPA, we were assisted by Louise Lorenzen, Alma Spring, and Lou Kallis. CBI's own archivist, Bruce Bruemmer, gave unstintingly of his knowledge and labor to advance our research and our searches of CBI materials. He also helped with computer systems and software programs. CBI assistant archivist Kevin Corbitt led us to advantageous sources as well. At the Massachusetts Institute of Technology,

Kathy Marquis and Helen Samuels led us through the myriad of records associated with MIT's IPTO-related projects. Catherine Johnsen provided us with materials in the Carnegie-Mellon University archives. At Lincoln Laboratory, Mary L. Murphy obtained access for us to view documents and videotapes associated with graphics developments there. Administrative matters at CBI and transcription of the taped interviews were carefully and cheerfully attended to by Judy Cilcain and Janice Jones. Diane Larson helped us with the preparation of the final draft.

Along the way, we had the assistance of a number of able graduate students, who made important contributions to the project: Collette LaFond, Deborah J. Shepard, Lance M. Smith, and especially Shawn P. Rounds and John P. Jackson.

Our association with the Johns Hopkins University Press was a rewarding and happy one. We are grateful to Bob Brugger for his many helpful comments and for shepherding the manuscript through the many procedures required at the press. The manuscript was exceptionally well edited by Miriam L. Kleiger, and we thank her for her fine work with the manuscript and her patience with us.

List of Frequently Used Acronyms

ACCAT	Advanced Command and Control Architectural Testbed
ACM	Association for Computing Machinery
AFIPS	American Federation of Information Processing Societies
AN/FSQ32	Army-Navy/Fixed Special Equipment computer; also known as SDC Q-32
ARPANET	Advanced Research Projects Agency Network
BBN	Bolt Beranek and Newman
C²	command and control (also C&C)
C³	command, control, and communication
CBI	Charles Babbage Institute
CCA	Computer Corporation of America
CINCPACFLT	Commander-in-Chief, Pacific Fleet, United States Navy
CMU	Carnegie-Mellon University
DDR&E	director of defense research and engineering
EDUCOM	Interuniversity Communications Council
FORTRAN	Formula Translation (programming language)
ICCC	International Conference of Computer Communications
IDA	Institute for Defense Analyses
IFIP	International Federation for Information Processing
ILLIAC	University of Illinois Automatic Computer
IMP	interface message processor
INWG	International Network Working Group
IPL	Information Processing Language (programming language)
IPTO	Information Processing Techniques Office; later ISTO (Information Science and Technology Office), which split into CSTO (Computer Systems Technology Office) and SISTO (Software and Intelligent Systems Technology Office)

ISI	Information Sciences Institute of the University of Southern California
LISP	List Processing (programming language)
MAC	Multiple Access Computer; or Machine Aided Cognition
MILNET	Military Network
MIT	Massachusetts Institute of Technology
MULTICS	Multiplexed Information and Computing Services
NABDS	National Archives Branch Depository, Suitland, Maryland
NPL	National Physical Laboratory, Teddington, England
NSF	National Science Foundation
ONR	Office of Naval Research
PARC	Palo Alto Research Center (of the Xerox Corporation)
Project MAC	See MAC
RISC	Reduced Instruction Set Computer
RLE	Research Laboratory of Electronics, MIT
SDC	Systems Development Corporation
SDS	Scientific Data Systems, later sold to Xerox and known as XDS
SRI	Stanford Research Institute (now SRI, International)
TCP/IP	Transmission Control Protocol/Internet Protocol
TIP	terminal interface message processor
UCB	University of California, Berkeley
UCLA	University of California, Los Angeles
UCSB	University of California, Santa Barbara
USC	University of Southern California

Transforming Computer Technology

Introduction

During World War II, scientists and engineers established a partnership with the military for the development of better defensive weapons systems, a partnership that had been growing since the beginning of the twentieth century. They contributed to the defense effort against the Axis powers by helping to define defense problems and design and implement new defense equipment. During the war, these scientists and engineers worked largely under military oversight, and the military came to like the situation. Every effort was made at war's end to retain the services of these professionals in fixed employment under military supervision. But the scientists and engineers preferred the prewar freedom of their academic and other laboratories, and left military facilities in droves. However, they did not wish to give up the new-found support for their research, so they sought some accommodation with the armed services, and defense agency leaders encouraged those returning to the civilian sector to maintain organized research programs in defense-related areas. To maintain the partnership with the technical community during the postwar period, the Department of Defense instituted broad research and development activities that went beyond the development of particular weapons systems.[1] Through this support, over the past five decades, the DOD has fostered the partnership and significantly affected agendas for research at universities and in industry.

The end of the Cold War brought into sharp focus the range and nature of these R&D activities; their justification on grounds of defense and national security, especially as they involve academic institutions; and the extent of the partnership in American society. Several recent studies in the history of technology published at this critical juncture of postwar history help us to understand better the effects of military and security concerns on academic programs during the cold war. Among the best analyses of the partnership from the perspective of the academy are Donald MacKenzie's *Inventing Accuracy,* David Noble's *Forces of Production,* and Stuart W. Leslie's *Cold War and*

American Science.[2] MacKenzie examined the history of nuclear missile guidance as a historical product and a social creation, focusing on the intense interservice struggle over the control of strategic weapons and the design of guidance systems at the three main guidance system design organizations in the United States, one of which was the Massachusetts Institute of Technology's Instrumentation Laboratory. Noble studied another MIT activity: air force-sponsored research into and development of numerical control technology. Both *Inventing Accuracy* and *Forces of Production* deal with only one of the several areas of activity that existed at MIT. Stuart Leslie, by contrast, not only looked at several laboratories and departments at MIT, he did the same for Stanford. *The Cold War and American Science* covers several decades from the 1930s to the 1970s, and reviews both scientific and engineering programs. All three of these books focus on the research and training at these educational institutions and the emphasis of these programs on defense R&D concerns.

In the present study, which examines the way in which computing research-support programs inside the DOD were organized, developed, and changed over the years since 1960, we illustrate the other side of the partnership between the DOD and academic scientists and engineers by analyzing the internal workings of one of the premier research-support agencies of the Department of Defense: the Information Processing Techniques Office of the Defense Advanced Research Projects Agency. We investigate IPTO's administrative activities and the research the office funded at a number of organizations, and chronicle the office's origins, development, and changes. The study includes an analysis of the management of the office, the origins and growth of representative IPTO programs, the interactions of IPTO staff members with the R&D community, and IPTO's military-related mission. The influence of IPTO programs, hence the DOD side of the partnership, has been charted and interpreted in terms of significant developments in computer science and engineering, notably, time sharing, networking, graphics, selected areas of artificial intelligence, and early parallel-processing systems.

To illustrate the impact of the IPTO-supported programs on research, we examine the major effects on computing research and education at several institutions involved with the DOD, primarily the Massachusetts Institute of Technology, Carnegie-Mellon University (CMU), and Stanford University. We show how this partnership between the DOD and researchers came about naturally as a result of postwar concerns. Thus, this examination broadens our knowledge of how R&D programs inside DOD came about, illustrates how civilian scientists and engineers played a major role in designing and administering these programs, and enlarges our understanding of the breadth of impact of

these programs on educational institutions, providing an even greater understanding of the partnership. This history also complements other studies of
postwar science and technology R&D, the government's role in R&D, the development of computing since 1945, and the interaction between defense organizations and the technical community.[3]

Since this study is about the activities of a major post-1945 R&D agency, our
analysis and conclusions offer an anchor point in the current debate about
the need for an enhanced government R&D policy and the place of a civilian
equivalent of DARPA.[4] A recent question in policy circles was, what lessons for
a civilian economy can be learned from the history of a DOD office such as
IPTO. A better understanding of IPTO, which is seen as having been very successful in transferring technology to the civilian sector, will aid policy makers
by focusing on the areas in which such agencies involved with fundamental
R&D have been successful, and on what contributed to that success. This history contributes to policy analysis by illustrating where IPTO was and was not
successful and how the agency managed its activities.

The Departments of the Navy and of the Air Force led in the post–World
War II promotion of R&D activity. Even before the cessation of hostilities in
1945, the Office of Naval Research (ONR) was planning for R&D programs after
the war. ONR's Naval Research Laboratory, for example, pursued its interests in
naval-related weaponry while considering new applications of atomic energy
in ships and in the DOD missile program; and ONR began a funding program
to assist scientists and engineers at universities and in industry in maintaining the flow of knowledge considered important to national security. With the
establishment of the National Science Foundation in 1950, ONR's extramural
role became muted and less funding was provided by Congress for the ONR
external funding program, but ONR continued to be an important funder of
research.[5] The ONR extramural funding program shifted toward problems related to the navy's mission of national defense, and away from general support
for basic and applied research for science and technology.

The U.S. Army Air Corps, renamed the U.S. Air Force shortly after the war,
had emerged from the war with an expanded defense mission, which included
the development of jet engine aircraft and ballistic missile systems. To accomplish this development, the air force mounted an extensive research program
that spread over many air force facilities, as well as organizations outside the
DOD. Air force personnel, like their counterparts in the navy, came to believe
that a broad R&D program that included research sponsored within organizations outside the DOD was needed in these new technological areas. In the late

1940s, the air force organized its own "think tank," the Rand Corporation. The Air Force Office of Scientific Research (AFOSR) was organized in 1950 to manage several air force R&D activities, including several research projects modeled after ONR. For example, to continue the involvement of academics in defense-related R&D, the air force, army, and navy stimulated the establishment of a number of electronics laboratories on campuses across the nation through the Joint Services Electronics Program (JSEP).[6]

Postwar support for research in computing was part of this broad pattern of DOD encouragement of R&D in the United States. Without this support, the enterprise would very likely have been much slower in starting and developing. During World War II, the military services had become leaders in the funding of experimental projects to design and construct computing machines because of a perceived need for faster and more powerful computing. Some of these projects have become legendary. The navy had supported Howard Aiken at Harvard, who led the design team for the Mark I and II computers; Jay Forrester at the Massachusetts Institute of Technology, who headed the design team for the Whirlwind; and some of the computer projects at Bell Laboratories under the direction of Ernest G. Andrews.[7] The army supported the ENIAC project at the University of Pennsylvania and the IBM modifications of electronic accounting machines for ballistic table calculation at the Aberdeen Proving Grounds.[8]

Immediately after the war, support continued for these and other machine designs. ONR sponsored a project at the Raytheon Corporation, which became the Raydac computer; the army funded EDVAC at the University of Pennsylvania, as well as the IAS machine at the Institute for Advanced Study in Princeton; and the navy supported Harvard's project on the Mark III computer. In one way or another, all these projects influenced industrial developments. The navy also facilitated the organization of Engineering Research Associates, a company organized to design and build data-processing equipment for navy intelligence functions.[9] In the 1950s, various groups with DOD contracts developed computers for missile-defense systems, jet fighter aircraft, command and control, logistic support on land and sea, and general management duties for the armed services and the DOD. When the space program accelerated after the launching of the Soviet satellite *Sputnik* in October 1957, digital computers became an integral part of that activity as well.[10] The more sophisticated the various military systems became, the greater the demands placed on their computing elements; and the greater those demands became, the more apparent were the shortcomings of the systems' computing elements. The need for a major research program for advanced computing technology useful in noncommercial computer systems became more and more evident. It was not until after the U.S. response to *Sputnik* that the momentum for such an R&D program took shape in the DOD.

The United States responded to the Soviet Union's launching of *Sputnik* in a number of ways. President Eisenhower appointed James A. Killian, president of MIT, as a presidential assistant for science, and Killian's first action was to create the President's Science Advisory Committee. With its aid, Eisenhower set about to separate the various space and ballistic missile programs, because of his belief that separation would facilitate an early successful launch of a satellite.[11] Eisenhower, well versed in the rivalry between the services, approved the organization of a new agency in the DOD to oversee the development of the U.S. space program and separate it into distinct military and civilian components. While this was its first task, the new agency's mission was soon to be more broadly designed, and it came to play a very significant role in further R&D support for computing.

The Defense Advanced Research Projects Agency

The new agency, the Advanced Research Projects Agency (ARPA), was the result of an effort to rationalize research and development at different levels within the DOD, and to stimulate new elements in frontier technology development in response to *Sputnik*. ARPA later became the Defense Advanced Research Projects Agency (in this study, we have used the acronym DARPA throughout), and in 1993 it was renamed ARPA. DARPA's mission throughout the period covered in this study was to prevent major technological surprises such as *Sputnik*, and to serve as the mechanism for high-risk R&D in cases where the technology was in its early stages and where the technical opportunities crossed the military department role and mission lines.[12] Along with the space program task, DARPA received a handful of additional presidential directives in the areas of nuclear test detection, missiles, satellites, and materials. And after it completed the space program task, the agency focused on these areas and served as a research arm of the Office of the Secretary of Defense. DARPA's objectives in these and subsequent areas were to maintain technical vigilance and to provide a quick response to technical developments abroad.

DARPA programs have been a source of research support for projects with possible multiservice-oriented needs. This research was carried out on the high-technology frontier, so as to give DOD operations a distinctive edge. DARPA has pursued its mission by stimulating R&D support programs, but it has had no research laboratories of its own. Instead, it has supported R&D programs in organizations outside the DOD and has participated in these R&D efforts through close cooperation between DARPA personnel and members of the research organizations.[13] While contributing substantially to the DOD mission throughout its history, DARPA has also contributed to the general economy by supporting the development of generic technologies useful well beyond military systems as the starting point of new products for the civilian market.[14]

DARPA's success resulted from the nature and range of its research-support programs and from its management style. Its programs focused on R&D at the frontier of high-technology systems for defense. Consider the range of its programs in the 1960s, which dealt with ballistic missile defense, nuclear test detection systems, special technologies for use in the Vietnam conflict, materials science, and information processing. In the first three areas cited, many, if not all, of the programs supported were classified, had a specific DOD mission, and rarely contributed beyond their defense mission. However, the materials science and information-processing programs, by deliberate design and occasional good fortune, contributed to both the military and civilian sectors of society. When DARPA directors referred to military and political success, they usually cited accomplishments in the nuclear monitoring and ballistic missile programs. When they wished to call attention to their agency's contributions to civilian society, they described the accomplishments of the information-processing and materials science programs. Indeed, the public's perception of DARPA's programmatic prowess is based almost entirely on the latter two programs.[15] (For a summary of the major elements of the DARPA program over time, see table 1.)

In all of these program areas, DARPA was devoted to the study of problems on the R&D frontier. More than that, DARPA involved itself consistently with high-risk projects. We need to make clear at the outset what DARPA meant by high-risk projects. The usual definition concerns the economic aspects of a project or program. When a firm invests in R&D for a new technology, it commits resources that are irretrievable if the project fails. The loss of these resources can affect the entire company adversely. For example, when IBM set out in the early 1960s to develop a family of compatible machines, it invested heavily in the System 360. If that system had not come to market or if customers had not accepted it, it is generally believed, IBM would not only have lost the market for 360s but would have had no new products to counter the developments of other firms.

DARPA used the term *high-risk* in a different sense. The DOD did not have

Table 1   DARPA Program Areas and Offices, Fiscal Years 1963, 1974, and 1988

Program Areas, Fiscal Year 1963	
Defender (Ballistic Missile Defense)	Propellents
VELA (Nuclear Test Detection Systems)	Energy Conversion
Climate Modification	Materials
Technical Studies	AGILE (Counterguerilla Warfare Studies)
C&C/Information Processing	

Program Areas, Fiscal Year 1974	
Strategic Technology	Nuclear Monitoring
Materials	Behavioral Sciences
Information Processing	Tactical Technology
Technical Studies	Management Support

Offices, Fiscal Year 1988	
Contracts Management Office	Program Management Office
Resource Management Office	Aero-Space Technology Office
Defense Sciences Office	Directed Energy Office
Information Science and Technology Office	Naval Technology Office
Nuclear Monitoring Research Office	Prototype Projects Office
Strategic Technology Office	Tactical Technology Office
Technical Assessment and Long-Range Planning Office	

Source: Barber Associates, *Advanced Research Projects Agency,* figs. v-1, ix-1; *Defense Advanced Research Projects Agency,* DARPA brochure (1988), 5.

an economic interest to preserve. The limitations it worked under were those imposed by congressional budgets. DARPA's high-risk projects were those with ambitious objectives that existing technology was not adequate to achieve. High-risk projects in DARPA dealt with the development of fundamental or enabling technologies with which to reach newly defined DOD objectives. It is important to note that when DARPA judged success, it applied technical rather than economic measurement standards. Because it could ignore cost, a number of high-risk projects were successful; hence DARPA was judged a success.

The other important factor in DARPA's success was its management style. In virtually all offices and programs, the agency had a lean administrative structure. Personnel used extensive field contact, sought fast turnaround times for proposals, and supported high-risk concepts and development projects. DARPA assembled an exceedingly capable staff, which accumulated a history of block grants and multiple-year contracts, employed a proposal evaluation mechanism that was largely internal rather than dependent on peer review,

Table 2   DARPA Directors, 1958–1987

	Term of Office	Educational Background
Roy W. Johnson	Feb. '58–Nov. '59	B.A., University of Michigan, 1927
Austin W. Betts	Dec. '59–Jan. '61	M.S., civil engineering, MIT, 1938
Jack P. Ruina	Feb. '61–Sept. '63	Ph.D., electrical engineering, Polytechnic Institute of Brooklyn, 1951
Robert L. Sproull	Sept. '63–June '65	Ph.D., physics, Cornell, 1943
Charles M. Herzfeld	June '65–Mar. '67	Ph.D., physics, University of Chicago, 1951
Eberhardt Rechtin	Nov. '67–Jan. '71	Ph.D., industrial engineering, California Institute of Technology, 1950
Stephen J. Lukasik	Jan. '71–Dec. '74	Ph.D., physics, MIT, 1956
George H. Heilmeier	Jan. '75–Dec. '77	Ph.D., electrical engineering, Princeton, 1962
Robert Fossum	Dec. '77–July '81	Ph.D., mathematical statistics, Oregon State University, 1969
Robert S. Cooper	July '81–July '85	Sc.D., electrical engineering, MIT, 1963
Robert C. Duncan	July '85–Apr. '87	Sc.D., instrumentation, MIT, 1960

and in some offices showed a consistent ability to fund new technological concepts that have had major effects in society.

This management style was a combination of the military services' approach to R&D support, and the talents and experience of the civilian directors of the agency. DOD officers are used to supporting R&D through large contracts, carefully monitored by DOD personnel who frequently participate in the determination of the requirements of the contract. DOD program directors participated in the research process to ensure that the results of projects were coordinated to influence larger systems development. DARPA followed the same pattern of management. To lead DARPA, successive secretaries of defense chose men who had a mix of experience in industry and government, along with advanced training in higher education in a technical area. (See table 2 for a list of DARPA directors, their dates of tenure, and their educational backgrounds.)

During its first eighteen months, as we noted above, DARPA was consumed with attention to the space program. It was staffed mostly by people from the Institute for Defense Analyses (IDA) in Alexandria, Virginia. By 1960, most of the space programs had been assigned to NASA or to specific military facilities, and DARPA had assumed other assignments consistent with its mission. These activities in ballistic missiles, solid propellant chemistry, materials science, and nuclear test detection had much more of the flavor of basic research than did DARPA's space mission. The Office of the Director of Defense Research and Engineering (DDR&E), created by DOD in 1958, took responsibility for the further development of many of these areas after the research stage. Hence, DARPA served the needs of the DDR&E. As a more conventional staffing pattern emerged, IDA staff members were gradually displaced by military officers and civil service personnel.[16]

Several developments, events, and explorations shaped the DOD response to the need for more R&D in information processing. The most important of these was perhaps a recognition inside the DOD of shortcomings in command and control systems. By 1960 the DOD recognized that it had a problem with respect to large amounts of information requiring timely analysis, and DARPA received the task of examining how to meet this need. *Command and control* denotes a set of activities associated, in military contexts, with rapidly changing environments. These activities include the collection of data about the environment, planning for options, decision making, and dissemination of the decisions.[17] Increasing the amount of strategic information that could be controlled would improve the command decisions needed in a rapidly changing environment; and computing technology had the potential for controlling greater amounts of information and presenting it in effective ways to aid decision making. The military services regarded computers as important components of modern warfare. Weapons systems and missile guidance systems used computers as the central control element. The capabilities of computer systems made them a natural adjunct to command and control problem analysis. While concerns for command and control problems had existed throughout military history, it was only in the late 1950s that the area received general recognition and the term *command and control* became common.[18] Earlier military concern with the processing of information and the use of the results in command decisions had focused on the problems of human relations.[19] In the early 1960s, however, the focus shifted to the informational aspects, and computer use in military systems expanded.[20] Military interest in computing technology was heightened by the merging of information technology, command techniques in the new military weapons environment, and missile control systems.

In early 1960, DARPA contracted with IDA to perform a survey of the information sciences and their relevance to defense concerns. The purpose of the study was to look at the current situation in the DOD and in the information sciences, and then recommend changes in DOD practice.[21] The survey noted five critical DOD problem areas related to R&D in information science: (1) pattern recognition and formation of concepts on the basis of data; (2) decision making; (3) communications; (4) control; and (5) information storage and retrieval, data handling, and data processing. The report described some research activities concerning information science which were already supported by agencies of the DOD. They ranged from mainstream applications of psychology, the social sciences, and biology, to specific projects in encoding of basic information, self-organizing systems, and heuristic and adaptive computers. No immediate action seems to have been taken on this report pending a new administration in the White House and in the secretary of defense's office.

Another event that shaped the DOD response to needs in information processing was the arrival in the White House of the Kennedy administration, with concerns about defense and a desire for new programs. In March 1961 President Kennedy, in a special message to Congress on defense spending, called for the improvement of command and control systems to make them "more flexible, more selective, more deliberate, better protected, and under ultimate civilian authority at all times."[22] In June 1961, the Office of the DDR&E assigned a Command and Control Project to DARPA.[23] By the end of June, DARPA had assigned IDA to do a "Digital Computer Application Study" to explore how to apply computing to command and control problems.[24] One of the report's recommendations was to expand research in computing. The areas of computing that seemed "potentially fruitful" included the development of improved techniques in problem (or algorithm) formulation, analysis, and programming, and the development of procedures and languages for communication between machines and their users. The report called for basic research in pattern recognition, concept formulation and recognition, problem solving, learning, and decision making, as well as research directed toward improving computer dependability.

Yet another factor in the DOD's response was the growing interest, in companies such as MITRE and Systems Development Corporation (SDC), in information-processing research, which led to proposals requesting DOD support. As a start in implementing the recommendations of the IDA study and in response to these proposals, DARPA gave SDC, in Santa Monica, California, a multi-million-dollar contract to pursue research on the conceptual aspects of command and control systems in August 1961. While a division of the Rand Corporation, SDC had been responsible for operator training and software development for the Semi-Automatic Ground Environment (SAGE) computerized air defense system of the 1950s. In 1957, SDC had become a separate organization. In November 1960, it submitted a proposal to DARPA for "Research into Information Processing for Command Systems."[25] It proposed to establish a command and control laboratory using the SAGE Q-32 computer that had been part of a canceled program for SAGE Super Combat Centers. SDC's proposal called for "conceptual and operational studies; surveys of related experiences and plans by military commands; research in organization of command, information flow, and system description; modeling and simulation of different command networks, and operation of a command laboratory for experiments and verification." [26] The principal justification for the SDC command and control research program in the DARPA program plan was that the "application of automation is threatening to usurp the commander's role, stultify command organization and response, and make more difficult the modernization and exercise of an integrated, strong, selective national command." [27]

The SDC proposal was typical of the period: the research it advocated had several overarching objectives and lacked focus. SDC researchers proposed five initial tasks. The first involved the creation of models of information flow and organization, and of information use in command situations. The second task concerned studies on interactions between humans and machines. Research in human-machine studies attempted to ascertain which duties were best performed by people and which by machines, the best techniques of presentation of data to commanders, and the impact of introducing automation on command structures. The third task was research in new areas of gaming, simulation, and information retrieval to assist commands, as well as the investigation of new needs. The fourth was the establishment and operation of a command system laboratory. The last task was simply "research," and called for SDC to "conduct research in the analytical techniques related to system definition and evaluation, information organization and languages, artificial intelligence, and man-machine and machine-machine relations." [28] These research tasks, and the language used to describe them, are typical of the command and control research activities supported by the military in the 1950s. We will see how the establishment of a computer research program in DARPA redirected this command and control activity by establishing DARPA programs that focused on computing research issues rather than directly on command and control issues. But before this change took place, DARPA funded the SDC project with an initial $6-million contract, and provided one of the SAGE Q-32 computers.

Since the Q-32, like the other computing machines of the period, did not come with a complement of software, much of the research to be done was in software and applications systems. In 1960, for many of the well-defined problems of interest to large organizations — accounting, inventory, logistics — existing computing methods were adequate. But a range of poorly defined new scientific and engineering problems — those being investigated by university research departments, the space and missile defense programs, the military's command and control proponents, and the nuclear weapons program —

stretched the capability and capacity of the installed computer base to the limit. Computing, as it was practiced circa 1960, was not equal to the task of information management when there was a need to analyze large amounts of data in a rapidly changing context. So it is not surprising that the SDC proposal focused on the overall objectives. SDC and DOD personnel knew that a great deal of work was needed to define the tactics to meet the objectives. There were guideposts, however.

The SAGE system, designed at MIT, had indicated in the 1950s what was possible in the handling of large amounts of changing data by means of faster and more capable computing. But as of 1960 not much progress had been made in exploiting these possibilities. Most computing R&D in 1960 focused on the problem of massive data processing, either for business or scientific research, where a large amount of processing was done to yield repetitive or individual results. For these uses, large new machines were under development; new languages for greater efficiency were in design; better data input/ output and display techniques were reaching the market. Advances in computing made at universities, where a computer system was seen as a specialized scientific instrument, tended to be separate from computing advances made in corporations, where the computer was seen as a large business machine. Moreover, these developments had minimal effect on computer use in command and control operations in the late 1950s. Bolder, more imaginative steps were needed. Machines needed to be more intelligent, able to interact with each other to gather relevant information, solve problems, anticipate data requirements, communicate effectively across distances, present information visually, and do all of this automatically. These steps required both imaginative individuals and substantial funding.

At the beginning of the Kennedy administration in 1961, Eugene Fubini became the DDR&E and Jack Ruina moved from a post in that office to become DARPA director. These two men were responsible for implementing the new administration's program for new defense systems, and sought ways to improve computing technology to control greater amounts of information and to present it in more effective ways to aid decision making. Instead of continuing to develop specific computing machines, as the DOD had done in the 1940s and 1950s, DARPA officials generalized their interest in computing into the Information Processing Techniques Office. In 1962, IPTO became the newest, and perhaps the most significant, of the DOD's activities in computing.

The Information Processing Techniques Office

From its founding in 1962 to the mid-1980s, DARPA's Information Processing Techniques Office provided substantial research support for bold and imaginative development of computer science and engineering.[29] It did so by exploiting the partnership between the defense and academic communities. When IPTO began in 1962, computing was a group of loosely related activities in fields such as theory, programming, language development, artificial intelligence, and systems architecture. Researchers created advanced architectural systems by designing new logic circuitry and developing new connections among component systems, better numerical techniques for the solution of problems, new input/output designs, and new concepts for more efficient use of the entire computer system. For convenience, we term all these computing activities in 1962 "mainstream computing." In the 1950s, several areas of mainstream computing made rapid advances. Systems became more reliable, more efficient, and less costly, and computer companies introduced a wide variety of models for business and for scientific computing.

Some researchers, however, were starting to view computers as partners in creative thinking: they sought to explore the interaction between people and computers, to create artificially intelligent systems, and to develop natural language (that is, human language) and machine translation processing systems. The university researchers were interested in smarter, more flexible, and more interactive systems. These objectives were of vital importance to two groups: the university computer researchers, who were interested in achieving better interaction with computers in order to make progress in many areas of computing research, and the DOD, which was interested in integrating computing into command and control.

IPTO's early program emerged from the goals and desires of these university researchers eager to investigate new computing techniques. Through the IPTO program, these researchers cooperated with the DOD to achieve a new kind of computing. Although the initial DOD focus prescribed for IPTO was on applications of computing to command and control, IPTO, as we noted above, altered the focus to a concern for the general nature of computing, and as a result many new techniques entered the practice of computing. In the end, the military also benefited from these techniques. Because these new techniques became useful not only within the military but also outside it, in many areas of computing, IPTO achieved a reputation for being able to target its funding to areas of computing that had broad implications for both the civilian and the military sectors.

From the outset, IPTO substantially affected computing by investing in selected problems to provide more capable, flexible, intelligent, and interactive computer systems. While contributing to the content of computing, IPTO's R&D focus changed the style of computing. Hence, when we think of IPTO's achievements, we think about those elements that involved changing both the content and the style of computing — time sharing, networking, graphics, artificial intelligence, parallel architectures, and very large scale integration (VLSI) components. This new style of computing was intended for use in computer science research and ultimately in military systems, and it has been effectively used in both.

IPTO set ambitious objectives for new computing technology and, to attain them, supported the R&D efforts of outside contractors. Following DARPA practices, IPTO employed a dynamic, highly effective, well-respected, and insightful approach to R&D funding. The characteristics of this approach resulted from several factors: the DOD's policies and practices with respect to R&D funding, the ample size of IPTO budgets, the quality of the people brought in to manage the office, and the response of the research community and industry to IPTO's ambitious objectives. Starting with a clear vision of its mission, IPTO never lost sight of its responsibility to support the objectives of the DOD. IPTO's reputation for accomplishments in computing and for its management style has been known for some time among the numerous members of the computing community; in this work, we provide a detailed narrative that fills in the details and relates the history of IPTO to the larger history of technology since World War II, making the history available to a wider audience.

As one of its prime practices, IPTO promoted an array of high-risk R&D projects to expand the frontiers of computer science and engineering for both civilian and DOD needs. At the outset, the office instituted research projects to investigate, among other things, modes of displaying images other than text, the partitioning of memory, symbolic processing to improve the machine's capacity to assist humans, the linking of machines to share resources, and new architectures. At the time, all of these projects were uncertain of success. Attempts were made to increase the intelligence of computers through natural language communication techniques, understanding of human problem solving, and new graphical interfacing techniques. During the 1960s, IPTO and its contractors made substantial progress toward more flexible and interactive computer systems. For example, by the end of IPTO's first decade, time sharing was part of every IPTO-sponsored center, and the ARPANET emerged as a functioning communication medium.

High-risk projects often require long-term support. Because of its place within the DOD, IPTO had significant budgets to expend on programmatic objectives and thus could sustain these projects. As we will see, IPTO's funding policies helped computer science to develop in the office's programmatic areas.

In their pursuit of ambitious objectives, IPTO directors and program managers took advantage of several elements in the DARPA management style to advance specific embryonic areas of the field of computer science and engineering. This combination of elements provided the context for IPTO's success. IPTO employed an amalgam of approaches used by other military agencies in the 1950s, and some new ones designed for the special circumstances in DARPA. Among these approaches were fast turnaround times for project approval, and extensive personal initiative on the part of program personnel. The strategy applied to program development and change remained consistent over time in spite of, and maybe as a result of, all the changes in staffing and funding.

Perhaps most important of all in the DARPA management credo were the agency's personnel policies. For IPTO, DARPA chose managers with foresight, experience, and ability, and gave them substantial freedom to design and manage their programs. Often the difference between programmatic success and failure came down to the person managing a program, and the way in which he or she managed it. The fact that IPTO recruited capable and technically able members of the research community and allowed these people, within the context of DARPA's mission, an unusual amount of freedom to promote research as *they* saw fit, seek advice as they felt the need, and manage the office as they thought the program required, was the deciding element in IPTO's success.[30]

The people who directed and managed IPTO programs have often been visionary and insightful about the needs of the research community. Their insight came to them both through their training and through their own prior research activities. First, they evaluated the potential and needs of the computing research community. Previous assessments of computing by other research agencies, such as NSF, had been done to enhance the computer as a tool in scientific and engineering research. IPTO, however, viewed need from the perspective of research in the computing community itself. Adopting an overarching objective of interactive computing, the office created programs to achieve specific objectives in software development: connectivity, display, decision aids, and increased capability. Second, it moved the results of research out into the industrial and military sectors, which had an effect on computing generally.

The IPTO cast of characters includes many people who had pursued significant research careers in computing before joining IPTO. Among the office's

Table 3   IPTO Directors, 1962–1986

	Term of Office	Educational Background
Joseph C. R. Licklider	Oct. '62–July '64	Ph.D., psychology, University of Rochester, 1942
Ivan E. Sutherland	July '64–June '66	Ph.D., electrical engineering, MIT, 1963
Robert W. Taylor	June '66–Mar. '69	M.A., psychology, University of Texas, 1964
Lawrence G. Roberts	Mar. '69–Sept. '73	Ph.D., electrical engineering, MIT, 1963
Joseph C. R. Licklider	Jan. '74–Aug. '75	Ph.D., psychology, University of Rochester, 1942
David C. Russell	Sept. '75–Aug. '79	Ph.D., physics, Tulane University, 1968
Robert E. Kahn	Aug. '79–Sept. '85	Ph.D., electrical engineering, Princeton, 1964
Saul Amarel	Sept. '85–Sept. '87[a]	D.Eng.Sc., electrical engineering, Columbia, 1955

[a]In 1986, IPTO became ISTO; Amarel remained its director.

directors (see table 3) were Joseph C. R. Licklider, a psychologist who specialized in the interaction between humans and machines; and Ivan E. Sutherland, an important figure in graphics research and a significant contributor to industrial developments, academic programs, and VLSI. Sutherland was succeeded by Robert W. Taylor, a psychologist whose interests in human-machine interactions were similar to Licklider's. Taylor recruited Lawrence G. Roberts, whose work in graphics in the early 1960s set the stage for major developments over the next two decades, although he went to IPTO to design and implement a network program. Licklider, Sutherland, and Roberts, who with Taylor directed IPTO from 1962 to 1975, came from the Cambridge, Massachusetts, community around MIT. DARPA selected David C. Russell, a physicist with military experience, to direct IPTO in 1975, and Russell was succeeded by Robert E. Kahn, an electrical engineer from Princeton with experience at MIT and Bolt Beranek and Newman (BBN; an engineering consulting firm), who had participated in the development of the ARPANET and was a key architect of the Internet. Saul Amarel of Rutgers University, an important researcher in artificial intelligence, came to IPTO in 1985 to assume the directorship from the departing Kahn. Russell, Kahn, and Amarel led IPTO from 1975 to 1986. IPTO capitalized on these men's remarkably expansive visions of what was possible in computer science and engineering. All of IPTO's directors were of the generation that had emerged in computer science as it was flexing its muscles, and were embued with a perception of limitless possibilities in computing. For Licklider and his successors, the pursuit of these visions became almost a missionary activity.[31] The activities of these men inside DOD are an elegant example of the partnership between the military and the civilian technical community in action.

After DARPA organized IPTO in 1962, the early IPTO directors, Licklider, Sutherland, Taylor, and Roberts, believed that they could influence the development of significant improvement in command and control systems by focusing on certain areas of their own interests in basic computing technology. They exerted such influence because they were themselves very able members of the research community. As noted above, at the time DARPA contained a disparate mix of programs ranging from missile development to nuclear test monitoring. These Cold War concerns were pressing policy areas, and this limited the interest that DARPA's directors could take in basic research programs such as computing; thus, the management of IPTO was left to Licklider and his successors. Since the latter had the confidence of the DARPA directors, they had substantial freedom in running IPTO. DARPA directors believed that the program was appropriate to DOD needs and that it was directed by the best leaders.[32] Therefore, the early program evolved in directions related to each of the IPTO directors' strengths. IPTO's approach to research support was to focus on a few centers. Licklider began this program with MIT and Carnegie-Mellon University. Over the years, other directors selected additional educational centers for specific areas of research: Stanford University, the University of California at Los Angeles, the University of Illinois, the University of Utah, the California Institute of Technology, the University of Southern California, the University of California at Berkeley, and Rutgers University, to name a few. The number of centers grew over the two and one-half decades of IPTO history discussed herein, but by 1986 the centers were a small percentage of all the institutions engaged in R&D for computer science and engineering. Similarly, IPTO sought the help of a few industrial R&D programs in the 1960s, and this group also grew in number over the next two decades.

The addition of new groups resulted in part from changes in the environment inside and outside the DOD. In the 1970s, in response to these changes, many of the dominating themes of the 1960s dropped from DARPA's agenda and were replaced by a new set of themes. DARPA abandoned counterinsurgency research, which had been stimulated by military interests during the Vietnam conflict. It also deemphasized the Defender Project, a program to study ballistic missile defense. In place of Defender, DARPA organized a new Strategic Technology program. The authors of a 1975 study of DARPA, commenting on the early 1970s, concluded, "In essence, the door was closed on the original 'Presidential issues' that had been the foundation of ARPA's work for a decade. No comparable mandates on broad new issues were assigned to the Agency."[33] DARPA's new assignments were in specific defense problem areas, with a new emphasis on joint service-DARPA projects. The Strategic Technology Office, the Tactical Technology Office, and the Information Processing Techniques Office became the core of the DARPA organizational structure, requiring a greater emphasis on application systems and attention to DOD needs, rather than the needs of basic research only.

During IPTO's first decade, DARPA's demands for applicability to military objectives were expressed quietly. After Stephen Lukasik became DARPA director in 1971, and the winds of change were beginning to be felt in the DOD, he paid more attention to the relevance of DARPA research programs to military needs. For example, the Nixon administration adjusted its priorities in R&D, insisting that military R&D programs tighten their focus to military missions. In addition, congressional demands for greater accountability from DOD, and a sharper focus on military mission research topics, stimulated closer examination of DARPA projects. Lukasik addressed these new policies with more concern for application, while trying to maintain the funding levels of the research program. Lukasik was always a strong promoter of research in IPTO.[34] His successors did not always have that luxury, as the DOD began to respond to greater congressional and executive branch demands for even more military relevance under constrained budgets.[35]

The inclusion of "Exploratory Development" as a budget category in the fiscal year (FY) 1971 IPTO budget signaled a shift to greater emphasis on development in the program.[36] Previously, IPTO had operated with only basic research funds. The DOD included in the category "Research" (known as category 6.1) all basic research and applied research directed toward expanding knowledge in several scientific areas. The Exploratory Development category, known as category 6.2, included studies, investigations, and exploratory development efforts, varying from applied research to sophisticated breadboard hardware, and was oriented to specific military problem areas.[37] The 6.2 funding category for IPTO reflected IPTO's role in the "increasingly critical defense requirements in communication, high speed computation, secure time-shared systems, and advanced signal processing."[38] Despite the differences between the funding categories, it was possible to have considerable overlap between them.

In the 1970s, command and control was still considered the key to effective utilization of the armed forces. In a new emphasis, DARPA directors constructed a strategy to fill the perceived gaps in the technology base by way of "a synergistic relationship among computer science, communications, and information sciences."[39] To advance the technology base, DARPA management at all levels believed it necessary to attend to four areas. First, IPTO program directors saw a need for still more research in fundamental computer science.

IPTO officials believed that basic research in areas such as intelligent systems, VLSI, and software technology would result in the ability to automatically interpret images and to abstract text material, and develop advanced tools and techniques for engineering real-time microprocessor software and systems.[40]

Second, more research in advanced network concepts for communications was also seen as necessary. For example, IPTO wished to extend the proven technology of the ARPANET to packet switching in a satellite communication environment, and to packet radio in order to provide a mobile, distributed, and secure network. In addition, IPTO sought teleconferencing and advanced automated message-handling capabilities.

Third, DARPA believed it necessary to develop an experimental system to evaluate new ideas for command and control technologies that emerged from IPTO research. To accomplish this the agency established "testbeds," the goal of which was to allow the military to evaluate new technology in a "try-before-buy" fashion. Such testbeds would bring the work of the computer scientist, the communications engineer, and the information scientist together to achieve the synergism that DARPA thought was lacking in the development of command and control systems. Another goal of testbeds was to make possible the assessment of a variety of competing technologies. Testing new technologies in a testbed would also provide feedback about the users' needs.

And fourth, DARPA wanted to stimulate work on advanced digital components. Faster, more complex semiconductor microprocessors would allow the inclusion of more instructions in the integrated circuits. Furthermore, if feature sizes of semiconductors below one micron could be achieved, microprocessors could become even faster and more sophisticated. IPTO proposed a program to design advanced VLSI architectures so as to make possible integrated circuits that would far surpass existing silicon-based circuits for achieving high-data-rate signal processing. Not only were components needed, but major new systems employing these components would have to be designed and tested.

In the period 1974–80, when this new strategy was taking shape at IPTO, three men directed the office: Licklider, for a second time; David Russell; and Robert Kahn. New leaders with greater concern for what near-term payoffs research would have for the DOD assumed direction of DARPA. George Heilmeier, DARPA director from the beginning of 1975 through 1977, pressed IPTO to be more mindful of the military applicability of research and exploratory development projects. He also strongly advocated the idea of testbeds, the experimental testing of results, to examine the relevance of new ideas for use in military systems. Heilmeier's successors also insisted on military applicability.

The later IPTO activities in testbeds bore the stamp of all of these men and resulted in systematic expansion of the IPTO program. To handle the more diverse program, IPTO's staff grew. This growth, too, affected the program's development and its influence. As the IPTO program evolved, the elements of basic R&D remained, but the program included more and more explicitly defense-oriented projects. Although we will not discuss it in this study, the political circumstances in the world of the past three decades led the DOD to demand new developments in computing that would help to increase the sophistication and speed of new military systems; and this was a challenge that IPTO was fully prepared to undertake.

In the 1980s, the DOD expressed concern not just for the development of new computing for defense purposes, but also for the continued strength of the defense industry. The department's concerns were consistent with national concerns of the later 1980s: concern for the standing of American high-technology industry, especially as it involved the defense industry, and concern for the competitiveness of the United States in world trade. Computing had by this time become so ubiquitous that the DOD wanted to secure its continued advancement and a quick passage of useful ideas from universities' research programs and IPTO testbeds to industry. The announcement by the Japanese of the Fifth Generation Computer Program rang alarm bells at the Pentagon in the early 1980s. When the Reagan administration began in early 1981, Richard DeLauer, deputy secretary of defense for science and technology, and Robert Cooper, DARPA director, among many others, wanted a new program to help maintain the U.S. position in computing, especially as it related to the defense industry. Robert Kahn and his colleagues in IPTO and other DARPA offices responded with the Strategic Computing (SC) program, which received congressional funding in 1984. Thus, IPTO's horizons expanded even further. IPTO coupled its university and testbed programs to programs in the defense industry. The number and variety of programs increased significantly, and many more projects were pursued by companies for specific applications. A complete list of these programs in the mid-1980s can be found in table 7 (in chapter 6, below).

The IPTO budget expanded to accommodate these new desires. The total IPTO budget in FY 1971, for example, was approximately $28 million, up from $14 million in FY 1965; $17 million went to category 6.1 research in information processing techniques, image processing, intelligent systems, speech understanding and the like. IPTO spent another $11 million on category 6.2 projects in distributed information systems, mostly for work on ARPANET and ILLIAC IV. The IPTO budget grew in the next four years, reaching more than $38 million in FY 1975. The budget contained a decrease in the amount for category 6.1 research (which was down to $13 million) and an increase in the distributed information systems area, or category 6.2, to $17 million. A new category 6.2 funding line contained $8 million for programs in command, control, and communications (C³) technology.[41] Distributed networks and ILLIAC IV now accounted for only $5.3 million; the rest went to considerations of secure systems, climate dynamics, software, and speech processing. In general terms, then, 60 percent of the IPTO budget in FY 1971 was allocated for basic research and 40 percent for exploratory development. By FY 1975, the numbers were approximately reversed: 43 percent for basic research and 57 percent for exploratory development. The split remained essentially the same for the FY 1980 budget (a total of $50 million): 42 percent and 58 percent, respectively.[42] In the first half of the 1980s, the division of funds between basic research and exploratory development had returned to an even distribution, depending on how we classify the activities in the Strategic Computing program. The total IPTO budget was $154 million in FY 1985 and $175 million in FY 1986. In FY 1988, the budget for the new Strategic Computing (SC) program grew to $226 million, accounting for more than 27 percent of DARPA's budget. Figure 1 shows a comparison of the trends in DARPA and IPTO budgets.

In 1985, when Saul Amarel assumed direction of IPTO, he identified several "exciting scientific/technological developments in [computer science] today that promise to accelerate further the impact of computing on defense and on our national economy." He based his analysis on the developments in networking and VLSI. The VLSI program spawned several novel multiprocessor architectures. These architectures, which were promising results of the SC program, provided the high performance needed for research in vision, speech, complex symbolic processing, and large scientific problems. He believed that AI was starting to have a noticeable impact on a number of industrial and defense problems.[43]

Like computer science today (circa 1996), the more recent IPTO program under Kahn, Amarel, and their successors was a complex, highly interrelated set of activities. It included intelligent systems; advanced digital structures and network concepts; distributed information systems; advanced command, control, and communications; and strategic computing. Some parts of the program, such as its intelligent systems component, had changed only their projects, not their objectives. This IPTO program contained an array of projects on satellites, semiconductors, machine representation and utilization

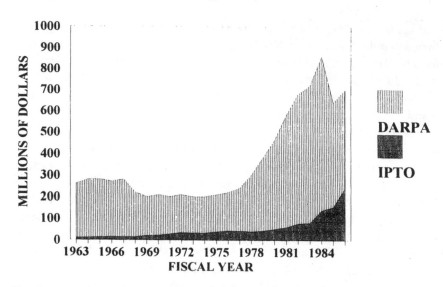

Fig. 1. DARPA AND IPTO annual budgets for FY 1963–86, in then-current dollars. DOD spending on advanced research projects declined slightly toward the end of the Vietnam War but rose sharply after 1977. Appropriations for IPTO rose steadily throughout the period. Data from Richard H. Van Atta, Seymour J. Deitchman, and Sidney G. Reed, DARPA *Technical Accomplishments,* vol. 3, IDA Paper P-2538 (Washington, D.C.: IDA, 1991), p. II-3, and *Congressional Record,* 86th–99th Cong., 1963–1986.

of knowledge, and generic hyperspeed computer applications. The program continued to focus on more capable, flexible, intelligent, and interactive computer systems.

In sum, program continuity exists throughout the more than two decades from 1962 to 1986, the period covered by this history. The basic technology program, although slowly adapted to meet new needs, continued on the research frontier of computing. IPTO was trying to accomplish its mission to help in the development of valuable DOD systems by adding testbeds and applications to the program. IPTO's attempt to make these applications functional in operating defense systems, and in the process strengthen an area of the defense industry, elevated it from a small DOD office to a player on the world stage of computing development. From Licklider's vision that helped to define the early IPTO program—a vision shared by his superiors Jack Ruina and Eugene Fubini—the program grew under his successors into one that shaped computing as a national activity.

The IPTO program resulted from the research outlooks of IPTO directors—their ability to envision, on the basis of knowledge of DOD needs, a radically different future for computing—as well as from the attention paid to the office's activities by DARPA directors, and the occasional interest taken by higher-level administrators in the DOD. The early research program focused on time sharing, networking, natural language programming, artificial intelligence, and graphics. Later, IPTO expanded this program to include several new themes, added in the mid-to-late 1970s in response to perceived new needs: VLSI design, distributed systems and software, strategic computing, and high-performance computing. These additions to the program reflected DARPA's belief in the mid-1970s that "the technology base for modern command and control is incomplete."[44] Moreover, IPTO added a communications dimension to this area of command and control because of the clear importance of communications to the research program of IPTO and DOD. IPTO was one group that contributed significantly to the information world in which we live, by envisioning an interactive computing world—and an appreciation of IPTO's history will lead to a better understanding of the nature of that world.

CHAPTER ONE

Managing for Technological Innovation

IPTO's emphasis on ambitious technical objectives and its years of nurturing the institutional framework for R&D in information processing resulted in some remarkable technical achievements. As a result, the office gained a formidable reputation with its associates and many outside observers, who are convinced that the achievements result from IPTO's special character, which they feel is worthy of being emulated by other government agencies involved in the support of R&D.[1] That special character, they argue, came from IPTO directors' ability to select technologies particularly ripe for development, their diplomatic facility in working with the research community, their skill in managing large contracts in a large program, and their standing in the research community, which they had earned by contributions made prior to their tenure at IPTO. These men are seen as the antithesis of the bureaucrats who direct many government programs. The successful aspects of the IPTO program have contributed to an enlarged view of this special character, and to the loyalty of a sizable group in the research community. The recent debate about organizing a civilian DARPA began because many believers in the efficacy of DARPA—a promoter of R&D—and, by inference, of IPTO, believe that this special character can be replicated in a civilian agency.

There are also critics of DARPA and IPTO, who point to IPTO projects that produced less desirable results after large expenditures of time and money as indicators that IPTO's achievements come at high cost.[2] But even while criticizing the high cost, the critics often acknowledge the significant effects that IPTO-sponsored R&D had on computing. Focusing only on the nature of the DOD's R&D mission, the critics seem not to acknowledge that to produce significant results from high-risk R&D it was necessary to conduct a range of projects, with the expectation of a low success rate. IPTO directors understood this balance between high risk and significant results, and were responsible for the design and promotion of ambitious objectives.

The IPTO programs were designed to focus on specific objectives, not to support general R&D in computing. It is important to recognize that the technical accomplishments that contributed to meeting these objectives were shaped as much by IPTO office management as they were by researchers' intentions. A strong correlation exists between the research experience of the managers and the selection of high-risk technological research projects, especially in the early years. Since IPTO often provided the initial stimulus for projects, and continually monitored projects so as to coordinate research and thus obtain integrated results, we focus first on the IPTO office: its managers, its program development strategies, its practices in budgeting and contracting, and its image of itself as seen through IPTO's and DARPA's interactions with its funder, the U.S. Congress. The remaining chapters will focus on IPTO programs and contractor projects, technical objectives and accomplishments, the range of projects—from the spectacularly successful to those abandoned for lack of sufficient progress—and on the way in which the relationship between managers and projects contributed to the development of generic technologies and to military defense systems.

People and Programs

Staffing IPTO: Office Directors

Outsiders frequently view government funding programs as bureaucratic and formal. From the outset, DARPA and IPTO tried to avoid bureaucracy and formality by operating with a lean management structure. Rather than describe the management structure, we craft our story around the people in the office. People made the difference, and with only a few people, it is possible to write of them as individuals. Considering the size of IPTO budgets and projects, the lean management structure was possible because of the capabilities of the people brought in to staff the office. The professional characteristics of individuals on the staff, including both educational and institutional affiliations, do not vary much over the history of IPTO.

When Eugene Fubini and Jack Ruina sought an inaugural director for IPTO, they turned to the Cambridge, Massachusetts, community, which has been a marvel in American history for more than three hundred years. For the past seventy-five years, Cambridge's scientific and technical community has dominated the activities of this geographical area, founding many companies based on the discoveries and inventions of workers in the laboratories of MIT and Harvard. In the 1930s, this community formed an association with the military community which defined what some historians have called the military-industrial complex in the United States. Forged mostly in the MIT Radiation Laboratory of wartime Cambridge, this community took a leading role in the production of defense systems for the Cold War world, with consequences for MIT, the New England economy, the nation, and the world.

By the 1950s, the researchers moved from one setting to another, forming interconnections mostly hidden from view, except to the people involved. The community in the 1950s, before IPTO, began with the laboratories in the universities. Recall the names of some of the principal laboratories: the Servomechanisms Laboratory that produced the Whirlwind computer; the Research Laboratory of Electronics, which led in information theory and many new electronic devices; the Instrumentation Laboratory, where many missile guidance systems were designed; and the Aiken Computational Laboratory. Throughout this history, we will report on many of the researchers from these laboratories who were supported by IPTO and who produced many new ideas that changed the style of computing.

When it was time to integrate what was learned in these laboratories into defense systems, the researchers founded a new set of organizations to act as intermediaries between the basic researchers and the industries that would build the new systems. Some of these organizations have become legendary in the history of science and technology in the United States — Lincoln Laboratory, Bolt Beranek and Newman, MITRE Corporation, and Air Force Cambridge Research Laboratories. The people who started or worked for these companies had received their training and early experience in the laboratories in and around Cambridge. Most of them maintained intimate connections with the DOD, many serving in various capacities in the DOD hierarchy. Defense companies — Raytheon, General Radio, Honeywell, Digital Equipment Corporation, and hundreds of others — formed the third part of the military-industrial complex that is the Cambridge community of interest to us.

As we will show, four of the six IPTO directors discussed in this history acquired experience in organizations at the middle level of this pyramid, at the systems integrators such as BBN and MITRE. A fifth came from another part of the defense community, the aerospace industry, and the sixth from the military. Thus, their experience prepared them for the mission given to IPTO: to stimulate or find the new ideas that could quickly be integrated into new defense systems — specifically, command and control systems.

Joseph Carl Robnett Licklider, a member of the Cambridge-Boston technical community who was already involved in research for the DOD, was the person whom Fubini and Ruina chose to lead DARPA's new office. A native of

Saint Louis, Missouri, Licklider had attended Washington University, where he earned an A.B. degree in 1937 with majors in physics, mathematics, and psychology, and an A.M. degree in psychology in 1938. In 1942 he received a doctorate in psychology from the University of Rochester. In many ways, Licklider was more a product of the Cambridge community than of the institutions from which he received his training. During World War II he had been a research associate and research fellow in the Psycho-Acoustic Laboratory at Harvard University, working on defense projects. After the war he stayed at Harvard for a few years and then joined the Psychology Department of MIT as an associate professor. During the 1950s Licklider was an associate professor in the Electrical Engineering Department at MIT, as well as a member of the Research Laboratory of Electronics and the Acoustics Laboratory. He participated in the Project Charles study of air defense which led to the establishment of Lincoln Laboratory, and was a group leader at the newly established laboratory. In addition, he served as a consultant to local laboratories and companies, served on the Air Force Scientific Advisory Board, and consulted directly with the DOD. In 1957, he became a vice-president at Bolt Beranek and Newman, which was engaged in studies in acoustics, psychoacoustics, human-computer interactions, and information systems.

A quiet, modest, self-possessed man, Licklider possessed important qualities characteristic of a good research leader. First, he had early acquired the ability to design research projects with expansive objectives. At BBN, Licklider had among other things directed a research project on human-computer interactions sponsored by the Air Force Office of Scientific Research. This research primarily involved study of the domain and parameters of systems theory, particularly with regard to the human factors aspects of the design of complex human-human and human-machine systems. Licklider and his associates were to study and outline a theory of human-machine systems which would help to identify the critical issues for further theory developments and provide a rationale for human factors recommendations.[3] In a letter of November 1960, Licklider wrote that a major part of the research had to do with the interaction of complex systems, and their simulation by computer. And in another part of this research, which was reflected in his later design of programs at IPTO, he divided human-human and human-machine system functions into their elementary operations.[4]

Second, he listened well and was ready to help young researchers by offering insight into the limitations of a concept and how the concept could be expanded. Moreover, he understood how to enhance a project's results so that it became part of a larger objective. And third, he demanded excellence of himself and of all around him. It was these last two qualities, along with his formulation of the IPTO program, that led to his colleagues' estimation of him as a visionary in technological development.

During his conversation with Fubini and Ruina about joining the DOD to lead IPTO, Licklider, using the knowledge of systems he had gained in his research for the air force, revealed his visionary talent when he asserted his belief that the problems of command and control were primarily problems of human-computer interaction. Fubini and Ruina enthusiastically agreed with him.[5] Licklider discussed what certain new computer developments that were on the horizon could do for the command and control problem if pursued. His views on this subject were adumbrated in a paper entitled "Man-Computer Symbiosis" which he wrote during the research at BBN.[6] The paper outlined a program for improving human capabilities through the use of computers. For example, Licklider described several ways to improve communication so as to provide a partnership in which both the person and the computer contributed to the overall goals in ways chosen on the basis of the specific strengths of each. In effect, the computer would help the human to think. Recognizing the limitations in this cooperative way of using computers, he identified a number of prerequisites for its realization. His paper listed research questions regarding computer memory requirements, programming for memory organization, natural and computer language compatibility, displays, and speech recognition. In each case, Licklider suggested possible research and the time required for significant results.

Through the 1960 IDA survey discussed in the Introduction, above, the DOD and DARPA had come to recognize that these areas should be attended to in order to advance command and control systems. The fact that these were all active research areas in the Cambridge community meant that there was a neat fit between research interests of academics at institutions such as MIT and the new system designs of interest to the defense community. No wonder Fubini and Ruina agreed with Licklider as to goals for research. They believed that IPTO was the right place for implementing programs to develop new interactive computer systems which would advance the capabilities of computers in ways Licklider described in this paper. Ruina offered Licklider the IPTO directorship, and Licklider started at DARPA in October 1962, although he had been involved in assessment of projects during the preceding few months.[7]

What principles would Licklider use as a guide? Licklider instilled into IPTO program management the criterion of excellence, a vision of the future of computing, a focus on a few centers of activity, and a concern for contributing to computing in order to serve both the military mission of DOD, and society generally. Shortly before he joined DARPA in 1962, Licklider wrote a letter to Charles Hutchinson, the director of the Air Force Office of Scientific Research, in which he noted that his main objective would be "to spend the Government's money very carefully and very wisely." Expenditures would have to satisfy at least three criteria:

1. The research must be excellent research as evaluated from a scientific or technical point of view.
2. The research must offer a good prospect of solving problems that are of interest to the Department of Defense.
3. The various sponsored efforts must fit together into one or more coherent programs that will provide a mechanism, not only for the execution of the research, but also for bringing to bear upon the operations in the Defense Department the applicable results of the research *and* the knowledge and methods that have been developed in the fields in which the research is carried out.[8]

Contrary to Hutchinson's advice to commit program money "in a hurry," Licklider felt it to be "of paramount importance to organize a coherent program and not to act until there has been an opportunity for careful investigation and deliberation (by me, as well as by advisors)."[9] Perhaps with more vision than he knew, Licklider set up a pattern in the IPTO program which his successors followed throughout IPTO's history: a convergence of objectives between academic and military interests, and a coherence in the initiated programs over time. To achieve these objectives, Licklider followed the DARPA style and adopted a lean administrative structure for IPTO. In fact, he managed two programs simultaneously—IPTO and the Behavioral Sciences Office, with a combined budget of $11 million—with the help of one secretary.

The elements of Licklider's approach to funding by IPTO were passed on through the next decades from office director to office director, and from program manager to program manager. For example, Taylor, the third IPTO director, had read Licklider's "Man-Computer Symbiosis" well before he went to DARPA and "heartily subscribed" to its goals. When he joined IPTO he assessed the IPTO programs, but as an adherent to the Licklider philosophy, he saw little reason to change the direction of them after he assumed the directorship. When Taylor assumed the directorship, Licklider's philosophy was firmly established:

What we were trying to do, on the one hand, was to select research problems that were ripe enough to have some hope of making progress on them, but on the other hand, to select them with the notion that if we succeeded it would make an order of magnitude difference between the way in which business was done then, versus what this new research finding might permit. . . . We were explicitly looking for things which, if they could succeed, would really be a large step beyond what technology could then permit.[10]

Taylor thought that IPTO's role was to fill a niche that "other agencies would be unable to fill because of their limited budgets." Believing that individual researchers were "already pretty well covered" led IPTO to focus on funding groups within institutions rather than on funding the research projects of individuals, as is the case with NSF and NIH.[11]

How was this philosophy passed on? When Ivan Sutherland joined IPTO, Licklider was still office director. The two men shared the work of the office so that Sutherland could more easily make the transition to his new post. Taylor succeeded Sutherland after a similar period as apprentice director. Similarly, toward the end of Taylor's directorship Lawrence Roberts worked closely with Taylor on a variety of areas, in a sort of apprenticeship. In the early 1970s, a deputy director's position was added to the office, and this post became a launching pad to the directorship. Both David Russell and Robert Kahn were deputy directors before they assumed the directorship.

Such a similarity of philosophical approach was further enhanced because the choice of a successor was often in the hands of IPTO directors; indeed, it was their responsibility. During one of Licklider's many visits to MIT while he was IPTO director, he learned of the work of the young superstar Ivan Sutherland. Sutherland, then a graduate student at MIT working in the area of graphics, was invited to one of Licklider's site-visit meetings held at MIT in early 1964. This conference was Licklider's first encounter with Sutherland. During the conference Sutherland asked a question of one of the speakers, which "was the kind of question that indicated that this unknown young fellow might have something interesting to say to this high-powered assemblage." Using an IPTO prerogative of reorganizing conference programs to take advantage of new ideas and information, Licklider asked Sutherland to give a presentation the next day. The presentation was a great success and impressed Licklider. When Licklider was preparing to leave IPTO in 1964, he recommended Sutherland to replace him. Licklider had some reservations because of Sutherland's youth, but Robert Sproull, Ruina's successor as DARPA director, saw no problem "if Sutherland was really as bright as he was said to be."[12]

At the time he joined IPTO, Sutherland was twenty-six years old, and was already serious and scholarly. Born in Hastings, Nebraska, he had attended

Carnegie-Mellon University and the California Institute of Technology (Caltech) before receiving his Ph.D. degree in electrical engineering from MIT in 1963. Claude Shannon was his thesis adviser at MIT, while Wesley Clark guided his work on the experimental transistorized computer the TX-2 at Lincoln Laboratory. (Marvin Minsky also served on his thesis committee.) While at Lincoln, Sutherland developed the famous Sketchpad system, the first interactive graphics system. He demonstrated Sketchpad to people in the IPTO-sponsored Project MAC at MIT just as the project was starting. Sutherland reported later that he had had several conversations with Licklider during these years but did not know him well.[13] But researchers in the world of interactive computing came to know the developer of Sketchpad very well.[14]

An army ROTC commitment left over from CMU days meant that when Sutherland finished his degree in 1963, he had to serve a stint in the army. He was assigned to an army project at the University of Michigan, where the project on side-looking radar was under way. Shortly afterward he was reassigned to the National Security Agency (NSA) for work on a display system. During this project he was asked to visit DARPA and discuss the possibility of assuming the IPTO directorship. "I felt at the time that it was probably too large a job for me to undertake at that age. I initially said no, and then thought about it for another six weeks. They twisted my arm a little harder, and I agreed to go."[15]

Sutherland, a hard-working and private individual, adopted the Licklider philosophy as his approach to contractor selection, not a surprising choice for a man of his age and lack of government experience. Licklider's approach, Sutherland noted later,

> turned out to be a remarkably wise thing. The principal thing I learned about that kind of research activity is that the caliber of people that you want to do research at that level are people who have ideas that you can either back or not, but they are quite difficult to influence. In the research business, the researchers themselves, I think, know what is important. What they will work on is what they think is interesting and important. . . . Good research comes from the researchers themselves rather than from the outside.[16]

Once Sutherland convinced himself that a research project was worth funding, all he needed to do was to convince his superior that the project "was sensible to do." While Sutherland needed to prepare justifications for his recommendations, he "always felt that the principal hurdle to get through was to convince the ARPA director that it was a sensible task." The two DARPA directors he had to "convince" during his tenure were Robert Sproull and Charles Herzfeld, both of whom understood the possible value of IPTO results for other programs in DARPA.[17]

Sutherland recruited as his deputy and possible successor Robert Taylor, a program officer from the National Aeronautics and Space Administration. Taylor, who was an ebullient and far-sighted individual, as evidenced in the many innovative projects he managed over the years,[18] was a native of Texas and educated at the University of Texas at Austin. During research on his master's degree, he worked with a group trying to build a model of how the auditory nervous system functions in localizing sound and isolating an intelligent signal against a noisy background.[19] The group was doing both basic and applied research in acoustics. During his work at Texas Taylor learned of Licklider's early work, but he did not meet Licklider until many years later, when he was at NASA and Licklider was at IPTO. In 1960 Taylor took a position as a systems design engineer at the Martin Company, now Martin Marietta, in Florida; he left the following year to work in human-machine systems research at ACF Electronics in Maryland. It was from that position that he moved in 1962 to NASA's Office of Research and Technology, which was responsible for explorations on the technological frontier in areas of concern to NASA and funded some work in computing. Through this effort Taylor came to know Licklider and later Sutherland.[20]

Taylor and Licklider met when Licklider organized meetings of an informal committee of government program officers whose operations in some way funded computing activities. Represented on the committee were the Army Research Office, the Air Force Office of Scientific Research, the Office of Naval Research, the NIH, NASA, and NSF. The group had no charter, no responsibilities, no budget, and no purpose. The members held several meetings in the early 1960s. To stretch their limited budgets as far as possible, members kept each other informed about the projects they were funding to avoid duplication. "We would discuss what was important, what was current, and what was going on. It was a wonderful way of getting information flow between the agencies."[21] Such liaisons are often formed among federal agencies. In the 1980s these liaisons took on a more formal character, as in the Federal Coordinating Council for Science, Engineering, and Technology (FCCSET), which prepared a study of high-performance computing.[22]

While Taylor served essentially as deputy director of IPTO, he and Sutherland shared all tasks. "We didn't divide up the work in any sense at all."[23] Taylor served in this capacity for some eighteen months. When Sutherland left in mid-1966, Taylor assumed the position of office director. At that time the office had several large contracts—Project MAC, ILLIAC IV, and SDC—along with a number of smaller ones in the areas of time sharing, artificial intelligence, graphics, and programming. Little maneuvering room existed. But as Kahn later said of him, Taylor "had incredibly good taste [technically] and intuition."[24] Taylor saw an opportunity to develop networking, and this became the major focus he added to the continuing activities during his tenure as director.

The next director, Lawrence Roberts, an office mate of Sutherland while the two were graduate students at MIT, had cut his computing teeth on the TX-0 (an early transistorized computer) and the TX-2 during graduate school. When the TX-0 was moved from Lincoln Laboratory to the MIT campus, Roberts obtained access to it as an assistant in the computation center. While using the TX-0, he taught himself the basics of computer design and operation. He did some support work for the machine as part of his duties, during which he wrote a program that recognized hand-written characters by using a neural net design. His dissertation dealt with a problem in computer perception of three-dimensional solids, and was written with the guidance of Peter Elias, an MIT professor specializing in information theory. During this period he worked first as a research assistant in the Research Laboratory of Electronics (RLE) and then as a staff associate at Lincoln Laboratory. At this time, he also developed an operating system and compiler for Lincoln Laboratory's TX-2 computer. By default, because of the departure from Lincoln of Wesley Clark and William Papian in 1964, he led the group associated with the machine. The computer research group working on the TX-2 was one of three groups in the laboratory's "design division." The other two groups investigated thin film memories and computer speech recognition, two areas Roberts brought into the IPTO program.

Roberts was well known to IPTO personnel before Taylor recruited him in late 1966 to join IPTO to manage a networking program. After his MIT colleague Sutherland became IPTO director, Roberts visited him seeking IPTO funds for a graphics project at Lincoln Laboratory. This request followed from a conversation between the two at the laboratory during one of Sutherland's visits. In March 1965 a contract was forthcoming, one of the first in Sutherland's new graphics initiative for IPTO.[25] Earlier, as one of the MIT contingent, Roberts had attended the November 1964 Homestead conference sponsored by the air force to consider the future of computing. Conversations with Licklider and others at the conference had convinced Roberts that "the next thing, really, was making all of this incompatible work compatible with some sort of networking." He concluded that something had to be done about communications. Following the meeting, Roberts began to shift his center of focus from graphics "to investigate how computers could work with each other and how to do inter-computer communications and computing." Shortly after this, he did an experiment on networking with Thomas Marill (it is described in chapter 4, below).[26] It was through Taylor's contact with various MIT laboratories that he knew Roberts.

At first, Roberts was reluctant to leave Lincoln Laboratory for IPTO and declined Taylor's invitation to join the IPTO staff. But as management of DARPA and Lincoln became increasingly insistent that he do so, he began to succumb to their arguments. Later he reported what had changed his mind.

> I was coming to the point of view . . . that this research that we [the computing group at Lincoln] were doing was not getting to the rest of the world; that no matter what we did we could not get it known. It was part of that concept of building the network: how do we build a network to get it distributed so that people could use whatever was done? I was feeling that we were now probably twenty years ahead of what anybody was going to use and still there was no path for them to pick up. . . . So I really was feeling a pull towards getting more into the real world, rather than remain in that sort of ivory tower. . . . In the course of all these discussions, people made me aware that I would then be exposed to a lot more, because everybody comes to ARPA.[27]

Accomplished in graphics and networking, fully cognizant of the benefits of time sharing and artificial intelligence, the dynamic and articulate Roberts was an ideal person to become director of IPTO. After Roberts joined IPTO, Taylor eventually increased the range of Roberts's responsibilities until Roberts could handle all aspects of the office and Taylor was ready to leave.[28]

In retrospect, the 1960s could be considered IPTO's glory years. Licklider had come to IPTO in 1962 with a vision of how to increase human-computer interaction and established a program of breadth and depth to realize this vision. His successors, Sutherland, Taylor, and Roberts, all subscribed to this vision and continued to promote it. When seen in light of their contributions to computing research, the directorships of all four men were successful ventures in government funding of R&D. Why then was it so difficult to find successors to Roberts and then Licklider, who returned to IPTO to serve another twenty months as director in 1974–75? Arguments have been made about salary penalties for those who left high-paying positions to work for the government. And some asserted that young "superstars," the type wanted for the office, were reluctant to interrupt a promising career to spend time at IPTO. This notion is supported by Roberts's example. He was coerced by the community to consider the post; but even he, as we saw in the quotation above, reported that he had a design in mind when he finally agreed to move to Washington.

It might be, however, that a small computer community saw IPTO's focus as very narrow, making the pool of candidates with the right background and of the right age to be director relatively small. For example, Seymour Papert implied this narrow focus in a brief comment on the attitude toward neural network research and artificial intelligence at the end of the 1960s. Members of MIT's AI community were determined, he asserted, to keep IPTO's focus on a particular type of AI research, and did so with their perceptron book, a book published in 1969 by Marvin Minsky and Seymour Papert which purported to prove that neural nets could never fulfill their promise of building models of the human mind. Only computer programs could do this, hence MIT's emphasis in AI.[29] Perhaps similar pressures existed with regard to the full range of IPTO programs. If so, the community must have exerted a concerted effort to maintain the IPTO program as it was, allowing for some additions, but by and large keeping it restricted. Any attempt to maintain a small group of activities in IPTO would of necessity result in a small pool of candidates because the number of workers in the small vineyard would be small; and the question of vetoes over choices outside the "IPTO community" should be asked. In contrast to the pool of candidates for director, there seemed to be no shortage of candidates for program manager; the positions were enthusiastically filled by many able graduate students and postdoctoral people, as we will see below. However, as seen in Sutherland's case, age was no barrier, either for program directors or program managers. Salary seems not to have been a barrier. Roberts remained on the Lincoln Laboratory payroll for a time—no loss there. Later program directors taken from industry were at least partly compensated for the change. We suspect that any screening was done unconsciously, if at all, and only outside forces could overcome the narrowness of perspective. This forcing function came first in the person of George Heilmeier, who was DARPA director from 1975 to 1977—a total of three years, but enough to change IPTO. In fact, as we shall see, all the elements for change existed inside IPTO. What made the change possible was the political environment in Washington.

By the end of Roberts's tenure in 1973, DARPA, IPTO, and the entire DOD had changed, and computer science had matured, so it was more difficult to sustain the IPTO philosophy. There was also heightened congressional interest in specific program content, increased industrial influence in DOD, and White House discontent with various DOD policies; and the general public attitude toward the military had changed from admiration to skepticism. The glorious days of the 1960s, with space, missile, and Great Society programs, were over, and the grim realities of the Vietnam War affected all planning and funding.

The relationship between Congress and the DOD had begun to change toward the end of the 1960s. An important concern for some in Congress was the question of whether DOD contracted only for R&D needed for its mission, or whether DOD contracts were a vehicle for support of universities' general programs, which had no relation to DOD's mission. Programs of the latter type were seen as the mission of NSF and NIH, not DOD. The changing relationship can be seen in the type of questions asked of DOD executives by congressmen and senators. During the Senate's appropriation hearings in 1968, Senator Mansfield asked John Foster, the director of defense research and engineering, about duplication of research and about the relation of DOD-sponsored research, particularly basic research, to that of other agencies.

> It was abundantly clear in his response that the Pentagon then believed all fields of science and technology were open to it, that it saw no inconsistency in funding basic research in fields already funded by civil agencies, and that all research projects it sponsored were somehow relevant to Defense needs. The Defense Department was adamant in its position that it must continue the full spectrum of research then being undertaken, even though by definition the outcome of much such research can neither be predicted nor its possible relevance to military science known.[30]

In 1969, Mansfield added a rider to the DOD military authorization bill (PL 91–121) for FY 1970: "None of the funds authorized to be appropriated by this Act may be used to carry out any research project or study unless such project or study has a direct or apparent relationship to a specific military function or operations."[31] Congress passed the Mansfield Amendment, which profoundly affected DOD's contracting activities thereafter.[32] Even though the amendment was short lived, members of Congress remained concerned about military support for research that was not geared to military needs, a concern that continues to the present.

Programs now went through closer scrutiny before approval and public presentation. Greater accountability was demanded of DARPA and IPTO personnel; there were more requests from DOD management for program and cost analyses to justify decisions. In 1970, the General Accounting Office had singled out some IPTO projects as questionable in terms of this legislation and as more properly supportable by the State Department than by military appropriations. DARPA asserted that none of its projects failed the test of the Mansfield Amendment.[33] Despite this assertion, the amendment caused some changes in IPTO. For example, IPTO programs began to be presented with more of the military's needs in mind. Simultaneously, the basic research program (category 6.1) remained static, and new program elements were included in the applied area (category 6.2). "For me," Roberts stated years later, "the Mansfield Amendment said that there were more 6.2 funds available." Roberts admitted to adjusting his presentations about projects to sound "more supportive of the military roles."[34] Such a change is also seen in the public presentations of IPTO and DARPA. Congressional hearings after 1971 contain more references to IPTO "testbed" programs involving the military, such as the packet radio and packet satellite projects.

The attitudes of the Nixon administration and the public toward defense programs, reactions in Congress to the DOD's funding of non-mission R&D, the shift of resources out of defense to more socially oriented programs, the ultimately successful attempts to extricate the country from the Vietnam War, and the perceptions of lessened flexibility inside IPTO all contributed to making many members of the computing community averse to joining IPTO. A successor had not been found before Roberts's exit date arrived in 1973. Finally, Stephen Lukasik, DARPA director, asked Licklider to return to IPTO for another term as director. Licklider arrived in January 1974.

Allan Blue, a program officer in IPTO from 1967 to 1977, couched his perceptions of changes at IPTO in these years in a description of the political and budgetary climates. He believed that, because of these changes, the leadership style needed for effective operation in the early days would not have been appropriate in his last years there. In the beginning, the aim was to find talented people, provide ample money for programs, and expect good results from managers. By 1977, Blue believed, "trying to do just good, basic research for the joy of it was extremely difficult, if not impossible, to get through the budget process."[35] According to Blue, directed research became the standard, and relevance tests were applied. Patrick Winston, director of the MIT Artificial Intelligence Laboratory, shared Blue's evaluation:

> In recent years [the 1980s], DARPA has not funded entire laboratories as centers of excellence, but has, instead, supported specific projects within laboratories. . . . During the 1970s, . . . there was tremendous pressure to produce stuff that looked like it had a short applications horizon. . . . the mid 1970s were days in which you had to find a way of explaining the work in applications terms. It wasn't enough to study reasoning; you had to talk about how it might be applied to ship maintenance or something.

Winston stated that he encouraged some of his colleagues in MIT's AI Laboratory to design projects that would meet these new IPTO criteria. "I was seeking to find intersections between what the laboratory was doing and what DARPA either was interested in or could be persuaded to be interested in. So in some cases, it was a matter of pointing out the potential applications of a piece of work."[36] Responding to congressional concerns about DOD support for external R&D, the DDR&E office was largely responsible for this changing emphasis in DARPA and IPTO R&D support programs. While this change in emphasis was taking place in IPTO, Licklider served his second term as IPTO director, from early 1974 through mid-1975.

After a stormy second term, Licklider was not succeeded by a member of the computing community. Instead, David Russell, a physicist who had received a Ph.D. degree in physics from Tulane University in 1968, was asked to take on the position. Russell's dissertation research had been a study of the energy levels of the element erbium using the Mössbauer effect technique. As an officer in the U.S. Army, Russell had been sent to Vietnam for a tour of duty. Later he received orders to report to DARPA for service in the Nuclear Monitoring Research Office. George Heilmeier, DARPA director, transferred Colonel Russell to IPTO in 1974 to serve as deputy director under Licklider; and the following year Russell succeeded Licklider as director.

The next office director was the dynamic and resourceful Robert Kahn, whose destiny it was to raise IPTO to new levels of importance. He joined IPTO as a program manager in 1972, having received his Ph.D. degree in electrical engineering from Princeton in 1964. Kahn oriented his dissertation on sampling and representation of signals toward applied mathematics and theory. While at Princeton, he was also part of the Bell Laboratories traffic group, which was responsible for the plant engineering of the Bell System. The group studied queuing theory, analysis of switching performance, and global engineering of the system. After graduation he joined the MIT faculty as an assistant professor of electrical engineering and pursued research in the Research Laboratory of Electronics (RLE) at MIT for two years. In 1966 Jack Wozencraft, professor of electrical engineering and a member of RLE, suggested that he take a leave to acquire some practical experience. Kahn decided to spend the leave at Bolt Beranek and Newman.[37]

Kahn accommodated easily to his new assignment, a trait he often exhibited during his career at IPTO, and began working on problems of networking computers. "I was largely involved in design of various techniques for networking, error control techniques, flow control techniques, and the like." He set out to expand the use of networks to link as many users as possible, as effectively as possible. He knew little about IPTO beyond its funding of Project MAC. Early in the next year, Jerome Elkind, the BBN division director to whom

Kahn reported, suggested that Kahn let the people at IPTO know what he was working on in networking. Kahn wrote to Roberts indicating his interest in networking and suggesting that this might be of interest to IPTO, and Roberts invited him to Washington to talk about his ideas, thus beginning the long relationship between Kahn and IPTO. "I found out at that point that [Roberts] was actually interested in creating this net. Having been a mathematician or theoretician, it really had not occurred to me at that point that I might ever get involved in something that could become real!" Kahn became peripherally involved in the design of specifications for what eventually became the ARPANET. However, he played a major role in BBN's obtaining the contract for the detailed design and the building of the ARPANET.[38]

During 1972, just before he joined IPTO, Kahn arranged for a public demonstration of the ARPANET. More than one thousand people attended the live, hands-on demonstration. While planning the demonstration, Kahn and Steve Levy were also planning a new commercial initiative for BBN—Telenet—to exploit the packet-switching technology developed for ARPANET. Levy became its first president. When Roberts left IPTO in 1973, he became Telenet's second president. A few months after the successful demonstration, Kahn joined IPTO as a technical program manager to run a proposed flexible automated manufacturing program, but the program was not funded by Congress and Kahn had to adjust to other duties.[39]

Of all the people to join IPTO, Kahn had perhaps the greatest prior preparation in IPTO matters. The development of the ARPANET had put him and his coworkers at BBN in touch with virtually every one of IPTO's contractors. For the next seven years he continued to work with this community. In 1979 Kahn became IPTO director and stimulated, planned, and oversaw a fivefold increase in the office budget which was driven by a number of new program directions initiated by the Reagan administration's defense buildup of the 1980s.

The last director of IPTO was Saul Amarel. In a sense, Amarel was the ideal candidate to direct IPTO in the mid-1980s. He had experience in both industry and the academy; and his research field—artificial intelligence, especially problem-solving systems—complemented IPTO programs for the increased exploitation of AI results in many areas associated with the 1980s defense program. When he arrived at IPTO, Amarel already had substantial and significant research experience in artificial intelligence behind him. He had obtained his degree from Columbia University in electrical engineering in 1955 and spent the next ten years directing the Computer Theory Research Group at the RCA Laboratories in New Jersey. Amarel focused his work on problem solving by machine. Supported by AFOSR, he investigated theorem proving in the propositional calculus; knowledge representation issues; and theory formation by machine. In 1969 he moved to Rutgers and organized the computer science department there. During sabbatical leaves, he worked with people at CMU and Stanford University. While on sabbatical from Rutgers in 1985 and working with Herbert Simon at CMU, Amarel was approached and agreed to head IPTO.[40] In the DARPA atmosphere of the 1960s, he might have become known as a great IPTO director. However, in the 1980s his success was limited by the expansionist tendencies of other DARPA offices that wanted to cash in on the enlarged computing program, an effect that was to continue later beyond the boundaries of DARPA and even DOD. In the end, IPTO was restructured in 1986 and became the Information Sciences and Technology Office (ISTO) under the direction of Amarel; IPTO per se ceased to exist.

IPTO had seven directors during its twenty-five-year span: Licklider, Sutherland, Taylor, Roberts, Russell, Kahn, and Amarel. Although it kept its lean administrative style, its staff gradually increased, and as its programs became more sophisticated, program managers were added to assist the directors.

Staffing IPTO: Program Managers

By the late 1960s the office staff had grown to accommodate changes in direction and increases in the IPTO budget. Program managers were hired to manage specific programs within the office, with the major selection criterion being their technical competence. In 1972 there were five professionals in IPTO — Roberts, Kahn, Stephen Crocker, Bruce Dolan, and John Perry. Allan Blue managed the office's business aspects. In the second half of the 1970s the number of professionals continued to increase, although the comings and goings make it difficult to cite a stable number at any given time. David Carlstrom managed the artificial intelligence program from 1974 to 1978; Vinton Cerf assumed responsibility for the interconnection of networks in 1976 and stayed until 1982; Floyd Hollister served in IPTO at the end of 1976 as manager for the Advanced Command and Control Architectural Testbed (ACCAT) and its related programs and departed in 1978; Duane Adams arrived at IPTO in late 1977 to work with programs in packet speech and very large-scale integration among others, and left in mid-1983; and Robert Engelmore came to IPTO in March 1979 and specialized in intelligent systems at IPTO until mid-1981. Cerf had a Ph.D. degree from UCLA; Hollister had a degree from the U.S. Navy Post-Graduate School; Engelmore had received his Ph.D. from CMU; and Adams had earned his Ph.D. at Stanford. Each man was relatively young when he joined IPTO.

Ronald B. Ohlander came to IPTO to manage the programs in artificial intelligence in 1981 and left in 1985; Paul Losleben arrived at IPTO in mid-1981 to manage the VLSI program and continued into the Strategic Computing era, leaving in late 1985; Stephen Squires came to IPTO in 1983 at the opening of Strategic Computing, had responsibility for software and architectures, and is still with DARPA;[41] and Robert Simpson Jr. came to IPTO in 1985 to manage a part of the AI program, which was growing under the Strategic Computing program, and left in 1990. Ohlander had received a Ph.D. degree in AI from CMU and an MBA from the University of Oklahoma and Simpson a Ph.D. degree in AI from Georgia Institute of Technology, both after military experience—in the navy and the air force, respectively.[42] Squires had an electrical engineering degree from Drexel University and had spent some time at Princeton, Harvard, and the National Security Agency before joining IPTO. Losleben had earned degrees in electrical engineering and technology and management from George Washington and American Universities, respectively. These people are representative of the more than three dozen program managers who served in IPTO between 1968 and 1986. In 1984 IPTO had eight program officers, including the director, managing thirty-eight programs in fifteen areas. Seen in light of the IPTO budget of $154 million, IPTO administration remained very lean throughout the office's history.

IPTO program managers needed to have other abilities as well as technical competence. The programs had large budgets, and projects at educational institutions were larger than those of most extramural basic research programs supported by other government agencies, such as NSF. Roberts discussed the criteria he used to select program managers who would enable IPTO to preserve a lean structure:

> I looked at how the military [managed] a lot of their [research] programs. I looked at their effectiveness. I found that they were able to hire a person of a certain caliber in the office; maybe a colonel or whatever. . . . that person was effectively in charge of a budget which might be $200,000 or something. He then doled that out in $10,000 pieces. . . . None of this was very effective, however, because he really only had three people working for him. So he accomplished virtually nothing, and spent a lot of time and effort doing it. Therefore, they could not get anybody who was better than a three-person manager to do that job, and his decisions were no better than that. . . . What I wanted were people who were really managers—like they would be in industry—substantial managers with substantial groups. To do that I had to give them the budget individually, not to their staff, not spreading it around to a lot of staff. So I took the staff down to a minimum

and had just two or three people in the office, and each person was handling ten million dollars or something, so that [each person] had a substantial capability. . . . the concept was that you really want to bring somebody in who can manage at that level.[43]

Personnel handling the processing of forms and the financial aspects of contracts in DARPA and the DOD made it possible for the young office directors and program managers to learn how to manage programs of this magnitude. They had to be able to insist on major objectives and be confident that contractors would carry out the objectives in a responsible and timely manner.

The program managers were usually participants in IPTO-sponsored projects before their arrival at IPTO, and some had held very responsible positions, even as graduate students. They came to IPTO with essentially the same philosophy as that held by the office directors and undertook their new duties at IPTO with enthusiasm.[44] They believed they were involved in a very significant undertaking, one that meant much to computer science and engineering, the institutions associated with IPTO, and ultimately the defense of the nation.

Program Development and Interaction with the Research Community

The impetus for program development came from four main sources. First and foremost was the substantial knowledge of computer science which directors and project managers brought to their new positions. Most IPTO directors were accomplished researchers in computer science. And IPTO directors, through extensive contacts with scientists at other laboratories, already had a broad view of computing before they came to the office. Moreover, for eighteen of the twenty-five years covered by this history, the directors came from the Cambridge-Boston area, where many innovative computer projects had occurred, including many for military systems, starting with the Whirlwind project at MIT and ranging through the R&D activities of MIT, Harvard, Lincoln Laboratory, the Instrumentation Laboratory, and the MITRE Corporation. Their own knowledge of research in computing proved invaluable in program development.

The second source of program development was the research community. IPTO worked very closely with the research community to generate research ideas and maintained extensive professional contact. Among Licklider's first actions as IPTO director was a tour of the country's major computing centers. From this tour he learned the limits of what researchers could and would do. He also identified researchers in new areas of investigation of interest to IPTO and took steps to bring them into the IPTO fold.[45] Taylor made similar visits to contract sites "once or twice a year, sometimes more often," [46] as did other directors. Other government agencies have also used site visits with varying frequency to stimulate programs. For example, the Office of Naval Research used site visits in the late 1940s. Mina Rees, who from 1946 to 1953 directed the ONR mathematics program, which included some computing activities, took a national tour as one of her first acts.[47] IPTO's site visits, however, were more frequent than those of other agencies and became an integral part of the office's continuing evaluation of progress and staff, a way to obtain ideas for new programs, and a mechanism for influencing subsequent research at the site. These visits took different forms. At times the office director would visit a group alone; at other times he was accompanied by other IPTO staff members or the DARPA director. IPTO occasionally engaged outside consultants to help in evaluations. For example, in the 1960s, Licklider twice assembled a group for site visits to the System Development Corporation's facilities in California. One of these visits was for a meeting on the topic of time sharing, in an effort to broaden the SDC staff's horizons about the subject.

The early IPTO program contracts illustrate the close connections between IPTO and the research community. Licklider's early program was representative both of his view of computing research needs and of that of the Cambridge computing community. (See table 4 for a summary of the early IPTO program areas and emphases, and table 5 for a summary of the later ones.) Led by this vision, expressed in his paper "Man-Computer Symbiosis," Licklider encouraged able researchers from Cambridge and elsewhere to propose projects and programs that would change the nature and style of computing. He took substantial interest in his two major projects: Project MAC, at MIT, which he initiated; and a project at SDC, which he had inherited. His involvement with MIT seems to have been mostly to encourage the activities in computing at that school and to offer Robert Fano, the first director of Project MAC, strong support for project expansion and change. SDC was another matter. Licklider took a much closer and more proprietary view of the work at SDC on that corporation's DARPA contract. He wrote critical reviews of results, took a hard line on renewal requests, sent in consultants to encourage changes, and refused to support some proposed activities.[48] Often Licklider and his successors received ideas that expanded the research program, such as the University of Utah proposal presented to Sutherland by David C. Evans which led to substantial IPTO-supported R&D efforts in graphics. These close connections between the office and the research community continued in the 1970s and are still strong today.[49]

Table 4   IPTO Program Areas and Specific Emphases,
Fiscal Years 1965 and 1971

Fiscal Year 1965
Research into command and control processes, particularly human-machine interaction
Computer languages
Advanced programming techniques
Time sharing
Advanced computer systems and their organization as related to defense problems and computer usage
Advanced software techniques

Fiscal Year 1971
Information processing techniques
Automatic programming
Picture processing
Intelligent systems
Speech understanding
Distributed information systems
Networks
ILLIAC IV
Climate dynamics

Source: Compiled from budget data presented by IPTO to House and Senate subcommittees on Department of Defense appropriations.

The third source of program development was various types of meetings of IPTO project research personnel, which began very early in the office's history. Licklider sought advice with some regularity, but informally:

> if I was going to be successful, I had to have some kind of system here. Maybe have one place interact with another, get these guys together frequently. . . . We would get our gang together, and there would be lots of discussion, and we would stay up late at night, and maybe drink a little alcohol and such. So I thought I had a plan at that level. I could talk about it to people at ARPA.[50]

Taylor saw wisdom in Licklider's approach, but he went further than assembling ad hoc groups of principal investigators (PIs) at professional society meetings and called special meetings of the PIs and their research teams. PI meetings began in the mid-1960s, dropped off in the mid-1970s, and were resumed in the 1980s and continue today under the auspices of ISTO. As an ex-

Table 5 IPTO Program Areas and Specific Emphases,
Fiscal Years 1975 and 1980

Fiscal Year 1975
Computer and communications sciences
Image understanding
Intelligent systems
Advanced memory technology
Speech understanding
Advanced C^3 Technology
Distributed information systems
Parallel-processing applications
Distributed networks
Climate dynamics
Secure systems
Software technology
Speech processing

Fiscal Year 1980
Computer and communications sciences
Intelligent systems
Advanced digital structures and network concepts
Integrated command and control technology
Distributed information systems
Advanced C^3 technology

Source: Compiled from budget data presented by IPTO to House and Senate subcommittees on Department of Defense appropriations.

ample, forty-six people attended the January 1973 meeting, six of whom were from IPTO. Project leaders from among the contractors made thirty-eight presentations about their projects.[51]

> I would ask over a period of a few days, each principal investigator to get up and give an hour or two description of his project, of the sort of description he'd like to hear from each of the other projects. I think those meetings were really important. Through those meetings these people, all of whom were pretty bright, got to know one another better. I got them to argue with one another, which was very healthy, I think, and helpful to me because I would get insights about strengths or weaknesses that otherwise might be hidden from me. Through these we could look for opportunities where one group might be able to help another in working on some technical problem, or making some contact with some resource for some kind of supplier.[52]

The meetings were a venue for established people in the field to exchange information. The entire group then discussed these matters intensely, which resulted not only in the formal transfer of research results and methods but also in camaraderie that aided informal exchange among the participants and is often cited as an important feature of the IPTO-sponsored research community. Today, however, the number of participants is in the hundreds, and the PI meeting is now more like a typical professional society meeting. "If you go to the PI meeting today [1989], it is a big, mass affair, where you hear 'big' people speak. Participants complain that the earlier sense of give-and-take is no longer possible with such a large group."[53] DARPA is now experimenting with new structures for PI meetings to recover the "small" feel of the earlier meetings.

Edward Feigenbaum believes that the early PI meetings served two important functions. First, they encouraged a broad exchange of information about projects of interest to IPTO, and brought together a number of the most creative researchers in the field to discuss results, plans, and problems. These researchers ranged across all the areas of IPTO support: time sharing, networking, graphics, and AI. Common problems were easily isolated; techniques were freely shared. Second, the meetings served as an evaluation of program progress, allowing IPTO staff to plan for subsequent years.[54]

Taylor also held the view that many of the innovative ideas emanated from the graduate students in these contractor projects, so Barry Wessler, an IPTO program manager, convened special meetings for the graduate students and ran them in the same fashion. Taylor intended this meeting to mirror the annual PI meeting. Vinton Cerf, later an IPTO program manager, remembered attending one of these meetings as a graduate student: he "met a lot of very smart people."[55] These meetings helped to build the newly developing discipline. They enabled the students to become part of the professional community early in their careers and provided connections that not only benefited IPTO but were helpful to the students for their entire professional lives. Taylor believed that both of these types of conferences were important to the health and quality of the program.

We have emphasized that meetings aided the community and IPTO in the transfer of research results and were excellent venues for monitoring the progress of the groups. Technical weaknesses in research programs became apparent as researchers sat around trying "to poke holes" in each other's presentations.[56] The presence of all the IPTO program managers and the office director at these meetings illustrated in a forceful way IPTO's interest in the research progress and results and a concern for making the results as useful as possible to other contractors as soon as possible. Moreover, if at any time an office director or program manager believed that "something was a key issue" on which "a lot was going to hinge, then he got right down into the nuts and bolts, into the details"[57]—a technique that is most clearly illustrated by Roberts in the networking and ILLIAC IV cases and by Kahn in the VLSI and Internet programs. These interactions between IPTO personnel and members of the community ran relatively smoothly because a great deal of respect and judgment existed on both sides. Severo Ornstein, a member of the research community, noted that he had "enormous respect for the people in charge of it [IPTO] because I thought they were *very* bright people. . . . These were people like ourselves, researchers."[58] Another member of the research community, Howard Frank, put it another way. "The philosophy [of IPTO] was more of stimulation of a group of people [rather] than telling them how to manage the work." The two sides functioned as a "group of independent professionals working towards a common mission."[59]

Throughout its history IPTO encouraged the organization of meetings on special topics, occasionally even arranging meetings. Reflecting his knowledge of the field, the office director or a program manager addressed the group early in the meeting to present the IPTO view on current problems or needs. Sutherland described ten significant problems in graphics he thought needed attention at a graphics meeting at the University of Michigan in 1965.[60] IPTO provided some support for a conference at the National Physical Laboratory in the United Kingdom in 1965 which promulgated time sharing. Many participants in IPTO time-sharing research projects in the United States attended, and IPTO funded their travel. The largest body of experience in the implementation and use of time-sharing systems, according to Donald Davies, the symposium organizer, was in the United States and "to a considerable extent in projects supported by ARPA." Papers were presented by Richard Mills and Jack Dennis of MIT; Jules Schwartz and Clark Weissman of SDC; Jerome Elkind of Bolt Beranek and Newman; J. Clifford Shaw of Rand; Roberts, then of Lincoln Laboratory; Alan Perlis of CMU; and Sutherland, representing IPTO.[61] During this trip Sutherland lectured on computer graphics at the universities of Cambridge, Manchester, Edinburgh, and London. He arranged for Roberts to visit these universities to discuss his IPTO-supported research.

As the number and size of programs grew and general PI meetings became unwieldy, individual program managers conducted PI meetings concentrated on their specific program area. This remains the pattern for IPTO's successors today. Robert Simpson, program manager for artificial intelligence research, did this repeatedly in the 1980s before groups separately assembled to discuss image understanding and case-based reasoning.[62] These meetings enhanced proposals, generated new, unsolicited proposals, and occasionally, as we shall see in subsequent chapters, created or modified IPTO programs. An example in the 1980s was the series of workshops on image understanding. During a workshop in February 1987, fourteen ISTO principal investigators from USC, SRI, MIT, Columbia, and Stanford; the Universities of Rochester, Massachusetts (at Amherst), and Maryland; and Hughes Aircraft Company, General Electric, Honeywell, and Advanced Decisions Systems presented project reviews. These presentations were followed by eighty technical reports.

The ARPANET had the effect of tightening the coupling between IPTO and the contractor community. Allen Newell and other PIs had frequent contact with IPTO through the network, sometimes involving requests that had short deadlines for response.[63] During Licklider's second term at IPTO (1974–75), he kept the community fully informed of what transpired in DARPA with messages sent across the network. For example, in an e-mail message to the community in April 1975, he noted

> a development in ARPA that concerns me greatly—and will, I think, also concern you. It is the continued and accelerating (as I perceive it) tendency, on the part of the ARPA front office, to devalue basic research and the effort to build an advanced science/technology base in favor of applied research and development aimed at directly solving on an ad hoc basis some of the pressing problems of the DOD.

Licklider went on to describe his interactions with the new DARPA director George Heilmeier. "The problem is that the frame of reference with which he enters the discussions is basically quite different from the frames of reference that are natural and comfortable, and familiar to most of us in IPTO—and, I think, to most of you." He noted the "fundamental axiom" that natural language processing, networking techniques, the use of AI in interpreting photographs, and large-capacity computer memories would all eventually be exceedingly important and useful to the military. This "axiom" was not uppermost in Heilmeier's mind, according to Licklider. He reported to the community that DARPA management's question was "who in DOD needs it and is willing to put up some money on it now."[64] Incidentally, the focus of these comments eloquently attests to Licklider's view of where his loyalties lay—not with the head office, but with the community.

The approach to the development of the ARPANET exemplifies another use of PI meetings, as well as the value of working groups in the definition of programs and projects. At the 1967 PI meeting, Roberts and the assembled researchers discussed the development of a plan for a networking project. They established working groups to design a standardized communications protocol and specify network requirements. These working groups discussed network plans throughout 1967 and were helpful in bringing the first few nodes into operation within two years.[65]

In addition to knowledge of computer science, knowledge from the research community, and a variety of meetings, the fourth source of program development was study groups organized by IPTO to investigate specific topics. These groups evaluated the state of a subfield, determined where substantial advances would be most advantageous, and suggested new program definitions. One of the most prominent of these task groups studied speech understanding in 1970.[66] Wanting to encourage further development in research on speech understanding, Roberts examined developments in computing power of machines and the results of research in artificial intelligence which might be applicable. He concluded that if work in speech understanding were pursued over the next five years, it would be possible to build machines with computing power sufficient to achieve results in this area.[67] In the spring of 1970, Roberts assembled a study group led by Allen Newell "to consider the feasibility of developing a system that would recognize speech in order to perform some task."[68] Not only did this group craft a five-year program to pursue speech research, but at IPTO's request it remained assembled, with the PIs added, as an evaluation committee for the duration of the program. It applied the criteria for progress outlined in the 1971 study report to the results of projects, and in 1976 prepared a follow-on report calling for continued research. However, for reasons of budgetary priority IPTO did not fund a follow-on until almost a decade later.[69]

In all these ways, directors and program managers consistently strove to interact with researchers sponsored by IPTO. Through meetings of various kinds, IPTO directors sought to have the researchers interact with each other and disseminate research results to other parts of the computing community. Their attitude toward the dissemination of research results within the professional community came from the IPTO staff's own close connections with the research community. The directors were researchers themselves, and they strongly desired to aid the community in its work in order to achieve the IPTO mission. We pointed out at the beginning of this chapter that IPTO directors had the confidence of their superiors. We turn now to a consideration of the interaction between IPTO directors and DARPA management over the years.

Interactions inside DARPA

Interactions between IPTO directors and DARPA directors varied, though it is not clear whether this variation was due to changes in DARPA or to the directors' individual personalities. For example, Jack Ruina, DARPA director from 1961 through 1963, reported that he did not see Licklider very much, "only once a month, at most." Computing was something peripheral in comparison to "the day's hot issues" — ballistic missile defense and nuclear test detection. The people directing these latter programs met with him perhaps "five times a day on different issues." Ruina's management style caused him to want to leave competent people alone, "especially those people I had confidence in."[70] Licklider remembered it similarly:

> As a manager, it seemed to me that he had spent long enough [with me] to decide he understood what I was trying to do, and that I would probably work hard to do that, and then just wanted to have a report periodically — was I on track or not? He had much bigger fish to fry; this was a small part of his life. I have told him since that I've felt that there was a kind of benign neglect.[71]

Two years later, DARPA director Charles M. Herzfeld's contacts with Sutherland and Taylor were more frequent. In Sutherland's case, some of this could be attributed to his youth. Herzfeld revealed later that the increased attention he gave to computing can be explained by his broader view of what computing could do for the other programs of DARPA and for DOD.

> I thought that my job vis-à-vis the computing program, when I was deputy director and director, was first of all to make sure to get the best office directors we could get, second, to help them do what they thought they needed to do, and third, to look for applications across the projects for computing, where you apply the new computing ideas, technology and its capabilities as widely as possible.[72]

He saw such a need when he was director of DARPA's Defender program. The Defender Project office sponsored anti-ballistic missile work that involved the development of new instruments — advanced radar, in particular, which took "data out of the air, literally, at an unprecedented data rate." They wanted "to get these data analyzed in as real time as possible, which drove us to the latest computing techniques." Nevertheless, even under Herzfeld, IPTO directors still had, in the words of Robert Taylor, "enormous freedom," and did not have to skew programs in the direction of other DARPA offices' needs.[73]

Sometimes DARPA directors were generous in adding extra funds to the IPTO budget for new programs previously unbudgeted. The front office looked kindly on IPTO when Taylor proposed to start the network developments that culminated in ARPANET. He took the network proposal to Herzfeld and gave him a number of reasons why he thought IPTO should fund the concept. Some of the reasons, such as the ability of a computer network to fail "softly"—that is, the processor continues to operate even though some part of it has failed—were in the interest of the Department of Defense, but they were not the basis on which Taylor made the argument, as we will see in chapter 4. Nor did Herzfeld ask specifically for a defense rationale. Instead, in the course of a single meeting, Herzfeld concluded that the concept was a good one and should be funded by DARPA. He took money from a lower-priority program and authorized Taylor to use close to five hundred thousand dollars in that fiscal year on the network project.[74] New funds for this activity were obtained from Congress in the following fiscal year. This was a common practice in DARPA: to begin an activity with "bootlegged" funds and appeal later to Congress to continue the activity.[75]

Herzfeld's attention to and interest in IPTO is typical of the DARPA directors in the 1960s and early 1970s. Largely from scientific and academic backgrounds, they at least appreciated the value of the computer as a tool in scientific and technical research. Jack Ruina (1961–63) was a principal agent in the organization of IPTO. As a former deputy secretary of the air force, he understood that the armed services were in need of help in evaluating both their computing needs and the technology to meet those needs. Although Ruina's successor, Robert L. Sproull (1963–65), was a physicist who had directed laboratories for atomic and solid state physics and the materials science center at Cornell, laboratories that were beginning to invest heavily in computing equipment for research, he divided activities in the DARPA front office with his deputy Charles M. Herzfeld, who eventually succeeded him (1965–67). Herzfeld was also a physicist, and had been at the National Bureau of Standards before joining DARPA's Defender program. As deputy director, Herzfeld had responsibility for IPTO and was very interested in the area. He also had a strong interest in how computing developments could serve defense needs.

Herzfeld was succeeded by Eberhardt Rechtin, an industrial engineer, whose interests were more in the exploratory development area than in basic research. He followed Sproull and Herzfeld's scheme of dividing responsibility for DARPA between himself and a deputy director. In this case the deputy director was Stephen J. Lukasik, a physicist trained at MIT, who had industrial (Westinghouse) and academic (Stevens Institute) experience. When he joined the DARPA Nuclear Monitoring Research Office it was across the corridor from IPTO. As he put it in a later interview, it was inevitable that there would be significant contact between members of the two offices. Moreover, he said that the IPTO people "happened to be some of the smartest people in the Agency," so others naturally gravitated to IPTO.[76] Lukasik assumed responsibility for category 6.1 research under Rechtin and developed more than the usual interest in IPTO R&D. When he was DARPA director (1967–70), he maintained that interest.[77]

While Ruina, Sproull, and Herzfeld appreciated the possible contributions that IPTO programs could make to defense system developments, the results emerging from the programs were insufficient to be transferred elsewhere in the DOD. This began to change in the late 1960s, when Rechtin received from the DDR&E the task of seeing that more results were transferred from DARPA to other defense agencies. Under Rechtin, and later Lukasik, this demand for greater transfer had little affect on IPTO until the late 1970s.

By the mid-1970s attitudes in the White House and Congress had changed, and a different type of person emerged in DOD to take over the department's R&D operations. George H. Heilmeier is representative of this group. He went from industry R&D to government at a time (1971) when the economy was causing scientific management personnel to ask more questions about the relevance of research to a company's interests. Heilmeier was used to asking about any potential short-term payoff from a research project. He valued university research, but like many in the DOD during the Nixon and Ford administrations, he believed that NSF and other federal agencies should support it. This led him to question the relevance for the DOD of academic research. His predecessors had asked similar questions, but much less often. Heilmeier wanted IPTO's programs to be evenly balanced between basic research and applications.[78] While he was trying to refocus IPTO to this view, he appointed Colonel David Russell as IPTO director. By the end of his tenure as DARPA director at the end of 1977, Heilmeier had made applications R&D more prominent in IPTO; and since that time IPTO and its successor organizations have been more applications-oriented.

Besides his orientation to applications, Heilmeier had a very different operating style than his predecessor, Stephen Lukasik. During Lukasik's directorship, Roberts, or any other office director, could go into Lukasik's office and converse about a program plan without having a formal review. Heilmeier, by contrast, took a formal approach to project review. He preferred items in writing, took his own counsel about whether to fund a project, and sustained more interest in short-term results.[79] This formal approach required that more memoranda be passed between IPTO and DARPA and lengthened the time it took to make evaluations and award contracts. After Heilmeier assumed control of DARPA, during Licklider's second term in IPTO, the office became a very different place: a place that was more intense and more hectic. Allan Blue, a program officer at IPTO at the time, noted that "Lick just was not up to the same speed as these guys [Heilmeier and his aides]." Heilmeier's requests for specific information from Licklider left Licklider convinced that DARPA had changed since his previous term as director. Blue remarked that he thought Licklider had been "pretty taken aback by this" but that he "eventually got into the system." [80]

In contrast, Robert Kahn thrived in the new environment. Moreover, he was an old friend of Heilmeier's; they had been in graduate school at Princeton together. Kahn's interactions with DARPA directors Fossum and Cooper were similar to those of Taylor with Herzfeld. When Kahn assumed the directorship of IPTO in 1979, he understood the new attitude in DOD toward R&D and believed he could use it to IPTO's advantage. He considered his first task to be the development of a new program that would help rebuild university R&D support, which had slipped during the 1970s. To achieve budget growth, he could then stress applications. With the support of DARPA director Robert Fossum, Kahn designed a VLSI architecture program that achieved these objectives. Kahn also had an advantage working with Fossum's successor, Robert Cooper. When Cooper took over DARPA in 1981, he "was looking for major new things to do." Together with Cooper, Kahn and his staff prepared the new program in Strategic Computing, noting that it would be an opportunity to involve industry to a greater extent in IPTO programs.[81] Kahn and Cooper orchestrated major increases in DARPA computing programs through Strategic Computing; many of these programs became the responsibility of IPTO. The budget expansions in Strategic Computing were devoted mostly to development and industrial programs, but funding for university programs also increased (in constant dollars) over levels seen in the mid-1970s.

In spite of these differences in style and emphasis, DARPA directors rarely overturned decisions from IPTO. It was in the interest of DARPA directors to strengthen the programs and proposals, and they took this side of their management responsibility seriously. Strengthening programs and proposals was done in the framework of budgeting and contracting cycles that remained relatively stable over the years. Stable budgets allowed programs to span many years, although DARPA worked mostly in five-year cycles. IPTO budgets increased slowly until the mid-1980s, when they rose dramatically.

Budgeting and Contracting

The DoD, IPTO, and the Budget Process

One of the distinctive features of IPTO over the years has been the size of its budget, which allowed IPTO to fund the institutions within its orbit consistently and copiously. Such long-term funding of centers of activity has often been cited as perhaps the most important reason, after personnel selection, for IPTO's success in the support of R&D. The capability to fund institutions with large contracts resulted from IPTO's being part of the DOD budget framework.

Although budget practices change over time, especially in complex organizations such as the DOD, a glance at the budget categories of 1970 is sufficient for our purposes. These categories are illustrative of DOD budgets throughout IPTO's history. The entire budget contains a group of titles that reveal in a general way how DOD spends money: for example, in categories defined as research and development, strategic forces, and logistics. The group of categories is known as the DOD Programming System. From the mid-1960s on, the major headings have been the following, with minor variations:

Program 1: Strategic Forces
Program 2: General Purpose Forces
Program 3: Specialized Forces
Program 4: Transportation
Program 5: Guard and Reserve Forces
Program 6: Research and Development
Program 7: Logistics
Program 8: Personnel Support
Program 9: Administration [82]

The Research and Development Program, category 6, the only category of significance for this study, is itself divided into six parts:

6.1. Research, which includes all basic research, and applied research that is directed toward expanding knowledge in the several scientific areas
6.2. Exploratory Development, which includes studies, investigations, and minor development efforts ranging from applied research to sophisticated breadboard hardware and is oriented to specific military problem areas
6.3. Advanced Development, which includes all projects for the development of hardware for experimental testing
6.4. Engineering Development, which includes development programs in

which items are engineered for military use but have not been approved for procurement or operation

6.5. Management and Support, which includes the overhead expense for the other subdivisions of research and development

6.6. Emergency Fund, which is available for use in any category at the discretion of the secretary of defense[83]

Category 6.1 contained research projects similar to those in basic research supported by NSF and ONR. In the 1960s, all funds used in the IPTO program were from that category. Substantial similarities exist between the category 6.2 projects and the applied research supported by NSF.

In 1962, IPTO began with a $9-million budget, all in category 6.1. By FY 1968, an increase in the range of programs to include parallel processing, graphics, and networking resulted in a larger budget of $19.7 million. In an effort to capitalize on the results of projects in the 1960s, IPTO's program in the early 1970s added to its basic research (6.1) programs activities in exploratory development (6.2). Congress appropriated $23 million for IPTO in the DARPA appropriation for FY 1970. The increase represented increases for Project MAC at MIT, computer-augmented management at SRI, heuristic programming and theory of computation at Stanford University, the ARPANET project at BBN, the experimental communications laboratory at the University of California at Santa Barbara, and an intended doubling of the effort on ILLIAC at the University of Illinois. When asked why the increase was needed, Eberhardt Rechtin, DARPA director, replied: "The increase . . . consists mainly of a funding increment for university contracts to permit improved planning and operating stability. In some instances, we were unable to fund for a full year's effort at certain universities in fiscal year 1968 and fiscal year 1969. Smaller increases occur in funding for the ILLIAC computer and the ARPA computer network."[84] By FY 1971, the 6.2 budget category for IPTO was $11 million. This money was used mostly to further ARPANET development and to complete the ILLIAC IV.

After FY 1971, increases in IPTO's budget came mostly in category 6.2, and by the end of the 1970s the 6.2 budget category in IPTO was considerably greater than the 6.1 category and supported a range of programs.[85] By FY 1980, the IPTO budget had reached almost $51 million. While IPTO budgets rose slowly over the years, the overall DARPA budget remained relatively constant. DARPA came out of this budget stalemate in the early 1980s, and IPTO took advantage of the increases in DARPA budgets. Kahn's proposal in 1982 for Strategic Computing is an example of a program that led to budget increases for both IPTO and DARPA. Kahn constructed the proposal to address a number of issues that concerned several DARPA offices, and in conjunction with DARPA director Robert Cooper he directly involved those other offices in the new program. In FY 1986, IPTO's budget climbed to more than $175 million.

The budget process in the early days of IPTO was relatively casual. In most years, DARPA directors acted on budget requests from IPTO personnel in a meeting that lasted some thirty to sixty minutes. As Taylor stated: "Never a half day!" When asked whether he had to defend his proposal against those of other offices, he replied:

> Not directly. No, not with other offices present. We'd go through each of our projects, line item by line item, with the director of ARPA, and the deputy director, and his program manager and tell what each research contract was about, and what their accomplishments had recently been, what their problems had recently been. Over the course of a year I'd try to get the director, and deputy director, and sometimes the program manager individually briefed on these projects by having people on the projects visit them in Washington. I'd take them in and they'd chat with the director of ARPA, the deputy director of ARPA, or both of them. Occasionally someone from the head ARPA office would visit one of our contractors. So I tried to make sure that the director of ARPA and his staff were not caught by surprise. One of the reasons that these budget meetings went as quickly as they did is because the director of ARPA and his staff were already familiar with most of these projects, and in some cases even familiar with the individuals who were working on them. That made the budget meetings run smoothly.[86]

In these years, the annual budget process occasionally began with a guideline figure from the DARPA director's office. Taylor noted that he built his budget by considering the number of proposals in hand, the commitments already made, and the quality of the research. Keeping the educational system in mind, especially the support of graduate students, he tried to preserve the multi-year commitments the office had made. No year contained huge changes. Over the three years Taylor was director of IPTO, the budget increased by a little less than one-third, from $15 million to $19.6 million.

Roberts described the planning and budgeting processes in the following way:

> I would work on preparing an order and think about what I was doing and what I wanted to do, and then I would work with them [DARPA Directors Herzfeld, Rechtin, Lukasik] on whether this would sell, whether they believed in it, whether they thought it was a good strategy. We had budget

reviews, and reviews of every project, and so they watched what was being done and how we were proceeding with it, and how we were getting the projects done.[87]

He went on to say that the DARPA directors he served with were happy to put more money into IPTO than into missiles. In other words, the idea of national significance was indirectly used as a selling objective to directors, but the selling objective used outside of DARPA was a program's significance for the military. At first sight, this might seem contradictory, but there was no contradiction in the minds of DARPA and IPTO personnel. A contribution to the country which included a contribution to the military is what IPTO directors sought.

At any one time, a program manager dealt with three annual budgets. Managers started with the year in which money was spent, added a plan for spending money in the following year, and estimated the money needed for the third year. This process might be more complex if the program had money left from a previous year. Consider the process in action in the 1980s. Program personnel worked with the guidance figures sent from the Office of Management and Budget (OMB) in September (say, September 1982). IPTO then prepared a detailed budget within these guidelines. OMB used IPTO's detailed budgets in preparing the president's annual budget for submission to Congress the following January (1983). This budget was for money to be spent after October of that year. Programs on which money would be spent in 1983–84 and which involved commitments for future years would need to be identified for the next several years. In the budget review documents, it was necessary to state how much money had been spent in the last completed fiscal year (1981–82) and to present the budget that had been passed for the current year (1982–83). Thus, DOD and OMB monitored at least five budget years simultaneously.[88] And the attention by IPTO to budgets increased considerably, and new support personnel were added to help with the task.

Contracting and Monitoring

IPTO directors adopted a set of practices for funding R&D which reflected the organization around them—DARPA and the DOD. Among the most interesting and effective characteristics of IPTO management were the nature and size of contracts, the time it took to award contracts, and decision making by program managers. In the beginning, the contracting process was relatively simple and straightforward, and the contracts written were brief. They specified in general ways the nature of the research to be pursued, any equipment to be purchased, and the duration of the contract. This type of contract was consistent with the emphasis on fundamental research in the early IPTO program.[89]

As with any government funding program, there were always more requests than could be funded. Many years later, Licklider discussed the number of requests to IPTO during his first term as director. "You can't imagine how many people, when they hear you have money, want to come to sell you something. Most of it was unbelievably irrelevant or low-grade."[90] As one of his main tasks, Gale Cleven, Licklider's deputy, visited with these people. "I just took advantage of Cleven and made him listen to all those things. He could be very charming; he was fantastically charming. His main job was just to make the visitors feel good and not give them any money, unless he could spot that they had something—which was one in 30 or 40!" Licklider was consistent in limiting the IPTO efforts to the "general area of interactive computing." One researcher came to see him about a project in the mathematics of digital system functions. "I didn't fund it, because, essentially, I would have clobbered myself, if I got to doing too many things I did not understand really well enough." In this instance Licklider knew the research well enough to know that it was good, but not well enough to know whether it was original. Instead of funding the project he put the researcher in contact with agencies that had money for mathematics.[91]

In interviews at the Charles Babbage Institute, all of the early IPTO directors commented that to reach IPTO objectives it was more effective and efficient to fund centers of activity. IPTO strove to develop a few centers of excellence in order to stimulate the computer science field in substantial ways, rather than to issue a large number of small contracts. This strategy was certainly used during the early years, when the emphasis was on research involving large-scale objectives and many researchers. Directors of IPTO also believed that it was more effective to manage a few large contracts than many small ones.[92] This attitude led to block grants that included many different activities, and grants to laboratories rather than individuals. Multiple-year contracts, which were also prevalent in other offices of DARPA at the time, were another feature of IPTO. Support continued for a sufficient period to sustain a coherent contractor activity, which might be a program with many parts such as MIT's Project MAC. IPTO contracts allowed adequate time for results and for assessment of high-risk areas, and they supported the continuation of a contractor's activity so long as the contract's intermediate results were satisfactory. The practice in multi-year contracts was to guarantee 100-percent funding for the three years of the contract; and even if full funding was not provided for subsequent years, that is, if the intermediate results were not satisfactory for whatever reason, the contractor was likely to receive 60 percent in the fourth year and 40 percent in the fifth year to cover closeout or to tide the contractor over until funds could be secured from other sources.[93] As the field of computer science matured and the interests of the DOD became more centered on applications, IPTO concluded that applications were more easily directed from a project focus. The shift to projects did not change the view about the size of contracts. The earlier centers of excellence, such as CMU and MIT, easily modified their proposals to fit this change to project support.[94] The same was true for the University of Utah, although its work continued to be confined to graphics. Stanford expanded its support from an emphasis on artificial intelligence to include semiconductor circuit design and production.

IPTO used two strategies in funding contracts: the cooperative strategy, which organized a number of contractors to work on a single objective (e.g., ARPANET, distributed computing, and security) and gave each a role to play in achieving the objective; and the competitive strategy, which put contractors in competition with each other. IPTO used the latter strategy in programs where several directions in research were to be tried because there was no single direction that would obviously produce the desired results. IPTO usually expected these competitive-strategy programs to have more than one five-year stage. If the program was to include a second stage, only some contractors would be involved in the follow-on stage. Examples are the speech-understanding program and the various computer architecture projects.[95] Each of these strategies included many high-risk projects, such as the ARPANET, ILLIAC IV, VLSI design, and Strategic Computing. The contracts for these projects included significant amounts of money to cover capital expenses as well as long-term personnel support. Contracts deviating from these strategies were occasionally made to accomplish a more limited objective, such as the development of the operating system TENEX. IPTO directors designed both strategies to facilitate the integration of results from individual projects.

To facilitate research, program officers kept the turnaround time for contracts short. People inside and outside IPTO remembered years later that it was often possible in the early years for IPTO to agree to a contract after hearing the details from the principal investigator and to present at least a letter of intent within a week.[96] For example, "we could get an idea in the morning and have the guy working under a Letter of Intent by 4:00 the next day."[97] Herzfeld remembers the process as even faster! "I used to say, ARPA was the only place in town where somebody could come into my office with a good idea and leave with a million dollars at the end of the day."[98] While the records show that such accounts are somewhat exaggerated, the process was relatively rapid.

The multi-million-dollar Project MAC is a case in point. MIT's official proposal was submitted on 14 January 1963. IPTO authorized MIT to initiate the project toward the end of February, and the project officially began on 1 March 1963. The actual ONR contract with MIT for Project MAC was dated 29 June 1963. A short implementation time was also seen when Keith Uncapher founded the Information Sciences Institute (ISI) at the University of Southern California in 1972. Approval of the ISI's multi-million-dollar, three-year contract took about thirty days.[99] Manager rather than peer review made possible this rapid evaluation of proposals and issuance of an intent to contract with an institution. This rapid turnaround of contracts was even more remarkable when one considers that DARPA did not issue contracts itself. DARPA offices used other government contracting units, almost always DOD units, as agents.

Increases in accountability required by Congress and the executive branch sometimes changed this rapid turnaround time into a lengthy negotiation process between IPTO and the contractor. In the 1970s, the first step of the proposal process was still the visit or early memorandum to IPTO to discuss project ideas. For large, multitask contracts, a visit to the organization might follow, or negotiations might be carried on by mail and telephone. During the visit or other interaction, IPTO and contractor personnel reviewed each aspect of the proposal. IPTO personnel would approve or disapprove each aspect. A September 1974 preliminary submission to IPTO from CMU serves as an example. The preliminary proposal was an outline of what the computer science department wanted IPTO to support in the following year. Licklider responded in early February with a twenty-seven-page criticism of the outline, suggesting what should be dropped, what added, and how the argument should be made in the formal submission.[100] The evaluation criteria in this and other cases were guided by how much of a technical advance seemed possible and how well the proposal would sell in DARPA and Congress.[101] The second step in the proposal process was the official submission and review at DARPA. This step usually involved close involvement of DARPA program management people, now that the technical people had evaluated the proposal. The third step was to secure the approval of the DARPA director.

During Taylor's and Roberts's directorships, the director could alter how IPTO's budget was spent by reducing or canceling contracts. These reductions and cancellations illustrate an aspect of the management process inside IPTO. Taylor canceled a University of Michigan contract involving graphics. Roberts, too, made his own reductions and cancellations. At one point he visited MIT and told researchers there that he was not interested in funding programs such as compiler development any longer, because they were only showing about a 10-percent improvement. He stopped funding work on operating systems after the TENEX system, described in the next chapter, "because it was just not worth building another one."[102]

As noted earlier, contracting was done through agents outside of DARPA. This enabled DARPA and IPTO to keep IPTO's own administration to a minimum. Other DOD agencies had ample facilities for issuing contracts, and IPTO used them. If the contract had obvious air force implications, as did contracts at Lincoln Laboratory or Rand, IPTO selected an air force agent. ONR had a single contract with MIT, so IPTO projects for MIT were sent to ONR. If IPTO wanted to write a contract fast, it often used the Defense Supply Service Washington (DSSW) to issue a contract. DSSW did no technical management; it just wrote contracts.[103]

In order not to burden the office, to relieve the office of responsibility for projects past the research stage, and to leverage its funds better, IPTO spun off mature projects to other DOD or government agencies for further development or use. For example, IPTO spun off ARPANET to the Defense Communication Agency, ILLIAC IV to NASA, and an expert system, the Automated Air Load Planning System (AALPS), to the army. Spinning off advanced systems to operating agencies ensured them a home and removed their maintenance costs from the IPTO budget for R&D.

As we emphasized in previous sections, IPTO managers paid close attention to the centers of activity they funded and the results produced by investigators in these centers. IPTO was interested in projects and their results. Principal investigators received frequent, often valuable, feedback from program managers. In contrast to the funding programs of other agencies such as NSF and NIH, a close connection existed between manager and PI. While involvement by IPTO staff in project decisions varied after contracts were awarded, the involvement of program directors and managers was a constant characteristic of IPTO activity.

An example of IPTO's monitoring in the middle 1960s was Roberts's involvement with the ILLIAC IV project. Roberts reports having been intimately involved in all contract decisions of the University of Illinois group:

> I was at every meeting. I sat in and was in charge of basically making the decisions with the University. . . . Because it was our money, I had an overview of the thing. . . . If there was any decision I could have made that disagreed with them, I would. . . . I was in all the project reviews and meetings with Burroughs with the Illinois people.[104]

Although Roberts disavowed involvement in technical decisions at the design level, he later discussed a major decision that he believes might have had significance far beyond the project. The ILLIAC IV design group debated whether to use thin films or semiconductors for memory system components. Roberts and the group decided in favor of semiconductors and against thin films.

Monitoring of IPTO by DARPA directors varied among the individual directors. As we saw above, many transactions between IPTO and DARPA directors were oral. But occasionally DARPA directors went on site visits to IPTO contractors. As time passed, the attention paid to IPTO by DARPA directors increased. When George Heilmeier became DARPA director in 1975, he felt that he had to read all proposals — an enormous task that, he believed, had not been done by his predecessors — to ensure that they met DOD needs. He sometimes sent them back to the office directors with questions.[105] This made more work for the program people, but as far as Heilmeier was concerned, the review improved the quality of the proposal and the outcome of the supported research.

In sum, then, the use of the DOD contract process for R&D allowed IPTO to target what were seen as highly significant areas in computer science, areas that appeared to offer the possibility of major advances in the field. One office director noted, using terminology taken from Thomas Kuhn's analysis of revolutions in science, that he looked for areas in which he could accomplish "paradigm shifts."[106] The tools employed by IPTO to promote such "shifts" included contracts providing large amounts of money; multi-year awards covering the work of a number of investigators; almost-certain renewal; and close cooperation between contractors and IPTO program personnel. As IPTO pursued its programs and made decisions about contracts, it maintained close contact not only with the researchers but also with the DARPA front office, and thus with Congress.

Relations with Congress

DARPA, including IPTO, did not normally seek publicity for its efforts. Rather, it shied away from publicity. "We sought no publicity whatsoever. . . . I think the theory was that once you get on the front page you are fair game and people begin to say, 'What the heck is this? What is an ARPA?' "[107] However, DARPA did not escape congressional oversight groups, regardless of its avoidance of publicity, because each year it participated in the DOD authorization and appropriation hearings. DARPA directors described to congressional committees the general program, and referred to specific offices when it would add strength to the DARPA appropriation request. In its early history, DARPA concentrated on the activities of its offices, including the Information Processing Techniques Office. The descriptions of IPTO activities contained primarily statements about how the results would help the DOD with problems of command and control. As DARPA's budget grew and the range of areas it studied increased in complexity and were described within a few overarching themes, the work of several offices involved information-processing research. In later years DARPA used these themes when making statements before Congress. All the statements from 1958 to the present show DARPA's interest in keeping Congress informed about its role in the DOD mission and—from 1962 on—the place of IPTO in DARPA, without garnering too much publicity to itself in the process.

In Licklider's first term as director IPTO received only small sums of money, by comparison with other DOD agencies, and had little visibility; and mentions of IPTO in congressional hearings were not very detailed. The DARPA testimony focused on concerns in ballistic missile and nuclear explosion detection systems. However, IPTO research results were quick in coming, and DARPA directors took advantage of them as they became useful. Throughout the rest of the 1960s, DARPA directors used the results of a number of IPTO projects as an important marketing tool with Congress.

Congressional views and procedures were important to the development, implementation, and growth of the IPTO program. Office directors understood that Congress, while it did not usually respond well to requests for funds to start new initiatives for which there was no prior exploration, would consider the results of preliminary studies as a criterion for providing more funding.[108] So IPTO started projects (e.g., the networking program) with funds reprogrammed from other activities until interesting results were obtained; it would then break out the program as a new area and expand the budget.

DARPA directors sold the IPTO program to Congress as an adjunct of other DARPA projects. When Rechtin was asked in 1970 how much of the IPTO effort was devoted to ballistic missile defense programs, he responded,

> Our research effort includes the investigation of new ways to organize computer hardware systems [the ILLIAC IV]. The objective of this research is to deal with a class of problems which involves massive amounts of computation in an extremely rapid time frame. Ballistic missile defense systems would be one beneficiary of a successful solution to this problem. Other uses would be in the fields of nuclear effects prediction, weather prediction, cryptoanalysis, simulation of fluid flow, seismic array processing, and large linear programming problems such as exist in the fields of economics and logistics.

He went on to say that, in particular, IPTO was concerned with the "future technology for collecting, processing, expediting and disseminating military information in the areas of administration, intelligence, operations, logistics and communications. As you know," Rechtin told Congress, "the DOD is the largest single user of computer systems in the world." Because of the DOD dependence on computing systems, it was important "to insure continued improvement in capability, reliability, and capacity" — another way of saying that IPTO was searching for more capable, flexible, intelligent, and interactive computer systems to satisfy defense needs.[109]

IPTO paid close attention to the DOD's increasing use of computers and to the problems, associated with the cost and compatibility of a large array of machines, that resulted from this rapidly increasing use. Rechtin reported that different users were buying different systems requiring different software. Users could not take advantage of each other's developments. In a statement delivered to a Senate committee in June 1969, Rechtin noted that "the ARPA computer network is addressing this question by electrically connecting a set of computers, permitting different users to use each other's machines and software whenever the latter's capabilities match the former's needs." This would mean that all users would not need to have the same data bases, software, or computing capability; and it would result in an increase in data transmission rates. Rechtin continued: "If this concept can be demonstrated as feasible, it could make a factor of 10 to 100 difference in effective computer capacity per dollar among the users." IPTO designed the ARPANET, Rechtin told Congress, to connect university programs in computer science so as to gain experience with network communications. The new knowledge would contribute to the DOD mission by connecting defense installations and saving money.[110]

A considerable amount of time was spent in congressional hearings discussing the ILLIAC IV project (see chapter 6, below). This offered IPTO many opportunities to emphasize the role of computer research in enhancing DOD efforts. The project was a computer science research program to explore a new form of computer architecture. It required substantial interaction with a user community comprising other government agencies, including the DOD meteorological community, the Army Ballistic Missile Defense Agency, and the Air Force Weapons Laboratory.[111] The ILLIAC IV would provide the DOD with capabilities to solve problems in such areas as the calculation of nuclear weapons effects, real-time processing of radar data, and long-range weather prediction.[112]

The ILLIAC IV project led DARPA occasionally to describe its interaction with industry and to cite the benefits to the government and the economy. For example, in the March 1970 hearings, when Lukasik was asked about the seeming lack of competitive bidding on the ILLIAC, he described the overall process in IPTO:

The competition came earlier in the game. Back in 1966, we ran a competitive design study that drew responsive bids from RCA, UNIVAC, and Burroughs. On the basis of this design study, we then elected to go with Burroughs. . . . We have negotiated an agreement with Burroughs that the Government having put in R&D money, can buy these at essentially only a 20 percent markup over the manufacturing costs. The normal procedure in the computer business is that there is a 100 percent markup between the cost to the manufacturer and the final cost. We have essentially gotten, we feel, very favorable terms for the Government on subsequent buys of this machine.[113]

The total contract to the University of Illinois up to 1969 was $29 million, of which $22.8 million was for the Burroughs subcontract. However, IPTO procured only one sixty-four-processor ILLIAC IV.

During Heilmeier's tenure as DARPA director, beginning in 1975, presentations to Congress began to change. First, DARPA moved away from an office-by-office approach and began using a more thematic approach, which was consistent with a more systematic and cross-cutting approach in DOD to the stimulation of technical innovation in order to replenish the technological base for defense systems. This meant that Congress learned of problems that DARPA thought were worth working on, rather than programmatic areas as defined by a budget or a subject area. For example, in the Senate hearings of February 1977, Heilmeier titled his statement "Technological Initiative and the National Security Issues of the 1980s." He boldly described "investment strategies" for advanced technology. His list of issues included space defense, antisubmarine warfare, passive surveillance, "really smart" weapons, threat-independent electronic warfare, and ballistic missile defense. Command, control, and communications played a significant role in the search for advanced technologies to contribute to the resolution of perceived weaknesses in U.S. defenses. While the problem description was thematic, the budget listing was still by office and program.[114] Five years later, however, budgets were also thematic.

Reagan administration appointees to the DOD believed that the mix of programs in DARPA was still insufficient for the country's defense needs. Richard DeLauer, new assistant secretary of defense for science and technology, and Robert Cooper, new DARPA director, submitted an adjustment budget in 1981 and a new budget in 1982 (for FY 1983) requesting large increases in DARPA programs; but Congress rejected both requests and kept funding levels static. Cooper began to work with the office directors to formulate new strategies that would obtain budget increases, especially for basic research. He decided that "the only way we could get money pumped into the basic research area was to start a major program, a big high visibility program that would take the technology that ARPA had been funding in the past and sort of aggregate it and aim it at major military problems and to pump the dollars up, not in the 6.1 budget, . . . but in the 6.2 budget." [115] At IPTO, Kahn responded to Cooper's request for a high-visibility program with a plan for a Strategic Computing program, composed of several testbeds that indeed aggregated research results and addressed major military problems. Congress agreed to fund the new program, which began in FY 1986.

On the whole, IPTO programs fared very well with Congress. The few negative effects on IPTO resulted from actions (such as the Mansfield Amendment) aimed at DOD programs generally. We might attribute the mostly favorable reception of IPTO and DARPA on Capitol Hill to the positive disposition of most members of Congress to the DOD and to the content of programs outlined in the testimony of DARPA directors.

In the 1960s, IPTO used its management practices in the support of research in many technical areas and selected projects to quicken the pace of research, thereby producing more interactive and flexible computer systems. It selected projects in artificial intelligence to produce more intelligent and capable systems. IPTO selected network research to promote the efficient sharing of resources and community building. While this is by no means a complete catalog of the office's program areas, it does represent the focus of IPTO in its early years. Through this research, the style of computing changed markedly. The stage was set both for further developments in these areas and for the integration of the new technology into the command, control and communications arena of the military.

IPTO's leadership role in computing can be appreciated by examining two levels of technical development. On the most fundamental level, there were the individual technical achievements that contributed to the advancement of the content of computer science and engineering. Contributions at the fundamental level included techniques for effective computer use, new architectures, and expert systems — results similar to those expected from any research proposal sponsored by any federal agency. The list of such contributions is a long one. The next level of IPTO's leadership in computing involved combining these contributions to create radically new systems. These developments at the systems level are the ones for which IPTO is best known — time-sharing systems, the ARPANET and the Internet, highly interactive graphics systems, parallel computer architectures, and infrastructure for VLSI design. What follows are descriptions and analyses of a selected set of examples to demonstrate the breadth and depth of the R&D areas of interest to IPTO; to provide more information on how advances in these areas were made possible through the personnel and management practices of IPTO; and to show how IPTO contributed to changing the style of computing.

CHAPTER TWO

Sharing Time

The principles underlying the system design of the modern electronic digital computer came into being during the decade 1935–45 and were incorporated into machines such as the electronic ENIAC, constructed at the University of Pennsylvania, and the Mark series of relay machines, built at Harvard University. These machines were cumbersome, and it was not always clear how to make them more capable and reliable. Computer researchers spent the next decade developing principles so as to produce a stable system and define mainstream computing. As we pointed out in the Introduction, the military played a significant part in this development. At first, however, as with all computing projects—EDVAC, UNIVAC, Whirlwind, EDSAC, and so on—those people and institutions supported by the military pursued the same goals as everyone else: to produce an effective and reliable system.

Between 1945 and 1962, the defense community invested significant resources in the further development of computing, and also in the incorporation of computers into defense systems. These efforts, particularly the work on Whirlwind and SAGE in the Cambridge-Boston community, set the context for subsequent investigations to change computing through projects supported by the DOD, DARPA, and IPTO. Both Whirlwind and SAGE involved computing on the leading edge of what was possible. To understand the vision instilled in IPTO by J.C.R. Licklider, who was a leading member of the Cambridge-Boston computing community in 1962, we first need to examine some earlier developments—the Whirlwind computer (which had a significant impact on mainstream computing), and SAGE (which incorporated new techniques of interacting with a computer into a production system)—and to discuss the state of mainstream computing in 1960. We can then begin to describe how IPTO changed computing, first through its support of time sharing, and later through the promotion of other R&D in computing.

The Background to the IPTO Vision

Whirlwind and SAGE

Project Whirlwind began at MIT during World War II with support from the U.S. Navy. The original goal of the project was the development of a generalized flight simulator to serve as a design aid for new aircraft and a trainer for pilots of the aircraft. The simulator would solve the equations of motion and aerodynamics of an aircraft and thereby simulate flight, and would use wind tunnel data to assume the characteristics of an aircraft that was still in its design stage.[1]

To develop the flight simulator, the MIT Servomechanisms Laboratory in 1944 organized a team under Jay W. Forrester, an assistant director of the laboratory. The simulator was to consist of two parts: a cockpit, and a computer large enough to solve the system of equations. The group first worked on an analog computer design, but by the end of 1945, after attending a conference on digital computing techniques and visiting the University of Pennsylvania to learn about ENIAC and EDVAC, Forrester decided to switch to a digital computer.[2] Eventually, building the computer became the primary goal of Project Whirlwind, and work on the design of the cockpit ceased. Forrester and his group transformed Whirlwind into a project to build a general-purpose computer, not a flight simulator or trainer.

The project requirements set by the military for Whirlwind placed it between general-purpose and special-purpose computer design. Forrester and the group working with him designed the Whirlwind computer with goals similar to those of other computer projects: speed and reliability. But in Whirlwind's case, the computer had to be fast in order to simulate the changing flight positions of the aircraft being modeled. The system needed to respond to external events as they occurred, a requirement common to "real-time" systems but not to other system designs. A simulation performed with a reliable system would last long enough to obtain useful results. The designers initially chose vacuum tubes for the memory component of the machine, but the tubes caused system reliability problems. The search for quick and reliable internal storage resulted in the monumental achievement of Forrester and the Whirlwind project: the development of magnetic core memory, which contributed as much to mainstream computing as to special-purpose computing.[3]

By the beginning of 1950, the Whirlwind computer was functioning. The project was consuming larger and larger sums of navy research funds, and the staff at the Office of Naval Research, which had been funding the project, became concerned not only about the cost but also about how the machine would be used. They were ready to withdraw their funds from the project because, at $1 million per year, it absorbed the lion's share of their research funds.[4] The close connection that the university researchers and military sponsors had developed in several wartime projects during World War II temporarily disappeared. As the defense budget shrank, it became difficult to sponsor high-cost research unless it was vital for the DOD. The impasse did not last long. As the Cold War created a new crisis situation, the Whirlwind machine found its vital mission. It became part of the air defense project, and was subject to no further concerns about cost.

Air defense became a central concern of the U.S. government after August 1949, when the Soviet Union exploded an atomic bomb. Officials of the Truman administration believed that the Soviets would soon have aircraft able to carry atomic bombs into the continental United States, and that the United States must prepare to defend itself. The concern for air defense was reinforced by the outbreak of the Korean War in June 1950.[5] The mobilization of scientists and engineers to combat the Soviet threat during and after that war continued the pattern established during World War II. Whirlwind became a key element in the response to this concern.

The air force turned to its scientific advisory board for advice on defensive systems to warn of any air attack. George Valley, a member of the MIT faculty, was serving on the scientific advisory board's electronics panel. He proposed that the board assemble a group of experts to address the air defense issue. The board formed the Air Defense Systems Engineering Committee, also known as the Valley committee.[6] After examining operating radar stations, the committee concluded that the air defense system left over from World War II was wholly inadequate for the current situation. It characterized the existing system as "lame, purblind, and idiot-like."[7] In the existing system, known simply as the "manual system," an operator watched a radar screen and estimated the altitude, speed, and direction of aircraft picked up by scanners.[8] The operator checked the tracked plane against known flight paths and other information. If he could not identify it, he guided interceptors into attack position. When the plane moved out of the range of an operator's radar screen, he had only a few moments in which to "hand over" or transfer the monitoring and control of the air battle to the appropriate operator in another sector. In a mass attack, the manual handling of air defense would present many problems. For example, the manual system would be unable to handle detection, tracking, identification, and interception for more than a few targets in the range of a single radar screen. The radar system did not provide adequate coverage for low-altitude intrusion. The system employed a "gap filler" radar to circumvent the lack of low-altitude surveillance. Because this "gap filler" radar only covered a few tens of miles, the system required more frequent hand-overs, thus further taxing the manual control system and its operators by reducing the time available to intercept an attacking bomber.[9]

In its report in October 1950, the Valley committee recommended upgrading the air defense system as soon as possible. The upgrade would include the use of computers to handle surveillance, control, and bookkeeping functions.[10] It divided the country into thirty-two sectors grouped into four geographical areas. Each sector would contain a direction center that had a computer and a system of hundreds of leased telephone circuits feeding into it. This new plan required a fast and reliable computer; and speed and reliability were the two design goals of the Whirlwind computer. Valley heard about the Whirlwind computer during a chance meeting with Jerome Wiesner, a friend and colleague at MIT who had participated in earlier meetings between MIT and ONR regarding Whirlwind.[11] After a demonstration of the computer and discussions with Forrester (and apparently without the knowledge that ONR was about to cut funding), Valley chose the Whirlwind computer for a test of the new air defense system.[12]

Within a few months of the Valley committee proposal, MIT began Project Charles, a short-term project to further study the air defense problem. A few months into Project Charles, in July 1951, MIT established Project Lincoln (renamed Lincoln Laboratory in April 1952) as a laboratory for air defense.

The implementation of the full air defense system began with a prototype, the Cape Cod system. The full system was originally named the Lincoln Transitional System (reflecting the desire to make the transition from the manual system); it was renamed SAGE (Semi-Automatic Ground Environment) in 1954. Whirlwind was useful for initial experimentation with air defense ideas and provided an elementary proof-of-principle for the new air defense design. However, the Valley committee and the Project Charles staff believed that SAGE required a new computer, which would serve as the prototype of the computers needed for the full SAGE system. In October 1952, the air force chose IBM to build the new computer, based on the Whirlwind design, that was later known as the AN/FSQ-7 (Army-Navy Fixed Special Equipment).[13] It is important to remember that as this decision was debated only a handful of one-of-a-kind computers functioned around the world. Several new, standard models appeared in the marketplace over the next three years.

The air force also issued a contract to the Rand Corporation to participate in software design and the implementation of the SAGE system, and to train personnel to operate SAGE. The air force had set up the Rand Corporation in 1948 as a nonprofit "think-tank" to conduct defense research. Through the contract with Rand, the project gained expertise in group behavior, specifically the interaction of people with machines. Additionally, Rand had knowledge about computing because of its experience with the JOHNNIAC computer. Lincoln Laboratory and Rand together developed the master program for all the direction center computers. The System Development Division within Rand then adapted the master program to the various installations around the country. It also improved training methods and developed training programs for each air defense station in order to exercise the system and maintain alert and knowledgeable crews.

SAGE had three major functions: to accept and correct target information from many types of detection and tracking radar equipment, to correlate and assimilate data on unidentified aircraft, and to direct defense weapons aimed at hostile aircraft.[14] A series of radar installations in a sector would feed information into the computer in order to produce an overall picture of the airspace of that sector. The telephone lines transmitted input data from the various radar installations into the computer for that sector. Information in SAGE entered computer memory through various mechanisms: directly from punched or magnetic tape, through telephone or teletype systems, and from radar data communicated along telephone lines. The computers received information about the thirty thousand scheduled flights over North America each day. In the design of the SAGE system, the high-speed digital computer would collect target reports from the radar network, transform them into a common coordinate system, perform automatic "track-while-scan" (tracking based on periodic radar reports at approximately ten-second intervals), and compute interceptor trajectories. The system then prepared "display materials" for automatic transmission to the proper users. The SAGE computers in each sector would be linked to other sectors across the country. This allowed rapid hand-over to another sector without operator intervention.[15]

Many operators used a SAGE computer in a single facility. To communicate with the computer they used an assortment of equipment, such as cathode ray tube (CRT) displays, keyboards, switches, and light guns (devices capable of sending signals coordinated with a display).[16] CRT consoles in the SAGE system displayed the tracks of radar-reported aircraft. Computer programs controlled the beam that created a track's image on the CRT surface by supplying the coordinates needed in drawing and refreshing the image.[17] The military personnel working on the SAGE system interacted with the display screens in front of them. For instance, an operator could request information about an unidentified image of a plane by selecting that image with a light gun; the system would respond by providing information about the specified image. If the operator determined that a plane was hostile, the system indicated which weapons were able to intercept it. The system provided the operators with the information they needed to make decisions, and provided them a way to send commands to the computer. Research conducted for SAGE improved the state of the art in graphical computer input and output techniques and tools.

From the beginning, the designers of the new air defense system wanted to complement rather than replace the operators in the manual system. The new system was *semiautomatic;* a human element remained an important part of it. The SAGE system required effective communication between human and computer. The operators performed the friend-or-foe identification function, assigned interceptors, and monitored engagements through voice communication with the interceptor pilots.[18] Human operators working without computers could not calculate quickly enough to counter an attack. Computers working without human operators were unable to make crucial decisions. The computers would keep the minds of the human operators "free to make only the necessary human judgements of battle—when and where to fight."[19] In Licklider's phraseology, this was a form of "man-computer symbiosis," in which the computer acted as an "intelligent" assistant.

The SAGE system incorporated, and pushed computing beyond, the accomplishments of Project Whirlwind. Both had graphic displays and on-line computer use, but SAGE required an advanced software system, many computers, and communication between computers. The use and subsequent improvement of voice-grade telephone lines for data transmission in SAGE were significant events for computer communications.

SAGE eased the way for the growth of on-line access to computers through computerized transaction processing systems. The on-line use of a computer had been demonstrated as early as 1940, but in the 1950s it had not become common practice.[20] In on-line transaction processing systems, people entered input at terminals, the central processor performed the necessary processing, and the system displayed the output at the terminal. Like SAGE, these systems allowed several users to make inquiries or update requests and receive the requested information after only a short delay. The system restricted the available services to those already programmed into the system. Although the operators specified which information the system displayed, they were unable to change the program. While the SAGE system allowed interactive input and output of data, it was only available interactively to the people running the air defense system. The functions provided by transaction processing were restricted and were used only for specific tasks. For example, the commercial airlines used transaction processing systems for flight reservations and seating assignments. Computerized transaction processing systems were useful in many applications, including air traffic control, process control, banking, and inventory control, as well as several military applications.[21]

Those computer professionals experienced with Whirlwind and SAGE, especially those around MIT, began to approach computing in a new way. They wanted a computer to be a partner in problem solving in an immediate, interactive, and responsive way. They wanted to engage a machine in conversation, with quick and intelligible responses available immediately; rather than in correspondence, with responses returned hours later. The responses they wanted were not listings of numbers that needed further analysis, but pictures, representing the numbers, and easily and quickly changed in response to changing requirements and new ideas. They wanted a machine to be a shared resource and depository of accumulated knowledge, rather than an individual tool not used in conjunction with others. This new way of using computers would make the conceptual jump from the computer as a fast calculator to the computer as a partner able to enhance creative human thinking. Many of the significant contributors to changes in computing over the next two decades — Licklider, Fernando Corbato, Charles Zraket, John McCarthy, Robert Everett, Kenneth Olsen, Marvin Minsky, and the like — and the people trained by them were members of this group seeking a new way of computing. To appreciate how far their thinking was from mainstream computing in the 1960s, we need to suspend our discussion of the new computing vision and examine computing as it was experienced by most users at this time.

Mainstream Computing, circa 1960

The art and science of computer design, construction, and use developed rapidly in the period 1946–60. Computer specialists were continually trying to adapt to major advances in technology and were striving to make the computer a useful tool for a growing number of applications. The principal technical concerns were reliability, memory, and speed. Persistent interruptions, inadequate memories, and cost hampered the diffusion of computers. By the mid-1950s developments in memory and reliable components, in part the products of Project Whirlwind, contributed to the spread and increased applicability of computers. Early machine designs contained tradeoffs between the speed with which the machine would operate and the size and accessibility of its internal, or primary, memory.[22] Early computer designs employed electronic components, but these components were not ideally suited to this new task. Components for achieving greater speed without otherwise affecting the operation of the machine had to be developed. New vacuum tube designs, and later the development of transistors, helped to solve some of the reliability problems. The invention of magnetic core memory greatly improved memory size and speed, and magnetic cores rapidly replaced the older acoustic delay line and magnetic drum storage systems. Consequently, reliability and speed increased and costs decreased, making wider diffusion possible.

The number of computers and their capabilities grew rapidly in the last half of the 1950s. While there were fewer than five thousand computers operating worldwide in 1959, the growth rate was significant: the number of computers had increased by more than 50 percent from 1958 to 1959, after having increased tenfold between 1955 and 1958.[23] The "magic word" in computing for 1959 was *transistor:* computers incorporating these new components were starting to appear on the market.[24] This new generation of computing equipment brought greater reliability, lower cost, and higher performance. Amid the constant change and growth in computing, business users asked manufacturers for cheaper computers in the small-scale and medium-scale range, random access storage at a reasonable cost, and assistance with developing applications.[25] Businesses such as banks and insurance companies used these systems to process volumes of data. In universities, computers were primarily doing calculations, replacing tedious hand calculations or slow mechanical calculations. These pervasive business and academic uses of computing, employing the newer, transistorized equipment, are what we are calling mainstream computing, in contrast to the special-purpose computing used in the SAGE system.

The software side of computing had also developed rapidly. Programming had improved dramatically when stored programs replaced the plugboards used in the 1940s to provide instructions to computers. However, programs still consisted of strings of digits. Metaprograms, known as assemblers, improved the accuracy and ease of producing these strings of digits, and compilers carried this process a step further. The latter were metaprograms that allowed the programmer to replace many machine language or assembler instructions with a single instruction, reducing the number of detailed machine instructions needed to write any given program. Compilers allowed more flexibility in coding and allowed programmers to use names to represent memory locations. In the 1950s, compilers such as A-2 and Speedcoding made it relatively easier for people to program computers.[26] By the end of the 1950s, advanced compilers allowed programmers to use languages such as FORTRAN to write instructions resembling mathematical formulas, and COBOL to write instructions resembling English. These higher-level (i.e., closer to human) languages gave people a way to express their instructions to the computer in a language they could better understand. Higher-level computer languages increased the number of people able to program and reduced the amount of training required for programmers. As more people were better able to use the machine by programming in languages better suited to humans, the demand for access to computing resources increased, along with demand for a wider variety of different programs. Consequently, programmers were in great demand, and programming became a major bottleneck in computing.

The capabilities of each particular system limited its users. Users had access only to the programs available on the computer to which they had physical access. The sharing of computing know-how was limited. People who worked with computers in the 1950s wanted to share data, programs, techniques, and knowledge about computing; and they recognized the need to overcome incompatibilities of hardware and software systems. Members of the emerging computer community recognized "a present and pressing need for the trading of mass information."[27] To share software for the same type of computer system, people copied it onto magnetic tape or cards and sent it through the mail. Most computers could not use programs written for a different computer. Programmers used contacts made with other programmers at conferences to receive information and share techniques. They formed cooperative organizations around particular types of computers in order to exchange information and programs and try to agree upon standards. For example, in 1955 users of IBM 704 computers formed SHARE, and users of UNIVAC 1103A computers formed USE.[28]

Computers were expensive, and even the largest and most powerful computers of the day had severe limitations in memory size and capability. While large-scale computers were the most expensive overall, the economics of the day favored them. For example, a computer that worked one thousand times as fast as a previous one would not cost one thousand times as much. However, obtaining the economy of large computers required that they be used to their full capacity, which most organizations found difficult.

The prevailing method of running programs was in part responsible for the difficulty of using computers to their full capacity. Many computers up to this time had been used by individual programmers who ran programs one at a time. Each programmer could use the computer only for a limited time, especially when it had to be shared with too many other users; and when time was scheduled, it was difficult to use it efficiently. Programmers had to sign up for sessions with the computer in advance. The time of the session was only rarely the best time to work on a problem, since the scheduled session might be at any time of the day or night. Because each user had time limits, the work of the programmer was under time pressure, and human "think time" suffered.

After writing a program, a programmer would typically go into the computer room and prepare the various machines required for a complete program run: one or more tape drives, a card reader, a card punch, a printer, and the computer. As the program went through translation into machine language, a listing of the program would appear on the printer and an "object deck" (the translated program ready to be executed) would be punched into cards. The programmer would carry the newly punched object deck to the card reader and start the reader, causing the program to begin execution. If the program executed as expected, the programmer collected the object deck and output, and left the machines for the next programmer.[29] Even if no gaps existed between programmers, the computer's processing unit was active for only a small part of the time.

Programmers, when writing programs, seldom produced error-free code. The programmer encountered many trivial problems during the first few runs of a newly written computer program. These arose from such minor things as punctuation or spelling errors in the code, or other syntactic errors. Although the programmer could sometimes analyze and correct these problems in a matter of a few seconds, fixing problems in the program was cumbersome. Changing the program consisted of punching replacement cards and manually placing them in the correct location in the program deck. Once the programmer corrected the obvious syntactic errors, the semantic problems needed to be debugged, leading to further delays.

Debugging logical problems was complex. Sometimes the programmer read "dumps" (listings of the contents of part of the computer's memory when a problem occurred). The programmer could also add statements to the program to trace which sections of the code were being executed or what values were assigned to variables during execution. Another method of debugging required the programmer to force a program to stop at a certain point in its execution and print the contents of various memory locations or print intermediate results. The cycle of analyzing errors, correcting them by re-coding portions of the program, punching the appropriate replacement cards, resubmitting the program deck to be rerun on the computer, and waiting for output would continue until the program was working properly. The burden of "debugging" a program was placed on the programmer; the computer was of little help. The programmer had to analyze the problem, attempt to correct it, and then either rerun the program or gather documentation for further analysis. Very few tools or utilities were available to make the programmers' interactions with the computer easy or more straightforward. Interactive, direct communication between the programmer and the computer was underdeveloped and inadequate.[30] Improving the debugging process was one of the early motives for developing time sharing.[31]

By the early 1960s, newer transistorized computers, with their greater processor speed relative to the speed of input and output devices, caused computer-room managers to become increasingly aware of the amount of time the processors were inactive. It was expensive to allow a costly computer to sit idle much of the time, as one program completed (i.e., was read in and executed, and the results printed) before the next was started. Using a computer one program at a time was an inefficient process. The idea of combining a group of programs together into a "batch" and running them continuously, without stopping to read in the next program from the card reader or wait for the output to print, was one strategy to increase computer efficiency.

Often the slowest and most time-consuming part of program processing was the time it took the mechanical peripheral equipment, including printers and card and paper tape readers and punches, to perform their function. The mid-1950s saw the origin of the "operating system" (often called the supervisory or monitor system) — a set of programs that controlled the operation of the hardware in order to use computer processing time more efficiently by overlapping the peripheral functioning, associated with input and output, with the computer processing. Magnetic tape read and write operations were significantly quicker than readers and printers, and magnetic tapes had enough capacity to hold several programs. The information from several input decks would be transferred to a magnetic tape on a separate machine. The computer then used the tape as input, and the programs would run consecutively, with minimal loss of time between programs. The output from each program, rather than going directly to a printer, was written to another magnetic tape and then transferred to a separate, special-purpose machine that would print from magnetic tape. This approach to computer use was called "batch processing."

In batch processing, the initial process of writing a program was unchanged. A programmer wrote a program on paper, often on specially designed coding forms, and sent the coding forms to a keypunch department where operators used keypunch machines to transfer the program onto cards (programs and data were stored on cards, paper tape, or magnetic tape, not on disks). The deck of cards was returned to the programmer, who then "desk checked" the program to see if the cards had been correctly punched.

The programmer in a batch system then submitted the card deck to the computer operations department, and at some later time received the printed output. The elapsed time between submitting the job and the appearance of the printed output was the "turn-around time." Turn-around time greater than the time required by the computer to process the job was inherent to a batch system: programs were not run immediately upon their submission to the computer center, nor were the results printed immediately at completion. Scheduled turn-around time would typically be two to three hours.[32] The delay in receiving the results of a run was often exacerbated by overloaded computer systems, so that turn-around on a single run would increase to a day or more.

In 1961, a data center manager proudly described the rationalized procedure for running programs in the batch style.

> When an engineer has completed a program, he sends his data sheet to data processing in a special gray envelope, used only by DP [the Data Processing Department]. The data sheet may be accompanied by a program deck, or we may already have the deck on file.
>
> A receiving clerk checks the sheet for legibility, then routes it to an IBM Card Punch Operator, who produces a data deck from it. Next stop is the set-up section, where the [data] deck is checked and placed in back of the proper program deck [i.e., the deck of punched cards containing the program for which the data is intended as input].
>
> The program and set-up decks are sent to a dispatcher who assigns several jobs to each tape, in most cases, and sends them into the [IBM] 709 [computer] room. There the cards are loaded onto one of two card-to-tape conversion units, and the information from them is reproduced on magnetic tape.

Finally, the job was ready to be run on the computer.

> The tape then goes to whichever of the 709's is open. The operator loads it onto a tape unit, starts the program and just monitors the time on each job.
>
> If there were any errors in programming, the 709 prints out a notice and the specific program is shunted off the machine.
>
> When all the jobs on the input tape have been run, the operator removes the computed data, on just one output tape, and places it on an off-line [i.e., not directly connected to the computer] Tape-to-720 II-Printer to produce a hard copy of the results at 500 lines a minute [of printed output].
>
> When a program calls for it, the 709 also produces a second output tape, from which a deck of output cards is punched.
>
> The computed data sheets, and any output cards, go into a gray envelope, and back to the originator by special courier.[33]

Thus, the batch method increased the efficiency of the computer but did not reduce delays and frustrations for the individual programmer, who, while waiting for turn-around, had to turn his or her attention to other tasks or other programs. Bouncing back and forth between problems led to further delays as programmers had to reacquaint themselves with each problem when the printout for it was returned. Programmers had little control in batch systems where "machine time" was the central concern. This attempt to rationalize the programming process worked poorly. Computer throughput increased, but the process failed to reduce the effort required to write and debug programs, and hence to solve problems.

The difficulties associated with the production of programs led to attempts to improve programmers' access to computing power. Some individuals, working from the batch environment, pursued improvements to batch systems to maximize the number of programs run on a computer and reduce turn-around time. People familiar with direct interaction with the computer, such as the community of programmers who had worked with the Whirlwind and SAGE systems, sought ways to maintain and improve computer interaction for individual programmers. This had the potential not only for improving machine efficiency but also for improving human efficiency in using the machine. Researchers exposed to the Whirlwind/SAGE form of interacting with computers, that is, to interactive computing, insisted on maintaining personal interaction with the most capable machines available at the time. MIT, and its associated laboratories involved with the development of Whirlwind and SAGE, formed the major center for interactive computing. Researchers there had computers available that provided quick response, on-line use for general purposes, and a variety of peripheral input/output equipment. Enthusiasts around MIT, including Licklider, pushed for a new concept — time sharing — as an economical way to allow the interactive use of the machine by several users, thus facilitating the task of programming. They had experienced the benefits of interactive computing and were unwilling to accept the batch method for their work. For example, Fernando Corbato, one of the people familiar with the use of the Whirlwind computer and later a leader in the development of time-sharing systems, pointed to the Whirlwind's effect on him and others in the MIT computing community. Comparing Whirlwind to "a big personal computer" with displays and typewriters, Corbato recalled: "many of us cut our teeth on Whirlwind . . . and one kind of knew what it meant to interact with a computer, and one still remembered," and, we might add, one wanted to continue to interact with similar computers in the future.[34]

Initial Developments in Time Sharing

Researchers in the late 1950s took two different approaches to the use of computers to solve the bottlenecks mentioned above. One approach emphasized having computers do more than one thing at a time; while the second emphasized the interactive use of computers by humans. The first approach resulted in multiprogramming; the second came to be called time sharing.

In 1955, Nathaniel Rochester of IBM described multiprogramming as a way for a computer to overlap computation with input and output processing.[35] Later, others expanded this concept to mean the ability to run more than one program in a single processor at the same time, which improved the efficiency of batch processing systems. Since multiprogramming addressed some of the technical problems involved in the sharing of computers, it applied equally to either batch or interactive operating modes.

In a multiprogramming environment, the system had two goals: to keep the computer as active as possible and to administer the computer. Multiprogramming operating systems allowed active programs to be independent of each other, a feature that was an advance over specialized systems such as SAGE, which required all the programs to work together for a single application. Multiprogramming removed programmatic constraints over which programs were able to run at the same time. It required minimal control by programmers (who would not have to specify the conditions under which their programs could run), enforced noninterference between active jobs, provided automatic supervision, and managed the allocation of computer resources among jobs. The executive program controlled the protection of programs—including the executive program itself—from each other. Additionally, it controlled allocation of resources, scheduling, and queuing.[36]

In the late 1950s, computing people used the terms *multiprogramming* and *time sharing* interchangeably.[37] In 1959 Maurice Wilkes, professor of computer technology at Cambridge University, described the ability of a processor to divide its time between a number of different tasks (what he called "time-sharing") as "in many ways the most important development in logical design since the introduction of stored program computers. . . . time sharing will enable the super-fast machines now being constructed to be fully loaded."[38] His view of time sharing had more to do with organizing computer time than with how people used computers, and thus differed from IPTO's definition of the term and from the definition that we use in this book. Time sharing soon took on a specific meaning, separate from multiprogramming, a meaning that included the interactive use of computers by people.

At the June 1959 International Conference on Information Processing held in Paris, a variation of multiprogramming, involving a human operator, was presented by Christopher Strachey of the National Research Development Corporation of London. While Strachey's emphasis was on multiprogramming, his presentation contained the idea of a person independently interacting with the computer to debug a program while the computer performed other tasks.[39] Thus, his proposed system went beyond multiprogramming.[40] Although he referred to his proposed system as "time-sharing," it involved only a single user.

At the MIT Computation Center, work was already under way on an interactive debugging scheme similar to Strachey's. The MIT Computation Center had been created in 1957 after IBM donated an IBM 704 computer to MIT, for the combined use of MIT, IBM, and a group of cooperating colleges throughout New England. The center began to experiment with interactive communication for programmers at the end of 1958, when IBM installed its "real-time package" on the IBM 704 computer at the center. This real-time package — a set of hardware changes — permitted the IBM 704 to respond to external events as they occurred. (IBM had developed the package at the request of aircraft manufacturers to allow IBM computers to accept wind tunnel input for testing airplane designs.)[41] The computation center planned to connect a modified flexowriter to its IBM 704.[42] The computer would then respond to a person's keystrokes on the flexowriter, thereby allowing the programmer to communicate directly with the computer. According to the MIT Computation Center's December 1958 progress report, this would "greatly simplify the work required to put various types of problems on the machine" because it would allow a person to "communicate directly to the high-speed elements, to add instructions, to change a program at will or to obtain additional answers from the machine on a time-sharing basis with other programs using the machine simultaneously."[43]

A way for programmers to interact with computers was especially important to MIT researchers because they were about to lose access to the technically obsolete Whirlwind computer, which had been available to them for research. Without the Whirlwind it was more likely that they would have to resort to the tedious batch processing system available through the computation center.

In a January 1959 memorandum, well before the Paris conference to which Strachey presented his concept, John McCarthy extended the time-sharing concept from the interactive use of a computer by a single programmer to the interactive use of the machine by several people. McCarthy had received a Ph.D. degree in mathematics from Princeton in 1951, and as an assistant professor at Dartmouth he had access to the IBM computer at MIT.[44] After spending the summer of 1958 at IBM, McCarthy in the fall of 1958 became an assistant professor of communications sciences at MIT, and a member of its computation center and Research Laboratory of Electronics.[45] He approached time sharing as a way to improve the interaction of programmers with a computer rather than a way to achieve more throughput from the computer.

McCarthy sent his January 1959 memorandum to Philip Morse, the head of the computation center and a professor of physics. The memo described how to convert the computation center's hoped-for IBM 709 computer into a time-sharing system. He believed his proposal was unique because of the nature of the active programs. The MIT system would "deal with a continuously changing population of programs, most of which are erroneous," whereas in previous attempts to share computers, such as SAGE, the "programs sharing the computer during any run are assumed to occupy prescribed areas of storage, to be debugged already, and to have been written together as a system." McCarthy viewed his proposal as an important development for computing: "I think the proposal points to the way all computers will be operated in the future, and we have a chance to pioneer a big step forward in the way computers are used."[46] McCarthy listed the equipment requirements for the system, including "interrogation and display devices," an interrupt mechanism (a way to allow an external source to interrupt the program currently running), and "an exchange to mediate between the computer and the external devices." He anticipated no problems with implementing such a system on the new transistorized IBM computer that MIT expected to receive in 1961.[47]

As is evident in this memorandum, program debugging was the intended primary application of early time sharing. A few months after Strachey presented his paper discussing time-shared debugging, a paper with a similar scheme describing the MIT system was presented by McCarthy and Herbert Teager at the September 1959 Association for Computing Machinery (ACM) Conference. Teager had received his doctorate in electrical engineering from MIT in 1955. Upon his return from naval service in 1958 he became an associate professor in the MIT Electrical Engineering Department and joined the staffs of the computation center and Research Laboratory of Electronics. He took charge of the time-sharing project in the computation center in 1959, when McCarthy decided to continue his research in artificial intelligence rather than develop time sharing.

McCarthy and Teager described their proposed system in terms of its effect on program debugging: "The system is being tried at MIT in order to determine the effectiveness of such an approach to program testing." They assumed that program debugging was best done in an interactive mode, so they explored time sharing as a way around the "wasted machine time" encountered during testing sessions. They would reclaim machine cycles for other work during these sessions "through a system of 'time sharing' the test with either a production run or other programs under test."[48] The system described in the McCarthy and Teager paper was in place at the computation center by January 1960.[49] This experiment with this limited form of time sharing was taking place just as MIT began to consider its future computing requirements.

A survey of computing needs conducted for the MIT Center for Communications Sciences concluded that planning for the center should fit into an overall MIT policy toward computing. The MIT administration, in an attempt to define its policy, formed an ad hoc faculty committee in 1960 to estimate campuswide computing requirements and make recommendations on how to meet them. In mid-1960, the committee of senior faculty members (Morse, Robert Fano, Albert Hill, and Jerome Wiesner) formed a working group, also known later as the Long Range Computation Study Group, consisting of representatives of various departments, laboratories, and centers around MIT interested in computing.[50] Teager was the chair of the working group; other original members who were committed to interactive computer use included McCarthy, Fernando Corbato (of the computation center), Jack Dennis (of the Electrical Engineering Department and the Research Laboratory of Electronics), Wesley Clark (of the Lincoln Laboratory), and Douglas Ross (of the Electronic Systems Laboratory).[51] By the fall of 1960, the working group reached its conclusion: MIT should plan its future computing around a large-scale, centralized computer that users would access by time sharing.

> MIT should obtain, within the next two to three years, an ultra large capacity computer, develop time shared remote input-output facilities complete with display and graphical input capability, and begin an intensive effort to develop advanced, user oriented programming languages for this system. This policy would seem the best possible way for MIT to obtain a research facility to multiply the intellectual effectiveness of the experimental and theoretical research workers and teachers in all fields. The policy would at the same time provide a necessary tool for frontier research in many fields involving physical simulation, information processing, and real time experimental work, which would otherwise be unfeasible.[52]

While agreeing on the goal of a centralized time-sharing computer, the working group disagreed on the best way of obtaining such a machine.[53] They considered several possible plans: building a machine, having one built to specifications, delaying altogether in hopes of a major breakthrough, obtaining a commercial computer with some modifications, or locating "a computer manufacturer in the course of completing a design and influenc[ing]" it to adapt the machine to MIT's needs.[54] Teager favored obtaining a commercial machine and recommended the IBM STRETCH computer over the Sperry Rand LARC and the Ferranti ATLAS. He wrote a report with this recommendation, which the rest of the working group rejected. They then wrote a separate report recommending working with an existing manufacturer to influence the design of a new computer.

The MIT administration (President Julius Stratton, Vice-President Carl Floe, and Dean of Engineering Gordon Brown) met with the ad hoc committee in May 1961 and agreed with the recommendation for obtaining such a computer. However, they observed that financing the machine would be a problem. The members of the committee were each to "investigate the possibility of getting funds for such a major undertaking."[55] The working group had projected that the purchase price of such a machine would be approximately $18 million, and that the cost of renting one would be $3 million per year. In either case, this was a "monumental project" that would require "a good deal of money from outside sources."[56]

Funding for computing resources at universities at the time came from a variety of places, but none of it was in the multi-million-dollar range specified in the MIT recommendation. The National Science Foundation funded projects for computers to advance research in scientific disciplines, but primarily through small grants.[57] Computer manufacturers offered educational discounts to help universities obtain computers for their mutual benefit. The budget for the entire MIT Computation Center, including the support it received from IBM, was less than $2 million per year.[58]

Undaunted by the magnitude of the need, some members of the group, including McCarthy, Corbato, Ross, and Dennis, visited computer manufacturers during the summer of 1961 to discuss the recommendations of the working group's second report. In October 1961, the group submitted to the MIT administration a final draft of a proposal requesting support for a large-scale time-sharing machine[59] In reviewing the specifications for the new computer, the working group noted that the manufacturers showed interest in the group's ideas.[60] However, no further activities related to securing a suitable computer system occurred until MIT's Project MAC began at the beginning of 1963, with support from IPTO. While the ad hoc committee had been studying computing requirements for MIT and planning the future with large time-sharing systems, a handful of projects to build small-scale time-sharing systems were under way.

In 1960, the MIT Computation Center had begun a time-sharing project:

CTSS (Compatible Time-Sharing System). This effort, led by Corbato, was mostly independent from Teager's time-sharing project, but based on the McCarthy-Teager ideas. After earning a Ph.D. in physics at MIT in 1956, Corbato had worked in the computation center as a research associate and as assistant director in charge of programming research. In 1960 he became an associate director of the center. Philip Morse, head of the computation center, described Corbato's efforts as focused on software while Teager's project focused on time-sharing hardware.[61]

The CTSS system allowed programs developed for the batch system to run while time-sharing users were using the same computer interactively. Thus, it was compatible with the existing batch system and did not require the majority of users, those using the batch facility, to convert their programs to time-sharing use. Corbato and two programmers at the computation center, Marjorie Merwin Daggett and Robert C. Daley, developed the system and held a demonstration of it in November 1961. No disk storage was available, so the system assigned a dedicated tape drive to each interactive user for his or her programs. As reported in the 1962 proceedings of the AFIPS Spring Joint Computer Conference, the time-sharing system at the computation center was able to support three interactive users.[62] Corbato expanded CTSS to allow up to sixteen simultaneous users by the end of 1962. The system was promising, but it served only a few users. The computer at the computation center provided computing to the entire campus; thus its use in time-sharing experiments was limited. While each of the experimental time-sharing systems allowed a few users, the computation center had to serve more than five hundred.[63]

Meanwhile, another small time-sharing system was under development at Bolt Beranek and Newman, a small Cambridge, Massachusetts, company engaged in research, consulting, and product development. McCarthy worked at BBN as a consultant one day a week. Edward Fredkin, who also worked at BBN, was another researcher interested in time sharing. Fredkin had spent 1956–57 as a project officer in the SAGE system, worked for Lincoln Laboratory in 1958, and moved to BBN in 1959.[64] While at BBN, he convinced McCarthy that time sharing was feasible on machines with limited memory sizes, such as the Digital Equipment Corporation (DEC) PDP-1, the prototype of which was at BBN. By using what Fredkin later described as the "trick" of swapping users in and out of a shared computer memory, multiple users of the small system were able to use the limited memory resource.[65] Fredkin then persuaded Ben Gurley, who had also worked at Lincoln Laboratory and was then the chief engineer at DEC, to make the necessary modifications to allow time sharing on the PDP-1 computer.[66] A high-speed magnetic drum memory was attached to the PDP-1. The drum transferred the contents of the core memory to a section of the drum while simultaneously transferring a different section of the drum into core memory. This allowed several active programs to share the limited core memory of the PDP-1. A time-sharing system with two typewriters attached was operating in September 1962. In 1963, when the system was described at the Spring Joint Computer Conference, researchers at BBN used the system in time-sharing mode for four hours a day with five typewriters attached.[67] However, the system was experimental and was not regularly used at BBN after it was demonstrated.[68] Nevertheless, one exceedingly important aspect of this project was that, as an employee of BBN at the time, Licklider was involved in it.

Also at about this time, at MIT, Jack Dennis was responsible for a separate PDP-1 time-sharing system developed in the Electrical Engineering Department. Dennis, an assistant professor in the department, had completed his doctorate at MIT in 1958. He was designing improvements to the TX-0 computer that Lincoln Laboratory had given to MIT, and following a suggestion by McCarthy, he included time sharing in his plans. When DEC provided a PDP-1 for the group in September 1961, Dennis switched his time-sharing plan to this new computer.[69] Through McCarthy and Fredkin, Dennis's group received information about the time-sharing system that BBN was implementing on its PDP-1. Although the BBN and Electrical Engineering Department systems were different, they both used the same specially designed swapping hardware to allow the limited memory to be shared.[70] The system eventually handled up to ten time-sharing users, but originally operated with just two typewriters, one display and light pen, and a tape reader and card punch machine.[71]

A group at Dartmouth College started a time-sharing project in 1962 partially in response to a suggestion by McCarthy.[72] Thomas Kurtz, director of the Dartmouth Computation Center, and John G. Kemeny, a faculty member in the Mathematics Department, along with some graduate students, designed and developed the BASIC language and the Dartmouth time-sharing system. The Dartmouth group designed the system to assist inexperienced computer programmers, particularly students, with the writing of small programs by providing a simple yet effective programming system that the novice could easily learn. The language interface for the system was BASIC, which later became widespread for simple programming. They developed the system on a General Electric (GE) 215 computer. The Dartmouth system could handle forty simultaneous users when it became available for campus use in September 1964. GE subsequently based its commercial time-sharing service on the Dartmouth system.[73]

The systems developed at the MIT Computation Center and Electrical Engineering Department, as well as the system at BBN, and to a lesser extent the Dartmouth system, were based on the idea of giving each programmer access to the machine for general-purpose computing. Other interactive systems developed in the early 1960s were special-purpose systems. They had less general goals for the programmer and focused more directly on the computer user who was not a programmer.

One of these special-purpose systems was the system developed at Rand in 1960. Rand had decided to provide a limited on-line service, similar to a quick, easy-to-use desk calculator, for its researchers. John Clifford Shaw, a former actuary, who had been at Rand since 1950, developed the system—named JOSS (JOHNNIAC Open Shop System)—on Rand's outdated JOHNNIAC computer.[74] A test version of JOSS was ready in 1963, and the system was available for general use within Rand in January 1964. Scientists and engineers at Rand used JOSS to solve a variety of small numerical problems. They liked the system because they did not have to learn how to code, use the operating system, or debug a program on a computer, nor did they have to use the services of a professional programmer.[75] The system was later rewritten for new hardware and remained in use throughout the 1960s.

Carnegie-Mellon University also wanted to improve the process of programming.[76] Researchers at CMU experimented in 1962 with a limited form of what they called "time-sharing."[77] A programmer would edit a program interactively and then submit it to run in a batch processing system. This approach was sometimes referred to as "instant batch," but it was more commonly known as "remote job entry" because the programmer submitted the program (or "job") to the batch system from nonlocal (remote) locations. Like most of the other early systems, the CMU system had some connection to the Whirlwind and SAGE projects. The leader of the project at CMU was Alan Perlis, who had received a Ph.D. degree in mathematics from MIT in 1950. In the early 1950s, Perlis worked on Project Whirlwind, developing programs for the Whirlwind computer.

IPTO and Time Sharing

IPTO marched into the time-sharing field in 1962, without fanfare but with much determination. MIT, having been involved with many advanced defense projects for decades, once again presented the ideas and concepts that would increase automated processing of the data in command and control situations, making the machine a greater aid in human decision making. This dream of the computer as an intelligent and able assistant was carried to Washington by Licklider, who served as the unofficial representative of the Cambridge community. He took it as his mission to accelerate the development of this dream, a dream that was already becoming reality in the MIT Computation Center and a few other computer locations in the Cambridge area. Licklider would lead the program to enlarge the concept to many more simultaneous users, and to spread its use beyond local borders. IPTO's first major venture was to ensure greater use of time sharing, increase the effectiveness of time-sharing in R&D, and encourage other groups to adopt the idea. Time sharing alone would not fulfill the potential evident to a number of researchers interested in computing. Licklider's larger, self-selected mission was to fulfill that potential by incorporating into IPTO's program work on a range of R&D problems that addressed the overall problem of providing more capable, flexible, intelligent, and interactive computer systems. These systems would eventually aid computing generally and the military specifically.

The concept of time sharing that IPTO supported was developed not by IPTO or Licklider, but by John McCarthy and others at MIT. Robert Fano and Fernando Corbato, along with several others, were anxious to carry out McCarthy's plan; so Licklider did not need to lavish any attention on the Cambridge computing community to arouse its interest. Licklider and the MIT group mutually subscribed to the same objective. Money for R&D was all that was needed. So Licklider turned his attention to spreading the concept, adjusted his focus from the East Coast to the West Coast, and set about to reorient the goals of IPTO's contract with the Systems Development Corporation. Licklider's encounter with SDC and his encouragement of others to investigate time sharing are addressed later in this chapter.

Not everyone was in favor of time sharing, or at least IPTO's version of it. Nor was time sharing automatically inserted into military programs. Later in this chapter the objections to and the limits of time sharing are discussed.

West Meets East

Toward the end of 1962 time-sharing research received a boost that guaranteed the spread of time sharing beyond the early experimental systems described above. The link between the MIT computing environment, as shaped by Whirlwind and SAGE, and the development of ideas and techniques about the IPTO program for interactive use of computers is strong, both institutionally and through the people who would disseminate the ideas and help develop other interactive techniques. The connections are perhaps strongest through the first IPTO director, J.C.R. Licklider, but continue with subsequent directors. In the early 1960s, projects funded by the IPTO shaped the overall development of time sharing. Before IPTO, various groups had sought funds simply to acquire computers and related equipment for time-sharing use by many different users in various scientific and engineering disciplines. IPTO funding made possible more than this. IPTO funding made cost constraints for time sharing irrelevant. Moreover, time sharing became a subject of research. IPTO leadership and IPTO-funded projects were largely responsible for spreading the idea of time sharing and making possible larger and more sophisticated time-sharing systems. In IPTO, time sharing—the seemingly simultaneous interactive use of a computer by several people working independently—became an element in achieving interactive computer use. Part of the reason for the spread of time sharing—specifically the Cambridge community's version of time sharing—was Licklider's conscious attempt to transfer these time-sharing ideas across the country. As the director of a multi-million-dollar R&D support program, Licklider played a pivotal role in the development of interactive computing; and time sharing played a major role in changing the way in which people used computers.

In 1960, as we pointed out above, Licklider had advocated interactive computer use in his very influential paper "Man-Computer Symbiosis,"[78] which outlined a program for improving human capabilities through the use of computers. Most relevant to the later IPTO program was the fact that in this paper he described several ways to improve communication so as to provide a partnership where both the person and the computer contributed to the overall goals in a manner based on the specific strengths of each. In effect, the computer would help the human to think. This conclusion was analogous to that reached by the SAGE development people, who said that command and control systems required both computers and people. Recognizing the existing limitations to this cooperative way of using computers, Licklider had identified a number of prerequisites for its realization. These prerequisites—research questions in memory hardware requirements, programming for memory organization, natural and computer language compatibility, displays, and speech recognition—were all active research areas in the Cambridge community and at a very few other sites; thus his review mostly reflected the research interests of the Cambridge community. It is not surprising that he carried this "symbiosis" idea with him when he became director of IPTO a few years later.

In part because of his training as a psychologist, the interaction between people and computers interested him. He had already experienced interactive computer use at MIT's Lincoln Laboratory, and he later used the BBN experimental time-sharing system and co-wrote an early paper describing the system.[79] While he was working in the Cambridge community, the limitations of the early experimental systems became clear to him. Licklider recalled of the BBN system, "It was such a weak little computer, there was nothing really to time-share."[80] Despite the limitations of the BBN system (or perhaps because of them), he had gained a vision of the future of computing. The successful experimental time-sharing and interactive systems showed that the underlying concepts of time sharing were feasible, and made a "true believer" out of Licklider, as it did others in the Cambridge computing community. He recalled, "I was one of the very few people, at that time, who had been sitting at a computer console four or five hours a day—or maybe even more. It was very compelling."[81] As the first director of IPTO, he was in a position to influence directly the further development of this compelling way to use a computer. Because of the heavy responsibilities of the DARPA director, Licklider had a free hand in organizing and managing both the command and control and behavioral sciences programs.

Licklider, among others, believed that a direct connection existed between research on command and control systems and the ability to use computer systems interactively.[82] While such an ability was useful for the military and for part of command and control research, Licklider understood that nonmilitary systems would benefit from the same techniques.[83] He believed that the computing problems inherent in command and control were the same as those in nonmilitary areas of computing. The idea that advanced data bases, displays, and computational ability would enhance effective decision making led him to shift the emphasis of DARPA's command and control research program from a broad range of command and control concerns including computing, to research oriented toward improving and changing the style of computing. With the limitations of the experimental time-sharing systems in mind, Licklider went about starting larger time-sharing projects and attempting to transfer time-sharing ideas across the country. He began this process by sharing with the Systems Development Corporation the Cambridge community's approach to time sharing.

When Licklider took over the command and control program, it consisted of a multi-million-dollar contract with SDC in Santa Monica, California. SDC, as mentioned above, had been the division of the Rand Corporation responsible for SAGE operator training and software development; in 1957 it had become a separate organization. In 1961 it received a multi-million-dollar contract from DARPA to establish a command and control laboratory based on a proposal submitted the previous year.[84] SDC created the Command Research Laboratory in October 1961. Its contract with DARPA called for SDC to conduct studies and to do R&D in the application of modern technology to military command.[85] By September 1962 it had completed the initial analysis of the problem and then separated the work into three primary areas of research: the functional analysis of command systems, the study of command processes, and the advancement of information processing technology.[86]

As part of its DARPA contract, SDC received an AN/FSQ-32 computer. This was the computer that the military had planned to use for the SAGE Super Combat Centers. These centers were to be an enhancement to the centers in the initial SAGE system, but the military canceled the program after the prototype machine was built.[87] The Q-32 was capable of on-line use, in that channels were built into it to allow the operation of teletype equipment; and SDC's plan took advantage of this on-line capability. SDC researchers intended to obtain teletype equipment as soon as the machine became operational.[88]

While the SDC plan had included on-line computer use before Licklider's involvement, it had not included time sharing. SDC's strength was on the command side of the command and control program. Consequently, the SDC project had been designed more for the objectives associated with command, and was, therefore, not directed toward the advancement of computing.

In October 1962 Licklider and Edward Fredkin, both of whom had been involved in the BBN time-sharing system, went to SDC to evaluate the work already done under this contract. Licklider was not sanguine that the results expected from SDC warranted the expenditure of IPTO funds. For example, he thought that the work in command functional analysis was not showing adequate results and suggested that if it did not "come to grips with definite problems or a workable set of variables . . . it should be dropped altogether."[89] The work in the economics of command system development he thought should be funded by SDC itself. Licklider believed that the research in human-machine interrogation techniques and command processes was the most promising for IPTO. He evaluated the SDC project in terms of his own—now IPTO's—ambitious objectives for advancing computer use, and decided to transform the SDC effort under this part of the contract into IPTO's West Coast time-sharing experiment. He believed that SDC, with the help and expertise of Fredkin, could produce a state-of-the-art time-sharing system in six months.[90]

With the support of DARPA's director, Jack Ruina, Licklider insisted that the SDC project turn its attention to making the Q-32 into a time-sharing machine and that a West Coast group of computer researchers be started around a shared system at SDC. In addition, IPTO and SDC agreed on an advisory group to assist in shaping the policies for the Q-32 computer facility as a national command and control research facility. SDC had no choice but to go along with Licklider's plan or lose the contract. In September 1962, SDC agreed to the plan, and Licklider renewed its one-year-old contract. The primary element of the renewal contract was Licklider's desire to have a time-sharing system available on the machine in six months.

To enhance the time-sharing effort and to ensure that some of the ideas of the MIT-Cambridge community were incorporated into the SDC system, IPTO contracted with Information International, a company founded by Fredkin.[91] In awarding the contract, Licklider expressed his belief that Fredkin and Ben Gurley (who had been involved with the BBN system and had left DEC to work at Information International) were the ones best able to transfer knowledge of time sharing to the West Coast. Licklider used Information International to fulfill what he referred to as the "hummingbird function" of transferring the "Cambridge know-how" into the "California program."[92] Besides preparing a survey of the state of the art of time sharing, Fredkin and his colleagues concentrated on software developments to make the SDC system a time-shared system as quickly as possible.

In November 1962, Fredkin helped Licklider organize a meeting at SDC's Santa Monica headquarters. Representatives of various time-sharing projects attended the meeting. In order to begin to share knowledge about the time-sharing developments with the people at SDC, the agenda included presentations on the existing time-sharing systems—MIT's CTSS and the BBN system—as well as on a handful of other systems about to go into operation: the MIT Electrical Engineering Department system, JOSS, and the CMU system. In addition, the group discussed the possibilities for future time-sharing systems.[93] The presenters included Corbato, Dennis, Shaw, Perlis, and McCarthy. This meeting introduced time sharing to Jules Schwartz, who was then a principal scientist at SDC. Schwartz had started working for Rand in 1954. The following year, as a Rand employee, he had worked on SAGE at Lincoln Laboratory, and he then joined SDC when it separated from Rand. While working at SDC he had developed an early implementation of an ALGOL-like programming language: JOVIAL (Jules' Own Version of the International Algebraic Language).[94] Schwartz credited the beginning of his understanding of time-sharing concepts to the talk given by Corbato about CTSS at this meeting.[95] In December 1962, SDC reassigned sixteen researchers to work with Schwartz on the time-sharing project for the Q-32.[96]

The SDC time-sharing project had three parts. First, SDC was to provide a usable time-sharing system on the Q-32 computer for programming and debugging. It was to be a general-purpose time-sharing system, using an approach similar to the CTSS approach. Second, SDC was to conduct research on time sharing itself. Analytical studies based on time-sharing simulations at SDC eventually caused changes in the way programs were processed by the system, for example, determining how much time each user was given before the system switched to the next user (time-slice) and how the system queues were set up to optimize throughput and response time.[97] And third, SDC was to increase the flexibility of display and input-output devices by providing more capability for the user. Licklider assigned this task to Fredkin. Fredkin achieved flexibility by using a PDP-1 computer (at SDC) as an intermediary between input-output devices and the Q-32 computer. This allowed a larger variety of input and output equipment than the Q-32 could handle directly.

By the middle of 1963, six months after being directed by Licklider to develop a time-sharing system, SDC was running a time-sharing operating system on the Q-32. Even though it was initially limited, it represented the current state of the art. It did not allow any batch jobs to run while the time-sharing system was active, it supported only eight teletype terminals, and it did not have disk storage available, although SDC added disks to the system within the first year. By November 1963, SDC's progress in time sharing had pleased Licklider. He praised the "outstanding achievement" of SDC in time sharing.[98]

Licklider promoted time sharing at other California institutions to reinforce the activity at SDC. He awarded contracts to Stanford Research Institute, the University of California at Los Angeles, the University of California at Berkeley, and Stanford University. All of the contracts specified the use (at least in part) of the SDC system. In some of the program plans, Licklider refers to this as the "California Network Group," the group of separate contractors who he hoped would become a community of researchers with common interests and goals.[99] Thus, Licklider transferred the MIT community's concern for time sharing and interactive computing to California, creating the basis for another time-sharing computing community.

Project MAC

At the same time as he was reorienting the SDC project, Licklider wanted to organize a research community around an existing machine. He chose to show that this was possible using some of the computing activities of MIT. In November 1962, he and Robert M. Fano, professor of electrical communications, discussed setting up a cross-disciplinary project at MIT with funding from IPTO. Fano, small in size but powerful in personality, who had also been a member of the earlier MIT ad hoc faculty committee, was an accomplished specialist in information theory. He had begun teaching at MIT in 1941, had received his doctorate in electrical engineering in 1947, and had become a professor in the Electrical Engineering Department in 1956. He had experience in MIT's Radiation Laboratory and its Research Laboratory of Electronics, and had been a group leader in Lincoln Laboratory in 1951–53; thus he was familiar with MIT's powerful administrators and knew many of the DOD representatives who provided liaison with MIT. He had been a member of the Valley committee for air defense which had led to the SAGE system. The mutual interests of Fano, representing MIT, and Licklider, representing IPTO, combined to form Project MAC. MAC was an acronym with two meanings: Multiple Access Computer, which was the main tool to be developed and used in the project; and Machine Aided Cognition, which was the overall goal of the project.[100] The goal of Project MAC was to stimulate the spread of time sharing, and to go beyond time sharing toward Licklider's dream of human-computer symbiosis. Fano directed Project MAC for its first five years. Licklider succeeded him in 1968, and Fredkin became director in 1971.

Although the MIT proposal for Project MAC was officially unsolicited, Licklider and Fano engaged in many conversations about the project before its submission. For example, shortly after Licklider arrived at DARPA, the Homestead conference on communications and computers we mentioned in chapter 1 took place. The conference was sponsored by the air force and organized by MITRE Corporation. During a "long" train ride to Washington, the two of them discussed what needed to be done in command and control research, and how the human-machine interaction could be employed to accomplish the mission of command and control. Nothing was resolved during the train ride, but over the Thanksgiving weekend that followed, Fano turned the issues over in his mind and resolved to design a project at MIT to pursue research in this area and solicit funds for it from IPTO.[101]

After receiving approval (from Charles Townes, MIT's provost) to proceed with planning, Fano prepared a two-page memorandum outlining the proposal. He began the memorandum by stating,

> The nation is facing many urgent information processing problems both military and civilian, such as those involved in command and control and information storage and retrieval. The number of people skilled in the techniques of information processing (not just programming) is insufficient to attack these problems both effectively and speedily.[102]

He pointed out that the techniques in information processing were still rudimentary because of the small number of people conducting research in the area. Thus, it was necessary to mount an educational program to train a larger cadre of people for research. The faculty and students in such a program would need more and better equipment to accomplish this goal, and Fano called for specific machine purchases. He noted the intimate connections between education, research, and personnel: "Adequate facilities are needed for research and education, research is needed to develop the facilities and to support education, and education is needed to provide the necessary personnel for the development of facilities and for the carrying out of basic research." [103] Such a program would benefit MIT students and faculty members by improving the educational environment. The IPTO program and the nation's defense would benefit from the research and development results obtained at MIT.

In December 1962, the MIT administration, under Provost Townes, appointed a committee for information processing to help set policy for the new project. Townes identified this project with the decision made the year before to use time-sharing techniques to expand the computer facilities available at MIT. [104] Project MAC, with a multi-million-dollar contract from Licklider, eventually fulfilled this MIT desire for a campuswide time-sharing computer system.

The handling of the proposal for Project MAC is a good example of the IPTO practices described in chapter 1. Licklider was deeply involved in the discussions about the content of the proposal before submission, in the objectives to be pursued in the project, and in the evaluation of the proposal—indeed, as far as we can tell, he was the only evaluator. The proposal received quick attention after arrival at his office, and within days of the proposal's arrival Jack Ruina, DARPA director, endorsed the approval order for a $3-million contract. Moreover, for the next two years Licklider remained in close contact with the activity, frequently visiting MIT, while he directed IPTO.

Licklider wanted Project MAC to maintain the momentum in time-sharing design and implementation and to formalize the transfer of the "Cambridge know-how" by "disseminating information and influencing the technical community." [105] Fano wanted to extend, speed up, and integrate work in human-machine interaction already in progress at MIT. In his proposal to IPTO, he pointed to several projects already under way at MIT. There was the research in the design, analysis, and evaluation of computer structures being conducted by a group led by Dennis in the Electrical Engineering Department. C. L. Miller led a group in the Civil Engineering Department in developing a system to solve geometric problems arising in the design of roads using an IBM 1620 with a graphical display unit. Douglas Ross, in the Electronic Systems Laboratory, developed an automated engineering design system, one of the first computer-aided design systems. Fernando Corbato, Marjorie Merwin Daggett, and Robert Daley were involved in the final stages of the development of the Compatible Time-Sharing System (CTSS) in the MIT Computation Center. In addition, Fano cited the work of Marvin Minsky in artificial intelligence, Herbert Teager in languages and devices, and Martin Greenberger on human-machine systems for use in business management. By providing a computing system with greater speed and memory, Project MAC would aid these projects immeasurably and integrate their capabilities into the MAC system to further the objectives of interactive computing, indeed, to work toward Licklider's dream of an intelligent assistant.

Both Fano and Licklider had the organization of MIT's Research Laboratory of Electronics in mind as the model for Project MAC.[106] Licklider believed that he "really had to have" a range of existing MIT activities, such as time sharing (CTSS), graphics (the Teager Tablet; the Kludge), and various engineering applications (Stress and Strudel), incorporated in Project MAC and associated with IPTO.[107] According to Fano these groups were more likely to cooperate as part of a project than leave their present organizations in MIT for a new department or laboratory.[108]

The organization of Project MAC reveals the close connection between the focus of Project MAC and IPTO plans. Licklider saw several advantages in the MIT proposal: "I wanted interactive computing; I wanted time-sharing. I wanted themes like: computers are as much for communication as they are for calculation. I wanted a summer study that would bring people from all over the industry, that was going to try to shape up this field and make it clear what we were doing."[109] He also wanted to be able to call on people in the Cambridge community for consulting help in projects elsewhere.[110] In short, he insisted that Project MAC be an R&D enterprise, not just a time-sharing project. Fano, too, saw this as the overarching reason for the effort.

As his justification of the MIT contract that established Project MAC, Licklider stated that "the cluster of concepts that surround time-sharing is expected in due course to have very profound effects" in computing.[111] He went on to describe how the computer might effect change in the university environment. For example, it might alter the way in which researchers interact with library holdings, and the conduct of research in the laboratory. The computer could also become an integrating mechanism in government and industry, making information more readily and quickly available, and might be an "amplifier of intelligence, entering into and extending the thought and decision processes of many individuals."[112] Then he took on the need of the DOD. IPTO's request to obtain a DARPA contract for MIT argued that the technical need in command and control was definite and urgent. According to Licklider's justification, by first supporting research on the broader questions in computing, this project offered "the possibility of a profound advance, which will be almost literally an advance in the way of thinking" about computing. "One can hardly overstate the technical need for such an advance, or its value to the Department of Defense and the nation."[113]

Within a few months of approval by IPTO in early 1963, MIT had organized a $3-million effort in software, languages, graphics, and related studies. Project MAC's "broad, long-term objective" was to build a complete system for user interaction, a system that had capabilities beyond those of the existing time-sharing systems and was available to a large number of users.[114] The Project MAC proposal noted that hitherto time-sharing systems had failed to make computing easy or economical, and that they lacked flexibility and quick response.[115]

CTSS, the time-sharing system developed by the MIT Computation Center, formed the software foundation for Project MAC. The project initially bought an IBM 7094, with the same modifications as were found in the computation center's computer. CTSS was already available and able to support twenty-four simultaneous users, and would work without modification on the new machine. As a result, Project MAC had a functioning time-sharing system up and running in November 1963, shortly after the computer was installed. The new machine was heavily used for time sharing, while the computation center's computer ran a heavy batch load on its system. Within a year of receiving its IPTO contract, Project MAC using CTSS had twenty-four remote stations at various places on the MIT campus, all linked by teletype to a central computer.[116] CTSS continued to be improved for the first few years of Project MAC and remained available to users at MIT until 1973.[117] While the initial CTSS system provided valuable experience and information, its organization and overall capacity were inadequate for meeting Project MAC's larger objective of a complete and widely available system for user interaction.

A crucial consideration in starting on the new system was the programming involved. The designers wanted to decrease the difficulty of a later upgrade to a new computer because they expected the first system to have a useful life of only two to three years. An upgrade would be easier if it did not require completely rewriting all the programs. In October 1963 Project MAC formed a special committee on program compatibility, which decided unanimously that all programming for the new computer installation would be written in a "basic language" that would be "sufficiently removed from machine language to hide the equipment configuration and special characteristics, and therefore have a probable life and usefulness extending beyond Project MAC."[118] Project

MAC staff members thus decided to write the new system in a higher-level language. By doing so, they believed, they could ensure that the effort invested in software for the new system could be capitalized upon when they later needed to upgrade the system hardware.

Fano believed that the MIT effort should also aim toward participation by people outside MIT in order to stimulate further activities nationally, a primary objective of Licklider. Following a suggestion of Licklider, Fano proposed to bring to MIT people from other universities and from industry for a summer study to develop "suitable specifications for the performance and structure of university computer systems, including type of consoles needed, character of the control program and of the programming languages."[119] Summer studies were a common feature of MIT departments and projects. People, often academics, with competence in a certain area were brought together during the summer months when they were relatively free to consult on a given military subject. For example, MIT conducted Project Lexington in 1948 to determine the technical feasibility of nuclear-powered flight. Earlier, we mentioned Project Charles, organized in 1951 to develop a set of recommendations for a projected laboratory for research into problems of air defense, the now-famous Lincoln Laboratory.[120] Fano hoped that a summer study in information processing would have the same kind of influence. All results would be broadly disseminated to universities and industry, and the time-sharing facilities would provide others with time-sharing experience.[121] Licklider believed that a summer study would also do much for dissemination of information about IPTO's program.

In the summer study highlighting time sharing, held in July and August 1963 on the MIT campus, more than fifty representatives from industry, government, and academia participated. In addition to sharing information regarding the MIT system, the summer study participants received technical notes from the SDC system and had access to the existing time-sharing system at SDC (through modems) and to CTSS in the MIT Computation Center.[122] These two systems represented the state of the art in general-purpose time sharing.

The dissemination of information continued to be an important part of Project MAC's goals. The project continued spreading the word about time sharing to its many visitors from other universities, industry, government agencies, and the news media, and its personnel gave hundreds of talks outside of MIT on Project MAC research.[123]

The IBM 7094 and the CTSS software used to set up Project MAC's initial time-sharing system comprised only the first step in the project's development

of better time-sharing systems. Project MAC took the same approach to obtaining a suitable computer that the earlier MIT working group had proposed:
close cooperation with a computer manufacturer. This is not surprising, since
some members of Project MAC had also been members of the earlier working
group. Project MAC staff members viewed the system they would develop as a
"pilot-plant operation" for all future interactive systems. Building the necessary hardware did not interest them. Ideally, a computer manufacturer would
provide a prototype that Project MAC would shape to meet the requirements
for multiple access computers. The manufacturer, they envisioned, would then
incorporate Project MAC's changes and market these new computers. In this
way, the new technology would become widely available to universities and
businesses. Project MAC sought proposals from manufacturers for the development of an advanced time-sharing system. Dennis, Corbato, and a few others
visited manufacturers to discuss plans for a new system. Dennis recalled considerable interest on the part of all the manufacturers visited, not necessarily
in the specifics of working with the researchers from MIT, but in keeping up
with what was going on at MIT and understanding Project MAC's approach to
time sharing.[124]

By February 1964, Project MAC personnel had participated in "detailed
technical discussions" with several computer manufacturers, including Digital Equipment Corporation, Control Data Corporation (CDC), IBM, and the
UNIVAC division of the Sperry Rand Corporation. In April, Project MAC had
its first discussion with people from the Computer Division of the General
Electric Company. The following month, GE made a presentation of its system to Project MAC personnel.[125] Fano and his colleagues recommended the
GE system as the base for its new time-sharing system.

Fano listed several technical reasons for choosing the GE computer.[126] The
configuration of the GE 635 corresponded to what the Project MAC people had
in mind. The machine was intended to be used in a memory-centered configuration. This permitted the effective use of several processors operating independently and simultaneously on the memory. The system that was planned
for initial delivery contained only one processor, but the system was easily
expanded by adding more processors. Multiprocessor systems were of interest because they could better utilize expensive memory, because the system's
capacity could be easily increased by adding another processor, and because
some processing could still be provided if one processor failed, an important
consideration when attempting to achieve the high level of reliability needed
for a widely accessible system.[127] GE, as a relative latecomer to computer manufacturing, struggled to make a dent in IBM's huge market share. The machine

Project MAC finally chose was the GE 635, part of GE's new line of computers, the 600 series. The computer division responsible for the 600 series was small compared to IBM, and GE was considered something of an underdog.[128]

Project MAC wanted to influence the design of the computer hardware, something its personnel believed it could do with GE but not with IBM. They believed that to fulfill the vision of developing the prototype for the future of computing, close cooperation between the manufacturer and Project MAC would be required. Fano reported to Julius Stratton, president of MIT, that IBM treated Project MAC only as a customer for its machines, not a design partner. Fano had the impression that IBM management never regarded Project MAC as anything more than a "prestige account."[129] While IBM was willing to work with Project MAC on some level, it was not willing to redesign its hardware to the project's specifications.

The Project MAC group's decision to use GE hardware caused concern at IBM. IBM was preoccupied with preparing its 360 family of processors, which was publicly announced in April 1964. While focused on its bread-and-butter commercial market (a major part of our definition of mainstream computing), it was trying to maintain its prestige accounts as well. When told of the likely Project MAC decision in favor of GE in May 1964, IBM responded quickly, convening a task force over the summer of 1964. IBM offered two counter-proposals, one in June and another in August.[130] Neither satisfied the Project MAC demands. Fearful that it would loose other prestige accounts when word of Project MAC's decision became known, IBM responded with a program to accelerate its development and marketing of a general purpose time-sharing system that could compete with the GE machine. Thus, IBM took steps to develop a general-purpose, large-scale time-sharing system like the one that Project MAC and GE were planning to build.

IBM's concern about losing other prestigious customers proved to be well founded. Bell Laboratories was looking for a new computer in order to implement a time-sharing operating system for its internal use. The staff evaluated the GE equipment and chose it for their planned time-sharing system. Bell then decided to cooperate with Project MAC and GE on the development of a new time-sharing system. In 1965 the three organizations announced their joint effort in a series of papers describing the planned Multiplexed Information and Computing Service (MULTICS). Each organization brought strengths to the project. Project MAC included some of the most experienced and knowledgeable personnel about time sharing, and had enough money, thanks to IPTO, to be able to purchase a computer. GE was in position to do the manufacturing, marketing, and maintenance. The computer users at Bell Laboratories were sophisticated and knowledgeable. They had experience in writing systems software and had a larger variety of users than either a university environment or a computer manufacturer provided.

Spreading Time Sharing through IPTO Projects

The SDC program and Project MAC were the two poles around which the early IPTO program revolved. Two other institutions also received contracts for time-sharing systems. In July 1962 the Carnegie Institute of Technology, later renamed Carnegie-Mellon University, submitted a proposal to DARPA. The strength of the Carnegie proposal was enhanced by the presence on the Carnegie faculty of four people working on various aspects of information processing problems: Herbert Simon and Allen Newell, who were working on problem solving and other areas of artificial intelligence; Alan Perlis, who was attempting to enhance the operations of the computing center; and Herbert Green, a psychologist friend of Licklider, who was directing work in cognitive science.[131] Licklider wanted to strengthen the work of these men at Carnegie. In August he submitted a program plan for a contract to be offered to Carnegie to aid Perlis's efforts to enhance the operations of the computing center. Later, when IPTO developed a "centers of excellence in information processing" program, Licklider identified Carnegie as one of the first two centers (MIT was the other).[132]

In the IPTO contracts to California institutions we can see another aspect of Licklider's strategy. With the contract in place at MIT and the contract modification complete at SDC, Licklider encouraged other computing organizations to use whichever of these facilities was nearest them. He carried out this strategy through the contracts IPTO issued at the time. For example, John McCarthy of Stanford University submitted an unsolicited proposal that included a request for funding to develop a time-sharing system at Stanford. IPTO refused to grant the request, arguing that "it seems inappropriate for ARPA to support another time-sharing system independent of the ones already being developed under ARPA contract." To encourage use of the SDC computer, the contract that IPTO gave McCarthy called for development of the LISP programming system for use on the Q-32 at SDC. Besides the LISP system, the contract called for the preparation and dissemination of "effective documentation, description and instruction material." The preparation of documentation and instructional material was another part of Licklider's plan for a loosely coupled organization of workers in the field, his "intergalactic network."[133]

The University of California at Berkeley, however, did receive a contract for time-sharing research in 1963. The Berkeley contract called for a model 33 Tele-typewriter and a leased line from Berkeley to Santa Monica. Harry Huskey, the principal investigator, intended to log onto the SDC Q-32 from time to time, evaluate what he found, and report to IPTO. Shortly after receiving the contract, Husky took a sabbatical to India and left the project in the hands of Edward Feigenbaum and David C. Evans, two young faculty members. The work became known as Project Genie. The university purchased a Scientific Data Systems Corporation computer, a model SDS 930, which was delivered in September 1964. Evans directed the team—which included Butler Lamp-son, Peter Deutsch and Mel Pirtle—that designed and implemented memory protection and wrote a new operating system for the SDS 930. The new system was operational in April 1965 and was demonstrated at the Fall Joint Computer Conference in Las Vegas in November.[134] The objectives of Project Genie were "to improve the computer as a research tool," and to focus on the problems peculiar to a smaller community of users employing a smaller system than either the Project MAC or SDC systems.[135] The modified SDS 930 time-sharing system served as the prototype for the SDS 940 machine introduced to the market in 1965.[136]

Licklider refused to expand the time-sharing development program beyond these institutions. Although DARPA in 1964 awarded to the University of Michigan a contract for research to improve the understanding of pattern-recognition and problem-solving processes through the Behavioral Sciences Program, and a year later would award a contract to that institution for work in the conversational use of computers,[137] in 1963 IPTO refused to support the development of a time-sharing system at the University of Michigan. As Lick-lider said in a letter to Robert Bartels, head of the Michigan Computing Center, "last fiscal year, we set up a system of contracts designed to advance the technology of time-sharing and close man-machine cooperation in information processing. . . . [which] essentially fills out the program that we planned and for which we have projected funding."[138] Licklider took a similar position with other universities and centers.

Nevertheless, IPTO occasionally saw a need to develop time-sharing systems as a part of work it funded on other problems. Later in the decade of the 1960s, for example, BBN received a contract to develop a paged time-sharing operating system for the DEC PDP-10. BBN claimed that it needed the new system to carry out its research in natural language communication between people and computers, because no existing system could meet the requirements of its research projects.[139] BBN developed the new operating system, TENEX, because the other operating systems then available at BBN—the DEC 10/50 and the Berkeley SDS 940 systems — were not of appropriate size and cost effectiveness for BBN's active and proposed projects. IPTO's program managers believed that a small number of people working on well-defined, independent tasks could produce the system quickly and efficiently. The contract did not aim for breakthroughs. TENEX was not a large-scale system, and the work was done by designers and developers experienced with time sharing. One member of the group worked full-time updating design documents that, in the opinion of the design group, "contributed significantly to the overall integrity of the system."[140] Working from knowledge of several other operating systems (the system for the DEC 10/50 at BBN, the Berkeley system for the SDS 940, MIT's CTSS, and MULTICS), the TENEX designers were able to draw up a list of specific design goals. In short, the goals were to produce a state-of-the-art machine using the K10 processor of the PDP-10 with good human engineering throughout the system, which would be maintainable, and modifiable. The project took only two years to complete, with the first stage of coding up and running for "nonsystem users" in only six months.

Beyond the Laboratory

While the GE collaboration with Bell Laboratories and Project MAC continued, other computer manufacturers, in addition to IBM, became interested in providing time sharing. Digital Equipment Corporation was the first computer manufacturer to announce publicly its intention to produce a commercial general-purpose multiple-console time-sharing system in 1964.[141] Other computer manufacturers were not far behind. IBM announced its planned time-sharing machine, the 360/56, in August 1965. Also in 1965, several other computer manufacturers, including Control Data Corporation, Scientific Data Systems, Sperry Rand, and Burroughs, announced time-sharing systems.[142] Richard Wright, in charge of IBM's time sharing, recalled that the computer industry was in a "state of agitation" because time-sharing appeared to represent the future for computing facilities.[143] Computer manufacturers sought to respond to the desire for time sharing. The annual Association for Computing Machinery (ACM) Conference in 1965 featured an all-manufacturer panel on time sharing, but it was clear that "the tremendous software effort needed to make these concepts efficiently and universally applicable [had] just begun."[144] Examining computing trends at the end of 1965, another observer noted that perhaps the most important trend was "the broad acceptance of time-sharing employing remote consoles as a technique to be used now rather than as an object of exploratory research."[145]

While commercial computer suppliers were interested in time sharing, their approach was different than the approach of MIT and Project MAC. Time sharing, as universities and research organizations developed it, did not fit the needs of commercial organizations. Universities and research organizations had many small, independent jobs that had simple input and output requirements and ran only a few times. In contrast, commercial organizations had large tasks that required several programs to work together and to process massive amounts of data. The same series of programs was rerun with new data, whereas in research environments new programs were written frequently. Commercial users were also content with a smaller variety of computer languages. Special-purpose systems, allowing quick results and offering one or two standard languages, were adequate for commercial use. The kind of applications useful in the commercial environment, such as order entry or account displays, did not require time sharing. Rather, they used transaction-processing systems that continued to improve through the 1960s. Perhaps the biggest difference between research and commercial environments was economic. Commercial environments needed to show that the new time-sharing systems both increased productivity and saved money. Only increased productivity was an objective of IPTO and its research community.

Computer service bureaus, which offered customers computer processing for a fee, played a significant role in acquainting commercial computer users with time sharing. These bureaus were used by small companies that could not afford their own computers, and were used by other companies to process specific types of applications, such as payroll, or to delay purchasing a new computer. Customers signed up for the service and then paid according to the amount of computing they used. Time sharing extended the services available from existing service bureaus. Registered customers used their telephone and a modem to access the time-sharing system.[146]

Commercial time-sharing service bureaus generally offered the limited type of service provided by systems such as JOSS and the Dartmouth time-sharing system. Early commercial time-sharing systems included KEYDATA, by Adams Associates, under the leadership of Charles Adams, who had worked on Whirlwind during the development of the SAGE system; and TELCOMP, by BBN, which reflected an attempt to commercialize BBN's early work in time sharing.[147] IBM offered QUIKTRAN, an on-line system based on FORTRAN.[148] GE offered a commercial time-sharing service based on the Dartmouth system, created by a different division of GE from the one participating in the MULTICS project.[149] These commercial time-sharing service bureaus grew rapidly.[150] By 1967, twenty separate companies were offering commercial time-sharing services.

But everything was not rosy in the computing world's reception of time sharing. *Computers and Automation* reviewed some disadvantages of time-sharing systems in a 1965 editorial that compared time-sharing systems both to stand-alone machines and to centralized batch processing. The editor noted five disadvantages of time sharing. First, the likelihood of a single failure of a time-sharing system was greater than that of several smaller machines failing simultaneously. Second, time sharing increased the demand for timely access, leading to increased pressure for computer time and availability. Third, scheduling done by computer programs showed "impersonal, mechanical arbitrariness" by comparison with scheduling done by humans; humans could be convinced that you really did have a high-priority task. Fourth, time-sharing system software was complex. As another mid-1960s observer noted, "the system devotes significant time to administering and keeping track of itself rather than providing service to the user." [151] Finally, the editorial questioned the cost of large-scale time sharing. Computing power was steadily dropping in price, "so it seems likely that soon more people can have more computing power for less cost by avoiding the unnecessary troubles of a time-shared computer system." [152] Whereas the researchers were enthusiastic about time sharing as an economical way to achieve interactive computing, to many in industry time sharing appeared uneconomical. To them it meant a loss of the full capacity of a computer due to the computer time used by the system to maintain itself and its time-sharing users.

Some critics questioned the soundness of time sharing as a research area. In a November 1965 letter to the editor of *Datamation,* Louis Fein, a computer industry consultant, expressed amazement at the way the computing community had "overwhelmingly and almost irreversibly committed itself emotionally and professionally to time-sharing." Referring to interest in time sharing as "mob behavior," Fein chastised the "promoter-scientists of time-sharing" for their lack of a scientific approach to time-sharing research. He pointedly referred to the MIT influence and prestige as part of the problem with time sharing.

> Many normally hard-nosed manufacturers, programmers, engineers, salesmen, analysts, scientists, users, educators, and stock market traders seem not at all to be concerned, as they are for other proposals, with whether the promoter-scientists of time-sharing are systematically, scientifically, and economically exploring it; or whether the probable outcome will be worth the investment; or whether time-sharing is good or bad and for what purposes. . . . It would be illuminating to obtain some understanding of the causes of this mob behavior. Is it the prestige of MIT and Project MAC? Had

Licklider at ARPA not funded Project MAC at MIT but at some other place would we not now have this irrational, bandwagon behavior?

Fein concluded that time sharing would work because the computing community and "especially Uncle Sam will continue to pour whatever time, energy, resources, and money it takes to make it so."[153] It did appear that time sharing was on shaky ground. As the 1960s drew to a close, large-scale time-sharing systems had severe problems. Yet even as the largest and most ambitious time-sharing projects appeared to be failing to meet their goals, interactive computing was becoming accepted for a wide range of uses.

Because throughout the 1960s batch systems continued to improve, not everyone saw time sharing as the principal approach to improving computer speed, reliability, and productivity. The advocates of time sharing sometimes overlooked the realities of batch processing. A critic of time sharing, Robert Patrick, pointed out in 1964 that the comparisons being made between time-sharing and batch systems were unfair because they represented only batch systems that lacked sufficient capacity. Such limited capacity would result in delays to the programmers no matter how a computer was used. From his experience with computer services, Patrick argued that better batch systems, involving short turn-around times, would be as useful to programmers as the time-sharing systems being discussed: "The turn-around time problem stems from the fact that our capacity is saturated. You can get good turnaround time if you have more computer capacity than you require. But if everybody immediately pounces on a console the minute the machine is turned on in the morning then some of them are going to have to wait."[154]

Time-sharing enthusiasts criticized batch systems for the hours and days involved in batch turn-around without giving much attention to what Patrick considered to be the root problem: instead of seeking new ways of using existing computing power, research should focus on solving the problem of the general scarcity of computing resources. Time-sharing systems also had problems when overloaded. The turn-around between keystrokes in time sharing would not be hours, but finding an available terminal could cause a delay of hours. Waiting for a slow time-sharing system to respond after entering a command left users feeling frustrated, sometimes to the point of finding something else to do until the system was less busy. Although enthusiastic about time sharing, Maurice Wilkes emphasized the problems of overloaded time-sharing systems in an assessment of the CTSS system after the Project MAC summer study. "A conventional computer system, when overloaded, may be a source of dissatisfaction, but it continues to do its job after a fashion. An on-line time-shared system, on the other hand, loses its point altogether when serious overloading sets in." [155] Patrick noted that if the number of users greatly exceeded the capacity of a time-sharing system, some users would have no access to the system, because only a limited number of people could be logged on at one time. In his view, time sharing was unnecessary because the major problem was the lack of capacity which caused delays in turn-around, rather than any inherent weakness of the batch approach. This consideration was a substantial concern for commercial users who needed to process large quantities of data in batch form, but was unimportant in research environments where the computer needs of researchers were the primary work required of the machine. [156]

It was apparent during the mid-1960s that computing power was becoming cheaper, but the consequences of the trend were unclear. Increased computer performance indicated to some people that smaller machines should be used directly, and to others that the larger time-sharing machines would have an even greater advantage in speed. The debate played itself out in trade publications. A 1965 article in *Computers and Automation* argued that the speed advantage in large-scale time-sharing systems was fleeting. "At best it is of temporary duration, since the speeds of small computers are rapidly rising." [157] The question of the "free-standing" computer versus the time-shared system was addressed again the following year in *Datamation,* with the time-sharing proponent concluding that "when it comes to general-purpose computing, the economies of scale are swinging increasingly in the direction of large machines. Computers offer such large economies that the large-scale computer utility seems imminent." [158] The advocate of small computers noted that "computing-power-per-dollar in small machines is increasing rapidly and will continue to do so," thus making the smaller machine an attractive alternative to large-scale time sharing. [159]

Consequently, time sharing had a more powerful rival than batch processing. Computer hardware continued to get better and cheaper, which made it possible to share computers among a limited number of people and still have enough capacity to get work done. Since smaller yet more powerful computers were available, some critics of time-sharing believed that it was a waste of research activities. [160] One of these critics, Severo Ornstein, who had worked at Lincoln Laboratory and on the SAGE system, later recalled, "A tremendous amount of effort was not directed at dealing with people's problems but with the problem of sharing resources. It derived from the notion that machines were always going to be big and expensive and that you therefore *had* to share them." [161] The new, smaller systems also deemphasized the need for large-scale time-sharing systems, though this was not obvious to IPTO in 1965 when it decided to fund the large-scale time-sharing project MULTICS at MIT.

Confronting Limits and Transforming Computing

Licklider's and IPTO's support, enthusiasm, and insistence on time sharing whetted programmers' appetites for the benefits of time sharing, but the technology lagged behind the expectations. Licklider believed that it was imperative to develop large-scale time-sharing systems as computers continued to get bigger (physically larger and more powerful). Nevertheless, moving time sharing into commercial markets and transferring technology from the early, relatively small time-sharing systems to large-scale systems turned out to be more difficult than anyone had imagined. The mechanisms for sharing computers between many users were complex and difficult to develop, and the path to large-scale time sharing was precarious.

As noted above, MIT, GE, and Bell Laboratories organized a project to develop a large-scale time-sharing system. Project MAC personnel would bring their extensive knowledge of time sharing to the project; and GE would bring a knowledge of manufacturing, marketing, and maintenance; while Bell personnel's experience writing software systems for a wide variety of users would be of inestimable value in ultimately reaching a wider group of users and thus making the system commercially viable. Project MAC personnel convinced the other partners in 1964 that it was possible to have a new time-sharing system available for general MIT use by the end of 1965.[162] By March 1966, the system was well behind schedule. Fano reported that he expected the system to be ready for general MIT use on 1 September 1967.[163] By December 1968, Project MAC had reported only that the MULTICS system was running and that the "path to reliable operation is clear." They had a new goal of a "solid, reliable, useful MULTICS System on the 645 computer by the end of September, 1969."[164] The other partners began to be discouraged. Moreover, GE withdrew its model 645 computer from the commercial market a year after its announcement, when it realized that "what we had on our hands was a research project and not a product,"[165] a perception that lessened the company's incentive to continue in the MULTICS project.

IPTO, which funded the Project MAC portion of MULTICS, conducted a project review at the end of 1968. Licklider, who by this time had become the director of Project MAC, after two years at IBM after leaving IPTO, felt that it was necessary to examine the negative impact MULTICS was having on the rest of Project MAC research, because the MULTICS project was late and it continued to use a significant part of the available resources.[166] By the end of December 1968, he and those involved with MAC and MULTICS at MIT considered their options for how to proceed. He noted, "There is some chance of convincing ourselves and the world that the Multics project has already been successful. It may be possible to bring Multics to a point, within say six months, within which it would be obvious to many that the operating system, itself, is indeed successful."[167] MULTICS was available for general use at MIT in October 1969, nearly four years later than originally planned.[168] By the end of 1971 the MULTICS system was able to support fifty-five simultaneous users with satisfactory response time, but it still had problems: its users at MIT experienced "an average of between one and two crashes [unexpected system failures] per 24 hour day."[169] In 1972, the IPTO director noted that "the Multics system effort which dominated MAC for years is now at a minimal level."[170] MULTICS no longer dominated Project MAC, but it still was not finished. In January 1973, "developing a demonstrably sound version of the Multics operating system" was still included in the Project MAC plans.[171]

Meanwhile, the other partners in the MULTICS project, GE and Bell Laboratories, were no longer part of the project. In early 1969, after three years of effort, Bell Laboratories withdrew from the project. Later that same year, Kenneth Thompson, Dennis Ritchie, and others at Bell Laboratories who had been working on the MULTICS system began developing the UNIX time-sharing system. UNIX is now a widely used operating system and is a de facto standard across a wide range of computers. Named as a play on "MULTICS," it was designed to provide the programmers at Bell Laboratories with a good programming environment. Because UNIX offered quick responses and had minimal system overhead, it was popular and used extensively in Bell Laboratories. MULTICS was big and ambitious; UNIX was more limited. As a product of Bell Laboratories, it remained noncommercial as it developed incrementally. Its popularity can be explained in part by its being simpler than large-scale time-sharing systems. It also gained popularity because university computer science departments could obtain it free of charge from Bell Laboratories. UNIX was also written in a higher-level language, making it relatively easy to transport to other hardware systems.

While GE intended the model 645 to give "top-of-the-line prestige luster" to the 600 line of computers, its experience with the 645 was a disaster. Weil testified: "As it came out, because of its lateness and its difficulty, it represented very little to General Electric except a drain on its resources."[172] In October 1970 Honeywell acquired GE's computer division, which included the MULTICS project, leaving Project MAC the only one of the original three partners to remain involved with MULTICS. Although the MULTICS effort had difficulties, it did survive into the late 1980s as a commercial offering of Honeywell Information Systems. Honeywell offered MULTICS commercially in 1973, and the Project MAC and Honeywell collaboration continued until 1977.[173] Honeywell continued to market the system until the 1980s and maintained a small but loyal group of customers.[174]

Some of the ambitions of time-sharing enthusiasts could not be accomplished through time sharing. Licklider summed up his thoughts in preparation for an internal project meeting on "The Future of Project MAC" in 1970. His disappointment with the MULTICS project is clear.

> After joining Project MAC [in 1967], I began to realize that the then-prevalent man-computer communication via typewriter (CTSS) was rather far removed from the fast-response graphic interaction of which my dreams were made and that the "on-line community" was not so much a community of users as an in-group of computer programmers. . . . It seems to me that the efforts of the 1960's fell somewhat short of what is needed and achievable in very close man-computer interaction . . . and quite short of the dream of an "on-line community." [175]

In Licklider's view the substantive use of the on-line computer facilities by nonprogrammers "now identifi[ed] itself with the realization of the concept of a network of multi-access computer systems." [176] While Licklider was disappointed with the progress toward symbiosis made by time sharing, he and others envisioned the larger-scale sharing that could be accomplished by a network of time-sharing computers.

Other manufacturers attempting large-scale time sharing also encountered serious problems and delays. IBM ran into severe software problems with its time-sharing system. Although the IBM 360/67 hardware was first delivered in the summer of 1966, IBM's software was delayed. By July 1966 the company doubled its estimate of the number of lines of code required to implement its time-sharing software (named TSS).[177] In January 1967, TSS continued to experience severe problems. IBM announced that the first release of TSS would be limited to "experimental, developmental, and educational" uses. It produced eight successive versions of TSS, most of which included the redesign of major components of the system. By 1969 the programming expenses of TSS were forecast at $57.5 million, with losses of $49 million for the entire 360/67 model.[178] Other IBM products offering time sharing eventually supplanted TSS.

Owing to the delay of IBM's release of TSS, other time-sharing operating systems were developed for the 360/67. These included the Michigan Terminal System, originally developed on a 360/50; and the Cambridge (later Conversational) Monitor System (CMS), developed by IBM's Cambridge Research Group (whose purpose was to test different operating systems). CMS provided terminal services for a single user in a separate "virtual machine," which mimicked the capabilities of the actual machine. The Control Program (CP) could establish up to twelve "virtual machines" supporting twelve users. IBM developed it on the 360/40 and later moved it to the 360/67 when the latter became available.[179] IBM did eventually offer large-scale time-sharing systems through the operating system VM/370 (which grew out of the CP/CMS work at the Cambridge Research Center) and through the Time-Sharing Option (TSO) on the OS/360 and OS/370 operating systems.

While large-scale systems had many difficulties, the basic idea of time sharing, and its realization in a number of small and more specialized systems, had become widespread throughout the computing world. By 1970 the term *time sharing* encompassed all on-line computing and transaction processing, as well as general-purpose time sharing.[180] It had a firm footing in commercial as well as research environments. For example, the airlines had mature systems, stock quotation systems such as NASDAQ were in use, the air traffic control system was functioning, and a variety of commercial on-line services existed. While large-scale general-purpose time sharing, such as MULTICS and IBM's TSS, were struggling, this expanded definition of time sharing made it appear that time sharing was overwhelmingly successful when, in fact, its success was much more modest.

IPTO, Time Sharing, and the Military

IPTO's (and DARPA's) prime objective was to develop and transfer new computing technology to military settings. To aid the transfer of technology from contractors to the military, IPTO needed either to sponsor or to oversee the development of specific applications of new computing system components. In the late 1960s, with this objective in mind, IPTO funded the development of the Advanced Development Prototype system (ADEPT) time-sharing system for the military. IPTO wanted a system that would demonstrate the potential of time sharing in an operational military environment.[181] SDC began work on ADEPT in January 1967.

Clark Weissman (head of SDC's Programming Systems staff), Alfred Verhaus (head of SDC's Data Base Systems staff), Eugene Jacobs (head of SDC's Programming Technology staff), and Clayton E. Fox led the project under the supervision of Jules Schwartz, who was then Technology Director at SDC. Programming of the system began in June 1967, and five months later ADEPT Release 1 was operating on an IBM 360/50; it was later transferred to an IBM

360/67. The basic ADEPT time-sharing system and its data management component were completed by March 1968. IPTO contracts accounted for most, if not all, of the computer programs used in this system.

ADEPT contained three major components: a time-sharing executive program, a time-sharing data management system (TDMS), and a programmer's package. The programmer's package was designed to use a JOVIAL compiler but could also handle FORTRAN and COBOL, and it allowed experienced programmers to write and debug programs on-line without interrupting computer service to other users. For less experienced users, the package included a terminal input interpreter known as TINT, which gave programmers the option of writing programs on-line in a JOVIAL-like language. TINT provided "tutorials" in programming commands and featured "step-by-step error checking suited to the inexperienced user."[182]

ADEPT's time-sharing executive program was designed to serve approximately ten concurrent user jobs.[183] The executive included the capacity for command and control managers to use conversational interaction between the computer and user, as well as "instantaneous response to multiple users," and file security that prevented both unauthorized access and accidental damage or destruction. Although users could work independently without affecting one another, they could also work together on the same problem or set of data.[184]

SDC held symposia on the ADEPT system in April and July 1968. The symposia were well attended by representatives from the DOD and other government agencies and featured "demonstrations showing a nonprogrammer querying an on-line data base on the status of Vietnam forces [which] aroused wide interest in ADEPT's capabilities."[185] In 1969 SDC offered TDMS to the public business market in the form of the Commercial Data Management System (CDMS). At that time, CDMS customers included Atlantic Richfield (which used the system for analyzing customer sales), Gulf and Western Industries (which used it for managing personnel files), Continental Can Corporation (which used it for developing a "container index"), and Texas Instruments (which used it for managing financial and personnel data). Despite SDC's high hopes for CDMS's success, which were based on the military's enthusiastic response to TDMS, CDMS failed in the public marketplace. Many users saw it as being not specific enough to their own companies' needs. Also, it ran on its own ADEPT operating system rather than on the popular IBM OS/360.[186]

SDC worked on ADEPT with IPTO support beginning in late 1967. IPTO initially contracted with SDC to develop a system "that would demonstrate in an operational environment the potentialities for automatic data handling introduced by time-sharing."[187] ADEPT was particularly suited for military uses. It allowed command and control managers using a "conversational" style to have "rapid, direct, on-line access to large stores of data from their own terminals, without programmer intermediaries."[188] ADEPT could be used as either a batch system or an interactive time-shared system.[189] Because it was intended for operation in a classified environment, it contained a security system that allowed users with different security clearances to work on differently classified materials at the same time on the same computer. ADEPT was perhaps the first system developed with security as a design parameter.[190]

The Time-Shared Data Management System component of the ADEPT system was influenced by SDC's experience developing time sharing for the Q-32 computer. An SDC document notes that "the SDC/ARPA Time-Sharing System, in wide use since June 1963, has provided extensive information on the scheduling and allocation processes involved in time-sharing."[191] TDMS allowed users to manipulate large amounts of data in a "conversational mode." As part of this "dialogue" the computer could ask the user to supply parameters, control information, file names, the operations to be performed, and the desired output report format. The user could ask the system to define a term, to comment or further explain a process, to indicate what next steps were permissible, to explain error messages, or to provide general tutorial help.[192]

ADEPT was at first used primarily at SDC and several government sites. For example, in 1968 the system was installed at the National Military Command System Support Center at the Pentagon and at the Air Force Command Post. The several field sites of the Air Force Command Post ran ADEPT more than eighty hours per week and provided two thousand terminal hours of time-sharing service monthly.[193] Other applications soon followed, such as the Nuclear Weapons Status System for the Defense Atomic Support Agency, and the Movements Requirement Generator (a strategic mobility model SDC developed for the Pentagon in 1968). ADEPT and TDMS also became the "underpinning" for the Tactical Information Processing and Interpretation (TIPI) system developed at SDC in 1968.[194] ADEPT and TDMS are important early examples of the integrative role played by IPTO in transferring several project research results to a single operational system. When requesting an increase in funds for hardware procurement for ILLIAC IV and ARPANET—IPTO projects to be examined in later chapters—Eberhardt Rechtin, DARPA director, pointed to the successful ADEPT system for the military under IPTO support.

Before the 1960s computers were rarely used interactively, and many people in the computing field saw little benefit in working directly with a computer. To these people it seemed wasteful of computer power for the computer to wait for human input, especially when computers filling rooms were very expensive and not very powerful—they were nowhere near as powerful as the computers found on many desktops in the 1990s. Time sharing was developed to promote interactive computer use and, as a result, stimulated important changes in the way computers were used. IPTO played a strong role in the promotion of time sharing. The successful implementation of time sharing in a variety of settings constitutes IPTO's first contribution to transforming the practice of computing, and it was a very important contribution.

Time sharing emerged in the early 1960s as a way for several people to access a single computer. The move toward time-sharing systems received a boost when Licklider became the first director of IPTO. While he held that office, he put his belief in the benefits of time sharing into action. IPTO helped to spread the techniques and enthusiasm for time sharing first to other IPTO-supported sites and later to the larger group of computer users. Time sharing represented an important contribution of the IPTO research community, involving an array of contractors across the country: MIT, BBN, the University of California at Berkeley, and SDC. General-purpose time-sharing systems were operating in 1963 under IPTO sponsorship at MIT and SDC. The system software consisted of executive routines that took care of scheduling, error recovery, input and output control, and basic commands; and of service routines that would be helpful to on-line programming, such as software to provide information about existing files, symbolic debugging aids, formatted input-output, on-line instructions, the ability to save and edit written routines, and tape-handling programs. Later, similar developments occurred wherever new general-purpose time-sharing systems were designed and put into operation.

The SDC time-sharing system was used by a variety of users from different intellectual and geographic areas and had many applications written for it. By the beginning of 1966 the SDC system was running seven days a week, fifteen hours a day, and had more than four hundred registered users, including representatives from thirty-five organizations outside of SDC, sixteen of which were military or government institutions. The remainder were universities and a few private organizations doing work of interest to IPTO.

Similar statistics could be given for the MIT system, but such statistics are unnecessary because the MIT system was intended to be more than a time-shared computer system. One of the goals of Project MAC from its inception was to spread the word on time-sharing, "disseminating information and influencing the technical community." Because the project was well financed and could hold classes, conduct studies, and give lectures, the ideas it promoted spread quickly to the larger community.

IPTO provided the financial means to expand the idea of time sharing and get more people working on the problems of large-scale, general-purpose time sharing. This allowed a larger number of users to become accustomed to the benefits of on-line computer interaction. One of the most visible large-scale time-sharing systems was MULTICS. The MULTICS project, which was partially supported by IPTO to build a prototype computer utility, made many improvements in the techniques of large-scale time sharing. The MULTICS system included the modular division of responsibility, dynamic reconfiguration, automatically managed multilevel memory, and protection of programs and data; and it was written in a high-level language.

Other time-sharing system projects with specific, limited goals were also supported by IPTO. Project Genie at the University of California at Berkeley modified a SDS 930 computer for time sharing. The commercial model, the SDS 940, reflected Project Genie's modifications. The Project Genie software was available to SDS users and later became part of the public domain. IPTO also supported the TENEX time-sharing system in 1969. The TENEX system ran on Digital Equipment Corporation PDP-10 computers, provided an environment for running interactive LISP programs, and was also used extensively by people on the ARPANET.

Commercial attention to time sharing came quickly. Most computer manufacturers had announced and were working on time-sharing systems by 1965. By the early 1970s, time-sharing systems were commercially available and were being used successfully. Relatively small machines with functioning time-sharing systems were readily available, particularly in research environments. For example, Scientific Data Systems successfully marketed a system based on research done at the University of California at Berkeley; and in 1972 DEC used TENEX as the basis for the TOPS-20 DEC operating system. General-purpose time-sharing was available on a wide range of computers, and time-sharing systems became an accepted part of the programming environment in commercial environments through the 1970s. While time sharing was successful and well received among many computer research groups, making large-scale time sharing successful in a commercial environment presented many problems. The large-scale time-sharing systems begun in optimism in the mid-1960s began to encounter major problems and became extremely complicated and unwieldy.

IPTO's support of time sharing clearly affected IBM, leading the company into a commercial time-sharing offering, although it was not fully prepared to make such a move. By the end of the 1960s time sharing was available on a wide range of computer systems, and it became an accepted part of the programming environment in commercial organizations through the 1970s.

The time-sharing work sponsored by IPTO increased the number of advocates of time sharing and the number of researchers familiar with both the research problems and the benefits of the technique. Programmers began to use shared on-line resources effectively and realized the cumulative positive effects of shared interactive tools. While productivity increases could not be precisely measured, the programming environment clearly improved. The impact of time sharing went beyond the technical development of large-scale time sharing. Time sharing affected the work of computer scientists and engineers in every development setting and made inroads into other areas of endeavor, such as scientific research and business systems. The sharing of programs and data, as well as the communication that went on among members of a time-sharing community, whetted the appetite of some researchers for a cross-country sharing of computers that had the flexibility provided by time-sharing systems.

The users and developers of the general-purpose time-sharing systems in use at Project MAC heralded the system for the sharing it allowed among its users. Fano and Corbato described their three years of experience with the CTSS system at Project MAC in a 1966 *Scientific American* article in which they noted that programmers and other users were building upon the work of others in a way not possible without time sharing. Because, in a time-sharing system, it was relatively easy to make programs available to other users, programmers began to design their programs with an eye to possible use by other people.[195] Early in 1966 Ivan Sutherland, who was just finishing his time as the director of IPTO, noted time sharing's effect on the sharing of programs and other files among users of the same system.[196] Communicating with other users of the same time-sharing system was possible through early electronic mail systems. CTSS had a mail command available to its users in 1965, which increased communication among its users.[197]

Most of IPTO's time-sharing developments were completed by the early 1970s. The time-sharing concept was proven for small- and medium-scale use, and computer manufacturers offered commercial models with time-sharing capability. Through the 1970s and 1980s, most programming switched over to time-sharing environments, and on-line computer use became common. In spite of early development difficulties, large-scale general-purpose time-sharing systems also became common. Interactive computing was accomplished with smaller machines using simpler time-sharing systems, as well as through transaction-processing systems. Eventually, single-user systems further reduced large-scale time sharing's direct role in providing interactive computing to most people.[198] Computing in the 1990s allows more choices in obtaining computing power, but time sharing continues to play a role. Combined with powerful and economical personal workstations and personal computers, time sharing has changed the computing environment. Its techniques still allow multiple users access to a machine when it makes sense to have a centralized or shared machine, such as an expensive supercomputer or a rarely accessed special device (e.g., a specialized printer or plotter). Time sharing led directly to evidence of the possible benefits of sharing—such as a sense of community, an improved programming environment, and remote access—but it could not, by itself, achieve widespread resource sharing. The next step forward toward this goal occurred when IPTO began connecting the time-sharing centers across the country through the ARPANET, a development described in a later chapter. Before considering IPTO's role in the development of a networking system for resource sharing, we turn to another substantial IPTO contribution to the increase in interactive computing: computer graphics.

CHAPTER THREE

Getting the Picture: The Growth
of Interactive Computer Graphics

Today, it is difficult to look anywhere without seeing computer graphics or the effect of their use in the preparation of visual materials. Graphics are the vehicle for video advertising, video games, films, and even art. Twenty years ago this was not the case. But twenty years ago the seeds of this remarkable change in visual images were firmly planted. These seeds were the result of the previous twenty years' research in interactive computer graphics, the story of which was dominated initially by the military role in support of research, and then by IPTO.[1] In the 1950s, even many of the researchers who began the investigation of interactive computer graphics believed that the applications of interactive graphics would be limited and specialized. They could not know that the pictures and processes they produced experimentally would lead to widespread use of computer graphics in business, industry, science, and the arts. However, these early pioneers created a vision that directed future developments toward improved computer use and the production of powerful images for a wide variety of applications.

Interactive computer graphics emerged as part of the search for new ways to improve communication between humans and computers. Early graphical output devices included pen plotters and line printers, which made it possible for understandable characters and lines produced with a computer to be seen on paper; but the devices could not be used interactively. Beginning in the late 1940s, researchers in graphic displays were primarily concerned with investigating the technical aspects of getting an image on a screen and interacting with a computer through that image, because pictures are inherently more informative than text. In the early days of computer graphics, any image—including text—produced with the aid of a computer was considered a graphic representation. Several early graphics programming developments grew out of work with the Whirlwind computer. For example, in 1949 Charles Adams and John Gilmore, staff members in the Whirlwind project, produced the first animated graphic image using Whirlwind: a bouncing ball. These researchers were using CRTs for output and believed that such devices would become vital to any computer system, in part because they had a perceptual appeal. As Norman Taylor, a leading designer in the SAGE project, also associated with the Whirlwind project, stated in a recollection of his early days at Lincoln Laboratory, "Displays attracted potential users—computer code did not."[2] CRTs also gained increasing acceptance because they made possible a more effective interaction with computers.

In the 1950s, several researchers investigated the possibilities of digitizing photographs to input a picture. For example, shortly after Whirlwind was built, pictures were scanned on a rotating drum with a photoelectronic cell and then processed in various ways using Whirlwind's central processor.[3] In 1957 Russell A. Kirsch, at the U.S. National Bureau of Standards, used the Standards Eastern Automatic Computer system to scan the first digitized picture—a picture of his son.[4] At that time, a computer could recognize densities on the basis of the scanned picture but could not easily define which dense areas were solid objects.

SAGE involved the first extensive use of CRTs. So, once again we must point to SAGE as an important system that paved the way for a transformation in computing—in this case, interactive computer graphics. Early CRT consoles had a display generator: an interface that connected a terminal to a computer and converted the information passed between them. The viewing screens were typically round and ranged from sixteen inches to twenty-four inches in diameter; the image display areas ranged from ten inches by ten inches, to sixteen by sixteen.[5] The CRT screens had a coating of phosphorescent material. Elements of phosphor were excited by an electron beam, and an image could be produced on the screen by exciting the phosphorescent material in a particular region. The energy transmitted to the phosphorescent material by the beam decayed relatively rapidly and had to be refreshed at short intervals. Beam control was such that small regions of phosphor—called picture elements, or pixels—could be excited at any location on the screen. The resolution of a display was determined by the size of the pixel, and the addressability of a display was the number of positions that could be programmed along each axis of the screen viewing area. As systems appeared with higher resolution and greater addressability, more control over the image on a screen became possible.

The SAGE system had interactive graphics consoles that used CRTs based on those first designed for Whirlwind, and included the first input device capable of sending signals coordinated with a display.[6] The device, called a "light gun," was built in late 1948 and early 1949 by the Whirlwind laboratory's technical director, Robert Everett. The light gun was devised to cause the display on a CRT to react. By 1955, as we described earlier, air force personnel working with SAGE were using light guns for data manipulation.

At MIT, Douglas T. Ross used the light gun as part of his computer-aided design system in which an operator composed a drawing on the screen using a light gun similar to the one used in SAGE. This work, done in the Electronic Systems Laboratory (ESL), was part of the development of the Automatically Programmed Tools (APT) language, the first language for the design and production of manufactured goods. In 1959, the Computer Applications Group of the ESL met with the Design Division of the Mechanical Engineering Department to discuss the possibilities of developing systems for design and manufacture.[7] In part, this meeting led to the beginning of MIT's Computer-Aided Design (CAD) Project, headed by Ross, and the development of the Automated Engineering Design (AED) language, a software-engineering language and system-building tool, which was later used extensively for interactive graphics.[8] "Under the AED project in the Electronic Systems Laboratory, John E. Ward and Douglas Ross sought ways of reducing [the] heavy, repetitive computational load of real-time graphics display."[9]

For computer systems to be more efficient and easier to use with graphics, the system had to be faster, and therefore the computational load needed to be smaller.[10] The need to reduce the computational load was the first major problem faced by computer graphics designers. One contribution to making the computational load smaller was making the technology of the light gun smaller. At Lincoln Laboratory, in 1957, Ben Gurley and C. E. Woodward developed a light pen, a smaller version of the light gun, facilitating the use of such pointing devices elsewhere.[11] To use the light gun, researchers developed a number of applications programs in the 1950s. In 1957, a "Moving Windows Display Program" was developed which allowed brain wave data to be stored and then shown graphically. This program was used with the light gun for feedback. Another program, "Scopewriter," emerged as a way of manipulating scientific symbols. With this program, researchers could take apart and recombine characters to make "drawings."

As a result of these early successes, researchers developed a desire to interact with computers in real-time, as easily and quickly as they interacted with people. They were convinced that graphical representation was an important part of such communication. Funders supported computer graphics research to improve the efficiency and power of computer communication. From these interests sprang a new vision of human-computer communication and a change in focus from punched cards and printouts to interactive computer graphics.

This early work in graphics developments concentrated on hardware and basic programming to produce a display on a screen. Several groups became interested in funding research in graphics in the 1950s and 1960s: NSF, NIH, NASA, various military groups, and eventually IPTO. NSF and NIH wanted to meet the needs of the scientific and engineering communities for computer instrumentation, including some graphics input/output devices. For these groups, standard-model computers were perceived to be adequate. NASA's concern was for effective transference of visual image data from space to earthbound laboratories. Screen displays were important to the military for systems such as SAGE, but the problem with data for command and control of the earthbound activities that interested the military was a more comprehensive and complex one. When it assumed responsibility for R&D in command and control, IPTO focused its support on projects to obtain high-resolution, high-speed interaction with graphics, and focused on problems in displays, input devices, and related programming. Researchers at MIT's Lincoln Laboratory (first with military support and then with funds from IPTO) and researchers funded by IPTO at the University of Utah were especially influential in the birth and development of interactive graphics. Much of the seminal technical work and problem setting on which today's elegant graphics is based had been done in these two institutions, and a very few others, by the late 1960s. Problems defined by researchers at these institutions included the representation of three-dimensional objects on a two-dimensional viewing surface and the invention of hardware prototypes and software models such as drawing tablets and graphical algorithms. IPTO-supported researchers had a greater concern for the aesthetics of computer images, that is, the picture itself. Thus, they set out on the quest for high resolution as they became increasingly aware of the overwhelming potential of interactive graphics.[12]

Becoming Interactive: Creating a Vision and Setting Problems

Licklider's Graphics Program

In graphics research as in time sharing, Licklider sought to capitalize on activities that he thought could be useful to enhancing command and control systems. IPTO's early support of graphics, although tentative, built on the foundations developed under DOD support. Licklider's overall objective was to foster computing developments intended to advance human-computer communication in general. Project MAC, one of the first IPTO contracts, drew on some of the early graphics work at MIT and Lincoln Laboratory and included work in graphics in its overall plan. Recall that the project had been funded by IPTO to be an interdisciplinary project that would "bring into being an interim time-sharing system and . . . advance to the greatest degree possible" several areas, including computer-assisted design.[13] Project MAC was using a new display for graphics, nicknamed the Kludge, which had been developed by the MIT Electronic Systems Laboratory as part of the CAD Project. As noted above, the air force funded the CAD Project through the ESL; and later the CAD Project received some support (in the form of office space and computer facilities) through Project MAC.[14] In contrast to the more general research and development work at Lincoln Laboratory, CAD Project graphics were intended to address product design and manufacture.[15] This display was the first console developed in the CAD Project, and it became part of the original MAC computer system.[16] The Kludge included hardware for rotating and scaling objects in three dimensions under user control. Because of this capacity, it became a useful tool in several fields. Using the Kludge, Steven A. Coons, professor of mechanical engineering, developed a surface-patch technique that was used by the Naval Architecture Department in designing ship hulls; Cyrus Levinthal, professor of biology, experimented with three-dimensional models of molecular structures; Kenneth Stevens, professor of electrical engineering, and his group developed vocal tract displays for speech analysis; and Michael Dertouzos, professor of computer science, used the display in his development of electronic-circuit design techniques.[17]

The Kludge was the predecessor of the Digital Equipment Corporation Model 340 display, a subsequently significant commercial product.[18] In addition to funding the graphics effort included in Project MAC, Licklider also issued a contract to the Applied Logic Corporation to improve graphics displays in order to help mathematicians developing theorem-proving programs. The lack of such displays was viewed as a stumbling block in making use of such programs.[19] IPTO's support played an important role in promoting the eventual proliferation of interactive computer graphics through applications of this kind.

Another objective of IPTO's early interactive computer graphics program was the development of drawing systems that would both enable a user to communicate with the machine and enable the machine to illustrate complex information in a way understandable to people. Soon after the early steps had been taken in producing input devices for use in drawing, this problem of human-machine interaction received attention. Since one of IPTO's missions was to make computers easier to use, the office supported projects aimed at developing the advanced programming required to improve interaction.[20]

Besides attending to graphics work at MIT, IPTO wished to capitalize on the early graphics work done at the Rand Corporation, supported by the air force. The Rand Tablet, completed in 1964 by a group managed by Thomas O. Ellis, was an alternative input device for interactive computer graphics. Developed in order to explore more efficient human-computer communication, it was the primary interactive communication device in the system. IPTO funded a graphics project, directed by Keith Uncapher, at the Rand Corporation, that resulted in both the tablet's construction and the creation of GRAIL, a Graphic Input Language. In the 1964 proposal for GRAIL, the intentions of the project were stated: "It will not *force* upon the user language devices which are inefficient for his purposes. The system's efficiency will be subordinate to the user's needs."[21] The original installation used the JOHNNIAC computer and a display with the tablet and stylus.[22] Later the tablet was connected to an IBM 360/40 and was used in programming.[23] The tablet aided real-time computer recognition of hand-printed text, could produce precise boxes from loosely drawn ones, and was used in making flow charts to assist in programming.[24] Rand researcher Thomas O. Ellis stated the developers' hopes for the future:

> The large amount of research being devoted to automated design will depend more and more on this kind of intimate contact with machines. Whether the research vehicle is oriented toward the design of automobiles, bridges, mathematical models or computer programs, it is resulting in a new way of life — an operational technique which can reduce design turnaround time.[25]

Today's CAD systems certainly fulfill this agenda.

The Rand Tablet had a wire screen below its surface. Each wire had a current passing through it which was affected by a magnetic field change produced by the movement of a stylus placed on or near the tablet. When the stylus was placed near the tablet, its position was shown on-screen by a point of light; and when the stylus was depressed on the tablet, the stylus switch closed and the system provided an "inking" response. By positioning the stylus and making ink appear on the screen, the user could produce finely detailed drawings. On-screen, there were several buttons representing system functions. There was also a list of command words that could be dragged to the "COMMAND" area of the screen with the stylus. If the word was dragged to another area of the screen, it would disappear. The tablet facilitated the use of flow chart symbols and text and had controls for doing the tasks—such as drawing, erasing, modifying, moving, connecting, page-turning, and referencing—necessary for organizing a program. The original tablet was ten inches by ten inches, and a larger (thirty-two inch by thirty-two inch) tablet was later constructed.

Three types of drawing tablets eventually emerged from various laboratories. One, like the Rand Tablet, had a mesh screen below its surface which sent signals to a receiver in the pen-like stylus. The second type had a pen-like device that emitted signals that were received by a grid in the tablet and caused an on-screen response when the pen was close to the tablet's surface. The third had a stylus used to draw on a pressure-sensitive surface. All three types are still in use. By 1969 a tablet developed by Sylvania was being used at Lincoln Laboratory with Rand Tablets in the time-shared system, and tablets had replaced light pens on the TX-2 consoles at Lincoln Laboratory.[26] At about the same time as the Rand GRAIL Project, researchers at MIT's Lincoln Laboratory were working on a character-reading input device that could relay information to the computer by comparing handwritten text to prototypical letters.[27]

One of the problems with using a tablet in a time-sharing system was that the response time of the computer was psychologically too slow for the user. Whereas a keyboard console was considered interactive if the user received some response from the system in a few fractions of a second and some computational response within several seconds, the system response to a tablet needed to be in milliseconds and computational response needed to occur in less than a second.[28] At Rand, IPTO-supported work to solve this problem and "couple" the user to his/her work so that the machine would seem to disappear continued during this period.[29] System Development Corporation, an offshoot of Rand, introduced a new buffer system for time-shared graphic tablet displays in 1969. At the time, typical buffers were external interfaces between tablet and computer. The SDC system included a buffer area internal to core memory, placing the data nearer to the user and reducing the interface time.[30]

Unlike Licklider, Sutherland came to IPTO with substantial research experience in graphics. In 1962, he had introduced Sketchpad, a system in which line drawings were made on a CRT interactively.[31] He had begun working on the system in 1961 at Lincoln Laboratory, while pursuing Ph.D. research under the direction of Claude Shannon, with funding from the military services. Sketchpad was developed on the TX-2 computer, which had an x-y point plotter display: each of the points it took to make up an image was plotted on the CRT; and the display was used interactively with a light pen (to designate the areas to be changed) and a control board with a series of buttons (for commands such

Fig. 2. *Left,* Sketchpad represented the state of the art in computer graphics until IPTO invested heavily in developing graphics capabilities in the late 1960s. Ivan Sutherland, second IPTO director, developed the Sketchpad program as a graduate student at MIT. Reprinted with permission of MIT Lincoln Laboratory, Lexington, Massachusetts. *Right,* Modern computer graphic representing human DNA. Photograph courtesy of Cray Research, Inc.

as "draw" and "move"), toggle switches, and analog knobs. A combination of switch settings, knob positions, button-pushes, and pen flicks were used for making and changing an image, for instance, erasing or moving something on the screen. A user could represent a straight line by designating the beginning and end points (called "rubberbanding"). When the user drew a rough curve on the screen, the computer could "correct it," showing the curve as smooth. The system was very effective in design of simple objects. For example, a 1963 demonstration film of Sketchpad showed the design of a bolt as it was sized and placed in position.[32] In his dissertation, Sutherland credited Welden Clark of Bolt Beranek and Newman with a similar program, developed on a PDP-

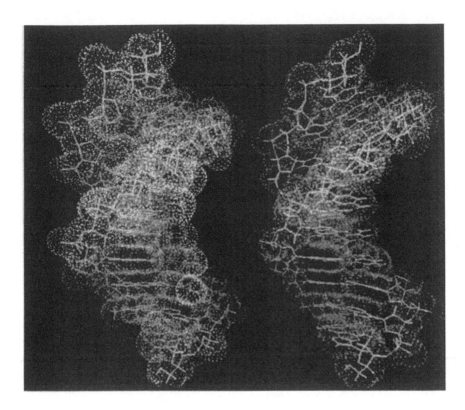

1 computer, that was used as a planning system for hospital design.[33] While Sketchpad had an experimental application, it was primarily an experiment in human-computer communication, fitting in well with the subsequent IPTO program.

One of the most important aspects of Sketchpad was its underlying data structure, which enabled the user to describe the relationship between the parts of objects to be pictured. Before this, a computer representation of an object had been a representation of a picture of the object rather than of the object itself. In Sketchpad, Sutherland introduced a clear distinction "between the model that was represented in the data structure and the picture that you saw on the screen."[34] The distinction made possible the representation of component parts of the object, and these could be constructed and manipulated interactively, as illustrated in figure 2a. He accomplished this representation through a technique he called a ring structure.

Sutherland explained Sketchpad in a paper presented at the 1963 Spring

Joint Computer Conference as part of an influential MIT symposium on computer-aided design. Owing in part to this symposium, Sketchpad became the most often cited development in early interactive computer graphics. Its capabilities continue to be the foundation of most graphics systems today (see fig. 2b), including the paint and computer-assisted design programs used with microcomputers.[35] Licklider evaluated Ivan Sutherland's Sketchpad program as one of the most significant developments in human-computer communication.[36]

Sutherland Introduces a Focused Graphics Program

Following Sutherland's arrival at IPTO, the office played a more important role in several developments that led to interactive graphics systems. Early graphics projects had attended to a few experimental applications, such as flow chart production, simple drafting and product design, and flight simulation. Shortly after his arrival at IPTO, however, Sutherland gave a contract to the University of Michigan for CONCOMP, a major project in graphics. The objective of this project was "to provide the theory and programs for a general network service system (GENESES) which, through the use of computer graphics in a conversational mode, will allow the System Engineer to specify specialized graphical problem-oriented language systems for the description and solution of diverse network problems."[37] CONCOMP continued until 1970, but Michigan did not evolve into an IPTO-supported center. Subsequently, IPTO did not consider the Michigan graphics project a success. In a 1966 letter to Taylor, Sutherland expressed concerns about the work being done at the University of Michigan, cautioning Taylor to "watch what happens there carefully";[38] and in 1968 Taylor did not renew the university's funding for graphics.[39]

In 1965, within a few months of his arrival at IPTO, Sutherland designed the Graphic Control and Display of Computer Processes program to

> couple the advanced computer drawing techniques now available with complex internal computation programs. Such coupling will make it possible to control complex computation by drawing and display the results of the computation as drawings. . . . The aim is to build tools necessary for controlling a wide variety of computations graphically.[40]

Under this program, IPTO funded a number of institutions, most notably Lincoln Laboratory. The programming concerns included the development of a unified list structure language (CORAL) at Lincoln Laboratory and the CAD Project's work on relating the language of graphics to more commonly used programming languages. Like Sutherland's "ring" structures, CORAL represented graphical models in computer memory as complex list structures.

In writing his justification for funding a contract to Lincoln Laboratory under this program, Sutherland (then the director of IPTO) stressed the importance of visual images to understanding: "In many cases, we may be entirely unable to think through a problem, or even recognize one, without an appropriate picture. . . . An ability to construct, manipulate and observe complex pictures of computed phenomena may radically change our ability to grasp complex ideas."[41] Lincoln Laboratory was to explore computers' ability to use drawings for control and display of computer processes, including a flow-chart compiling system capable of producing machine code from hand-drawn flow charts. While the CAD Project focused upon mechanical design, Lincoln Laboratory graphics researchers emphasized electrical and logic problems.[42]

Like Licklider, Sutherland wanted to capitalize on the work of his colleagues in graphics, but he wanted to do so within his overall program plan. For example, he wanted to support further graphics research by Roberts, his former office mate at MIT who had taken a position at Lincoln Laboratory. Roberts had combined matrices and perspective geometry to achieve a program for perspective transformations done in four dimensions. Roberts's program could recognize constructs of simple planes as objects and could draw other views of the objects. This work led to his dissertation in 1963, the results of which are still widely used. On the early work on graphics done at Lincoln Laboratory, Roberts later commented, "We kept on developing the technology in the graphics area to the point where we couldn't quite see how anybody was going to use it for the next decade — in fact they didn't much — because what happened was . . . we were . . . far ahead of what the computers were economically able to support."[43] At this time, he also explored the process of digital photograph transmission, in an effort to reduce the amount of data one needed to transmit faithful reproductions, making a further contribution to graphics, and wrote a program to use when scanning and digitizing a photograph.[44] Part of this work concerned an investigation of three-dimensional graphics, and this led Roberts to design the first hidden-line program, a program to remove boundary lines of an object which are unseen when viewing a real object from one perspective. All of these results became important in his later IPTO-supported work.

In his plan for the Graphic Control and Display program, Sutherland revealed part of his growing vision of the uses of computer graphics — and that of other Lincoln researchers he had worked with earlier — by suggesting the possibility of understanding social structures through images produced from numerical data and solutions to programming problems derived from pictures. He also pointed to what he saw as an important shift in computer graphics which went beyond the representation of objects toward the representation of ideas.

In a week-long course on computer graphics offered at the University of Michigan in June 1965, Sutherland gave a lecture identifying ten key unsolved problems in graphics.[45] (This course was one of the first university short courses on computer graphics, and was taught by several people who were connected to IPTO-funded projects.) The problems Sutherland discussed in his presentation involved hardware characteristics and cost, problems of technique, coupling problems, and problems in describing motion, halftone capability, structure of drawings, hidden lines, program instrumentation, logical arrangement, and working with abstractions. He noted the lack of a complete graphical console at reasonable cost, and pointed out that researchers had still to develop languages that included motion capability in their semantic structure. He called for extending Lawrence Roberts's work from three-dimensional solid objects to arbitrary curved surfaces. He also pointed to the difficulty inherent in presenting information in a "sensible or artistic or logical arrangement" when the terms had not been clearly defined. This course and the Graphic Control and Display program were turning points in IPTO's graphics program.

Computer graphics, as exemplified in Sketchpad, were initially two-dimensional line drawings. An interest in depicting objects in three-dimensional form soon emerged. The Sketchpad III technology was an attempt to achieve interactive three-dimensional graphics with the added capability of simultaneous on-screen presentation of frontal, side, plan (overhead), and oblique views of the object being drawn. Sketchpad III was developed by Timothy E. Johnson on the basis of the data structure and related utility programs in the original Sketchpad program. Johnson's work was part of the MIT CAD Project, and was supported by ESL and Project MAC as master's thesis activity.

In 1966, with IPTO support, Roberts developed the Lincoln WAND, which employed a four-dimensional perspective transformation program to draw in three dimensions.[46] Four transmitters were placed on the edges of a two-dimensional grid, and the WAND contained an ultrasonic four-channel receiver that picked up pulses from the transmitters as it was moved in the air closer or farther from them, thus providing a third dimension. The WAND was developed to be used with the TX-2 computer. The computer calculated the x-, y-, and z-dimensions from the four distances.

The mouse was another input device that later became important for interactive graphics. Douglas Englebart of Stanford Research Institute created it in 1964 as part of a project funded by IPTO and NASA to construct one of the early workstations. SRI researchers tried several ways of manipulating images and text on-screen while simultaneously using a keyboard. However, in SRI comparison studies the mouse appeared more efficient and effective than other devices, including a joystick and a light pen. The mouse was presented at the 1968 Fall Joint Computer Conference during a multimedia show conducted by Englebart, other SRI staff members, and several volunteers. Englebart and his colleagues demonstrated split screen graphics and text on the system and superimposed image and text. They also demonstrated the use of "windows" by "cutting a hole" in the screen and seeing an image in that space. Englebart used the mouse to select and change things on the screen while the audience watched the cursor follow the movements of the mouse.[47]

With only a few years of support in the second half of the 1960s, IPTO researchers had accumulated an impressive list of graphics innovations, especially in the much-needed areas of input devices, displays, and programs. Robert Taylor summed up these accomplishments at the end of 1967 and speculated on their future development:

1. Computer algorithms developed and tested for the efficient production of gray-scale pictures of solid objects with hidden area elimination. Should lead to real-time production of computer generated television type images for man-machine interaction.

2. With the rapid improvement of programming tools (languages and executives), it became possible this year for a system to be developed which permits a user to design integrated circuit masks in all their stages, completely in graphic terms on a display. This development required only 100 man-hours whereas five years ago such a job would have required several thousand man-hours.

3. Conic drawing display hardware designed and put into operation permitting the generation of any conic section without dependence upon costly straight line approximation.

4. Inexpensive storage tube display developed with full vector-text capability. Unit operates over standard phone line as a medium speed remote graphics console providing a good compromise between expensive dynamic displays and limited text-only displays.

5. Initial models of 32" × 32" Rand tablet delivered, providing 4096 line resolution, sufficient for tracing large maps and drawings directly onto the computer.[48]

Following these accomplishments, several of the problems Sutherland delineated in his 1966 paper on "ten unsolved problems in computer graphics" became the focus of future developments in the field. Sutherland had pointed to the importance of making graphics inexpensive, friendly (he called this "a problem of technique"), and available in various tonal values and colors. Several of the examples he used in referencing the problem of technique were later solved by designating certain functions to be performed, and listing those functions for selection on a screen menu. He described the motion language developed by Kenneth Knowlton at Bell Laboratories and used to produce animated films. In 1966, Kenneth C. Knowlton and Leon D. Harmon had scanned several photographs in their "studies in perception" series at Bell Laboratories, using a system that assigned picture tonal values to black-and-white micropatterns that were then used to reproduce the picture on a printer. The motion language drew upon laws of physics that are still used in producing animated computer graphics. Sutherland envisioned being able to produce such animations on-screen, interactively. He saw graphics as destined to make the unseen visible: he called for what later would be known as scientific visualization, and pointed to the power of the technology not only to reproduce visual reality but also to aid in representing new images and ideas. Sutherland was later to contribute to the solutions to these problems at the University of Utah, where IPTO helped to establish an important program for interactive computer graphics.

As the initial developments in computer graphics progressed, researchers defined programming problems on the basis of an underlying assumption about human perception and cognition. They assumed that graphical outputs representing images could communicate more effectively than other forms of computerized displays. Pictorial representation was simply more efficient and direct in conveying some information than were columns of numbers, and verbal descriptions. This basic assumption led to a vision for interactive graphics which included a search for realism in imagery, so that users could communicate with the machine in a faster, clearer, and richer way.

As researchers attempted to develop programs that would allow them to draw shapes with greater complexity, the differences between constructing abstractions in a computer and perceiving believable objects on-screen became increasingly apparent. An important issue to be addressed here was how to give two-dimensional information the appearance of being projected into three-dimensional space. The issue was resolved by the development of "wire frame" drawings that were simple outlines of three-dimensional objects. As researchers attempted to make drawings more complex than a wire frame shape, their awareness of the difficulties of creating believable objects on-screen grew. This awareness contributed to the solution of visual representation problems.

Once wire frame images had been produced, the problem of hiding lines that, from the viewer's perspective, would be behind the object's front surfaces had to be solved. Eliminating such lines would make the objects that were portrayed appear more three-dimensional and less transparent. As we noted earlier, Roberts's investigations at Lincoln Laboratory in 1963 into the possibilities of bandwidth reduction in photograph transmission and three-dimensional graphics had led to the first hidden-line program. And in an article on a display that mounted on the researcher's head, Sutherland pointed out that "in order to make truly realistic pictures of solid three-dimensional objects, it is necessary to solve the 'hidden line problem'. . . . We have concluded that showing opaque objects with hidden lines removed is beyond our present capacity."[49] More work was needed before the hidden-line problem for complex and multisurfaced images was solved. University of Utah researchers did much of this work under IPTO sponsorship. Work on the hidden-line problem continued through the 1970s as raster technology, and therefore, shaded, three-dimensional figures, became more prevalent.

Another problem concerned how to get the part of the image not on the computer screen out of the computational load. This problem interested Sutherland as well as others in the field. After leaving IPTO in 1966, Sutherland went to Howard Aiken's Computational Laboratory at Harvard University and worked on a problem that arose from the parts of images outside of view or off the screen. Parts of the image that went beyond the edges of the screen would appear on the opposite side of the screen and overlap the image. There were two solutions to the problem: interrupt the display of the beam for the period of the overlap, or predetermine the coordinates of the part of the picture to fit into the window (called "clipping"). Sutherland thought clipping to be better, because it saved time by not requiring the machine to draw unseen parts. He presented his clipping divider algorithm in 1968, the result of a project funded through IPTO and the Office of Naval Research. Using the clipping divider, Sutherland then developed a head-mounted display that would respond to movement and could be used for flight simulation. This project was also supported in part by IPTO, following some preliminary experiments at Lincoln Laboratory.[50] The system included two tiny CRTs placed so that each eye of a viewer could see each screen. Stereo images were shown on each screen so that the viewers would see an image in three dimensions. Images were stored in geometric form and shown as wire frames. Sonic transmitters and sensors tracked the movement of the viewer's head so that, as movement occurred, the image changed accordingly.

After the development of wire frames and clipping, and the solution of the hidden-line problem, the next step in achieving realism was to cover the wire frames with surfaces to give the appearance of solid objects. Steven Coons tackled this in the early 1960s as part of the CAD project. While on the Mechanical Engineering Department faculty at MIT, he created a technique for making "surface patches": polygonal forms that could have various planar and curved surfaces and could be attached to graphic wire frames. By placing these polygonal patches side by side, the entire surface of an object could be modeled on the computer. Later, Coons was instrumental in introducing programmers at Harvard and Utah to his matrix mathematics, aiding their understanding of three-dimensional surfaces.[51]

As research in computer graphics advanced, the quality of the viewing screen became an even greater concern for research and applications. During the 1960s, developing a standard of quality for displays was an increasingly important issue for the field, especially for the developers in industry. The criteria that these researchers employed were excellent visual representation and real-time operability. To meet these goals, designers had to consider display resolution capabilities and the development of display processors to efficiently handle information communicated between the computer and the display. Throughout the 1960s, the hardware of display monitors increased in sophistication, resolution improved, and processor alternatives developed.

Many of the improvements in displays came from commercial manufacturers. For example, in response to the problem of the need to continually refresh the image on a CRT, Robert Anderson of Tektronix had developed the direct view storage tube (DVST) in 1961 for oscilloscopes.[52] In 1964 his co-worker C. Norman Winningstad had the idea that the tube could also improve computer displays, and convinced Tektronix to begin to produce what became its first in a popular series of storage tube computer displays. This display had several advantages. It could hold an image for a longer period of time without being refreshed; it was relatively inexpensive; and it was very dependable. However, it was slow, and parts of an image could not be changed without erasing the entire picture. In spite of the shortcomings of these various devices, they served the research community well. By the late 1960s some companies, such as IBM, CDC, and DEC, had display generators that could produce images on several screens at once.[53] As display hardware became more sophisticated, displays previously connected to mainframe computers were connected to minicomputers that took over some of the tasks of the main computer, including image refreshment.[54] By 1967, DEC and Bunker-Ramo had such systems on the market.

Researchers used several CRT displays experimentally during the development of interactive computer graphics, and some were later commercially marketed. From the TX model consoles built at Lincoln Laboratory, DEC produced an interactive graphics display console, the Type 30, and then the commercial DEC 338.[55] In the late 1960s IBM also produced more sophisticated vector displays, which were used by graphics production companies.[56] These displays had a high addressability (4,000 by 4,000 points) and could handle real-time animation of three-dimensional wire frames. Their cost ranged from $50,000 to $1 million. While most displays had 1,024 rows and 1,024 columns, researchers at Bell Laboratories built a device that could "draw" with a resolution of 32,000 lines and 26,000 columns with a fine laser beam. They used this device to make high-resolution circuit masks (patterns for making computer chips), a growing application at the time.

Lowering the cost of displays was the focus of much of the graphics work by the companies. In fact, more attention was given to the cost of displays than to the cost of the hardware and memory needed to support the displays.[57] Eventually, raster displays became an inexpensive and important part of interactive graphics systems. A raster-scanned refresh image was reproduced by an electron beam moving along lines from the top to the bottom of the screen thirty times per minute, a process similar to image production on television screens. The Rand Corporation produced one of the first raster display interactive graphics systems[58] and initiated the Video Graphics System (VGS) project to develop a computer graphics display system based on buffered television components. Ellis described the system at the 1968 Fall Joint Computer Conference as a general-purpose graphics system.[59] The Rand VGS used video displays and a rotating magnetic disk that provided a fast rate of response between user and machine. The availability in the mid-1970s of inexpensive raster displays based on television technology was a vital step in the growing use of computer graphics.[60] In the mid-1970s, raster displays of television technology with solid state memory were less expensive, more easily made, and more readily available than other types of display.

Solutions to some of the few problems that raster-scan technology presented were found in the late 1960s. One problem was that early raster displays did not have the resolution of the vector displays. Another problem was the need to design raster display processors that could do some of the work previously done by the main computer. To shorten computer response time, raster displays required a buffer to hold picture information. The conversion of information into the proper sequence for scanning (top left to bottom right) required a set of instructions, a program. This involved the development of "frame buffers," which passed picture information to the display and simultaneously retrieved new picture information from the computer for use. Frame buffers included processing elements to draw lines and simple shapes. Raster displays also required the development of digital-to-analog converters for video display television technology. Scan conversion costs had previously been too high for the inexpensive raster technology to be used efficiently.[61] Costs did not decrease until the 1970s, when researchers developed several solutions to problems in scan conversion.

In the first decade of IPTO's graphics program, support focused on needs in graphics displays, input devices, and related programming, although R&D in these areas was not exclusive to IPTO-supported laboratories. However, with direction from Sutherland, Taylor, and Roberts, successive IPTO directors, two of whom — Sutherland and Roberts — came from a background in graphics research, new displays appeared in IPTO-supported laboratories and eventually became available as commercial models. A similar comment applies to new input devices from IPTO-supported laboratories and commercial development. Programmers attended to two areas: the need for programming related to the functioning of these new devices, and applications of interactive computer graphics systems. In the late 1960s a new raster display technology became available through IPTO-supported activities at Rand: it was to revolutionize display quality.

With suitable graphics hardware available, researchers could concentrate on programming the display produced on the screen. This area was of most concern to IPTO in achieving its objectives of developing more capable and intelligent computer systems. Graphics advanced in the 1960s from the search for novel ways to put images into a computer and display the images on a computer monitor, to the development of ambitious systems using time sharing for designing products, an integrative effect of IPTO support.[62] Developers of graphics systems were particularly interested in means of communication with the machine, which researchers saw as dependent on the production of synthetic images. They found previous technology to be ineffective for their purposes. The areas of study that were particularly important to IPTO-supported researchers were the production of synthetic images using computers, and the relationship between human perception and algorithms for data processing. The establishment of an IPTO center of excellence for graphics research at the University of Utah set the stage for substantial progress in this area, a subject to which we now turn.

The University of Utah: An IPTO Center of Excellence

It is through the Utah program that the influence of IPTO funding on the development of interactive graphics can be most clearly seen. In 1966 the University of Utah hired David C. Evans, who had participated in Berkeley's Project Genie on time sharing, as head of its computer science department. Evans decided to begin a graphics program there. Shortly after his arrival at Utah, he applied to IPTO for funds to do graphics research. And Robert Taylor, after succeeding Sutherland as director of IPTO, decided to expand the contract so that Utah could become a center of excellence in graphics. In an interview, Taylor noted the conscious attempt at IPTO to "build a center of excellence in graphics" at Utah.[63] According to Taylor, the University of Utah program

> began with the radical notion that one could produce pictures of dynamic three-dimensional computer modeled objects that were: (1) shaded realistically like photographs rather than mere line drawings (2) opaque (i.e., with obscured parts removed) (3) in perspective and color (4) in real-time motion (i.e., at 30 frames per second).[64]

In 1968 Sutherland joined Evans at Utah; he remained there until 1973. As we noted, Sutherland had left IPTO in 1966 for a position at Harvard, where he pursued graphics research for the next two years. He later ascribed to Evans concepts that helped to shape Utah's program and the future of graphics software development:

> The man who figured out the basic techniques that made modern graphics possible was David Evans. . . . He knew that incremental computing was a powerful way to get more results for less work. He suggested to a collection of his students, and to me, that we make use of all of the information that we'd computed for one pixel in doing computations for the adjacent pixel—that we try never to recompute *ab initio*, but rather to compute on the basis of nearby changes.[65]

This formula provided an important avenue for achieving more representational graphics because it could be used to produce subtle changes in an image by making each pixel appear only slightly different from those near it, and would not require the memory and time necessary to do entirely new calculations for each pixel.

Over the next two decades the Utah group exceeded IPTO's expectations, and Utah became the source of important contributions to graphics for research and applications. Examples of advances made at Utah are the creation and manipulation of three-dimensional models in a computer database, the generation of synthetic surfaces for three-dimensional objects, the development of algorithms for hidden-surface removal and shading, and the development of CAD/CAM vector-graphics displays. The Utah group used advanced graphics systems in the development of early flight simulation programs useful to the military.[66]

In March 1968, IPTO principal investigators met at the University of Utah. Graphics received a considerable amount of attention at the meeting, and in a trip report, N. Addison Ball of the National Security Agency summarized what he had heard at the meeting about work at IPTO project sites which had focused upon several of the ten problems in computer graphics which Sutherland had delineated in his 1966 paper. The summary mentioned a description by Rand's Uncapher and Ellis of video work "to provide flexible, low-cost graphics stations"; a presentation, by Ivan Sutherland, of work done previously at Harvard for the use of color on a conventional display; Lincoln Laboratory's William R. Sutherland's interest in simplifying display programming; a presentation by Thomas Stockham (also at Lincoln Laboratory) about picture processing; and Evans's desire to provide interactive graphics to users who were not computer-oriented.[67]

The University of Utah focused on both sides of the main mission in computer graphics: the production of synthetic images using computers, and the relationship between human perception and algorithms for data processing.[68] Work by students at the University of Utah produced several breakthroughs in these areas. For example, faculty members and students focused upon the task of developing a series of increasingly complex display processors that could interpret data directly. In 1970 Sutherland reported that "specialized display processors designed to handle . . . graphical computations are now beginning to reach the market."[69] Also in that year, G. S. Watkins, then a student at the University of Utah, developed a system to integrate an algorithm used for "scan line coherence" and Sutherland's clipping techniques for hidden-surface calculations. IPTO funding in 1972 made it possible to install hardware at the University of Utah for the use of Watkins's system;[70] later another student, Franklin C. Crow, developed an algorithm to improve scan conversion which would convert a limited number of vectors for a raster scan display. In 1972 Martin Newell, a faculty member at Utah, presented a hidden-surface algorithm that could be used with highly complex and transparent pictures. His work on curved surfaces was foundational for all the three-dimensional graphics development that followed. Another influential faculty researcher in the Utah group was Ronald D. Resch, who had worked on topological design since 1961 (when he had begun studying wadded pieces of paper). He drew upon the work of architects such as Buckminster Fuller to aid in understanding the construction of surfaces.[71]

During Sutherland's stay at Utah, he and Evans started Evans and Sutherland, a graphics systems production company. This was "principally a product-oriented and privately funded" company that developed early flight simulation graphics systems.[72] An important aspect of their flight simulations was their focused attempt at realism. Flight simulation research and development was supported, in part, because it was assumed that flight simulation would result in fuel savings when pilots were trained. Whereas IPTO's funding for the University of Utah program was slightly more than $10 million in the period 1968–75, the estimated fuel savings in the same period were $190 million.[73] Researchers soon realized that the graphic representations also provided greater detail and wider fields of view. Although their company was not funded by IPTO, Evans and Sutherland received IPTO funding for research at the university, and that research in turn contributed to the development of the company. For example, the company used the "continuous tone" technology that had been developed at the University of Utah with the help of students.[74]

In addition to systems specifically designed for flight simulation, the company produced display terminals and computer-assisted design systems. The first system it produced was the Line Drawing System–1, which used a vector graphics display and added a refresh buffer. The refresh buffer held the information required to redraw the vectors as the lighted phosphors faded on screen. The system could rotate, scale, and translate three-dimensional objects in real time and was similar to the computer-assisted design system produced by Adage, Inc., the year before.[75] Computer-assisted design systems of this type, such as the one produced by the MIT CAD Project, were basically tool programs used to create images. In contrast, flight simulation systems were developed for a specific application (to train pilots), but their needs drove the development of the tool programs.

In 1974, after leaving Utah for the California Institute of Technology, Sutherland, together with Robert Sproull and Robert Schumaker, published a paper titled "A Characterization of Ten Hidden Surface Algorithms." The purpose of the paper was to compare the available hidden-surface algorithms and synthesize a greater understanding of the problem. The authors concluded that there were two "underlying principles" on which the algorithms were based: sorting and coherence. All of the algorithms used sorting to determine which lines and surfaces should be hidden, and all used various types of coherence to reduce the number of calculations required for sorting.[76] The authors categorized each algorithm as belonging to one of three types. The first type dealt with hidden lines as a problem of "object-space": that is, in terms of whether the lines were visible in the object environment. This type of algorithm determined what was hidden by "testing edges against object volumes."[77] Roberts's 1963 algorithm, developed to be used with vector displays, was of this type. The second type of algorithm concerned "image-space" and had been developed for use with raster displays; it checked to find out what was supposed to be visible at any given raster point (pixel) location on the screen.[78] The third type of algorithm took into account both object-space and image-space. The researchers mentioned in conjunction with algorithms of the third type were Schumaker, Newell, Warnock, Romney, Bouknight, and Watkins. Much of the important work in this third area had been done at the University of Utah and at General Electric.

In their paper, Sutherland, Sproull, and Schumaker gave the word *environment* a specific technical meaning: the several objects "existing" in the computer, and the space in which they existed. An environment could include, for example, many "primitives" — three-dimensional shapes such as spheres, cones, cylinders, and boxes — that could be used as building blocks for making more complex objects. This change in wording indicated shifts in conceptualization as new possibilities for graphics were recognized. The authors pointed out the problems associated with the conceptualization of a single "object-coordinate system" when trying to represent many objects in a realistic field.[79]

The representation of three-dimensional space, as well as of shape and surface, was an important topic of study at Utah. The theory of linear perspective, based on a three-coordinate system, had been developed during the Renaissance and had been almost universally adopted in Western painting. It was a way of representing three-dimensional space on a two-dimensional surface. The difficulty in computer graphics was to calculate a point on the computer screen that corresponded to a point on an object perceived in the world by the viewer. When producing static imagery, the task was not so difficult because the coordinate system was based on a fixed observer. However, there was difficulty in producing smooth transitions in perspective when animated graphic objects were seen from a changing eye position programmed into the graphic sequence.

The University of Utah program attracted several graduate students who made contributions to the representation of images in three-dimensional space, particularly to the construction, lighting, and texturing of surfaces. In 1974 one of these students, Edwin Catmull, developed a technique for curved surface representation based on algebraic and mathematical formulas for "bivariate patches" associated with Coons's earlier work on surface patches. The technique was a shift from the previous model using polygon rendering, or the programming and drawing of each line of a shape. Catmull also combined an animated hand he had developed with the animated head created by his fellow student Fred Parkes to produce the first human simulation. Catmull's dissertation research involved design of a z-buffer system for hidden surfaces, a special-purpose random access memory and software which calculated line and shape removal for each pixel separately. Working from the models of visual perception and the understanding of photography developed by then–faculty member Thomas G. Stockham Jr., a former staff member of Lincoln Laboratory, several students developed solutions to surface problems. In 1971 Henri Gourand improved on Coons's work on surfaces by making the facets invisible and developing a technique for smooth shading. Phong Bui-Tong added highlights to that technique. Franklin C. Crow developed an algorithm for transparencies and worked on anti-aliasing programming.[80] Aliasing is the stairstep effect produced by pixels on the edge of a curved surface. Anti-aliasing is produced by averaging the colors of the edge and the background. Martin Newell developed an alternative to the existing practice of using long lists of numerical data to produce objects; he described a way to use short lists to describe the object, and procedural subprograms to do object construction.[81]

The work of University of Utah student James F. Blinn and others resulted in the development of ways to map patterns and textures onto mathematically defined surfaces.[82] The continuous search for realism was at the heart of Blinn's dissertation, and was the foundation for the work of Robert L. Cook at Cornell. The work of both of these men improved on previous illumination models by representing each pixel as three-dimensional where previous models had treated only the entire image as three-dimensional.[83] "These models accurately account for the directional distribution of reflected light on a wavelength basis as a function of surface roughness and slope, the material's properties, and the reflection geometry."[84] Like Catmull and several other students from the University of Utah program, Blinn went to work at the New York Institute of Technology (NYIT), where he more fully developed his textured-surface procedures. He then went to NASA's Jet Propulsion Laboratory, where he produced the graphics images for the famous Voyager film. NASA originally saw this film as a public relations device, but educators soon saw it as an educational tool. Blinn's work aided understanding through visualization. In order to produce pictures with increasing surface realism, Blinn incorporated fractal geometry, a mathematical model for many complex forms found in nature.[85] When applied to computer graphics, fractals produced naturalistic forms and rich detail. Capabilities for producing these rich surfaces were included later in microcomputer graphics systems.

By the early 1970s, IPTO considered the graphics work done at the University of Utah as part of a larger IPTO interest in picture processing (discussed below in the chapter on artificial intelligence). According to a 1973 IPTO request for the establishment of a new DARPA contract for Purdue University which the program manager John Perry prepared for Roberts,

> The processing of pictures is a task of fundamental importance to the DOD. Pictures are, for example, basic ingredients of the intelligence estimates which guide strategic planning. Reconnaissance imagery similarly dictates day-to-day tactical decisions, and real-time image transmission is assuming ever greater importance with increasing use of remotely piloted vehicles. . . . Digital techniques are rapidly becoming dominant in all these functions. . . . For these reasons, IPTO has organized a substantial program of basic research [in] digital picture processing. In contrast to the applications-oriented developments sponsored by operational agencies, this program seeks to develop understanding of digital images and their transformations as a foundation for later practical use by DOD.[86]

Concurrent IPTO picture-processing projects were under way at MIT, on visual and audio signal processing theory; at the University of Southern California, on image-coding algorithms for data compression and image enhancement and restoration; and at Purdue University, on models of digital images for coding, transmission, and enhancement.[87]

Lest the reader get the impression that the Utah group worked in isolation, we readily acknowledge that researchers at other sites influenced the work at Utah, and vice versa. For example, among the several innovations from other sites which dealt with surface problems was Richard Reisenfeld's development of "B-spline" theory, done under the guidance of Steven Coons at MIT and thought by those at the University of Utah to be the best available approach to solving curve and surface problems interactively.[88] Reisenfeld's work is still foundational in three-dimensional graphics development. At about the same time that Evans and Sutherland began their work together, General Electric initiated a program to develop flight simulators for NASA. GE's work resulted in the development of a system, first put into use in 1968, for the NASA Manned Spacecraft Center. It had "the first real-time solution to the hidden-surface problem."[89] William Newman's work also aided the Utah graphics group. While at Harvard, Newman wrote a command language for interactive graphics which facilitated entering, modifying, moving, viewing, and deleting images.

In sum, the University of Utah program had a far-reaching effect on the development of interactive computer graphics. The program helped the field not only to solve persistent problems but also to involve a greater number of people and generate ideas for applications. In fact, one of the most influential aspects of the University of Utah program was the effect it had on the growth of the professional community. Utah graduates went on to populate the profession and continued to promote its advancement through projects of their own at other universities, research centers, and companies. For example, Catmull's method was too expensive to apply in many centers. However, NYIT had the memory space it required. Catmull went there and helped to develop the system using his z-buffer. He later became head of graphics at Lucasfilm, Ltd., the film studio that produced graphics for several feature films such as the *Star Trek* series; and then president of Pixar, a company that developed graphics software.[90] Blinn, at the Jet Propulsion Laboratory, continued to work on and solve the problems associated with realism in computer generated pictures, such as the problem of creating a sense of continuous tone in raster scan images.[91] Newell went to Xerox's Palo Alto Research Center (Xerox PARC). Sproull, who was at Xerox PARC with William Newman in the middle 1970s, later took a position at CMU. After working for several years designing real-time imaging systems for ship and flight simulators for Evans and Sutherland, John Warnock became principal scientist at Xerox PARC, working on interactive graphics problems and graphics standards; in 1982 he co-founded Adobe Systems, another graphics software company.

The influence of Utah on the growth of the field of interactive computer graphics can also be seen in the later demographics of the field. Many Utah graduates went on to other institutions to guide their own students and in the process received IPTO grants to support them. For example, Franklin Crow went to the University of Texas at Austin and then to Ohio State University, receiving funds from IPTO and other sources to continue his investigation of algorithms for creating three-dimensional effects.[92] Henry Fuchs went to the University of Texas at Dallas and then joined the faculty at the University of North Carolina at Chapel Hill, where he has become a leader not only in interactive three-dimensional graphics but in VLSI architecture and medical imaging.[93]

Growth of the Professional Field

To pick up a thread we started above, the growth of the field of interactive computer graphics has been promoted by the increasing number of applications of computer graphics techniques. Early in the development of computer graphics, specific experimental applications and a few general uses emerged. Some of the earliest applications included the creation of histographs, bar graphs, and charts (including flow charts) for information representation and other business uses. General Motors began work in computer-assisted design and manufacture early, and its DAC-1 system became an important contribution to the design of GM cars and trucks. In the middle 1960s, corporations such as Lockheed, McDonnell-Douglas, and Boeing began to use computer graphics for plane and missile design: Lockheed developed CADAM for use with IBM hardware, and McDonnell-Douglas developed CADD, Unigraphics, and Fastdraw.[94] Other early applications included scientific displays involving pictorial graphing of information, such as meteorological maps. Johns Hopkins University and Bell Laboratories undertook early work in scientific visualization.[95] Military applications included the projection of computer-generated images of territories at the Combat Operations Center of the North American Air Defense Command (NORAD); the projection surface was twelve feet by sixteen feet, and the images could be projected in seven colors.[96] These developments in applications, combined with the principles and techniques developed in IPTO-supported projects described previously, set the stage for the explosive growth of computer graphics following the introduction of new display hardware after 1968.

While hardware and software costs were a research funding issue in the early years, they became a broader concern as computers moved out of the laboratory and into more people's lives. Several larger computer corporations that could afford the investments necessary, such as DEC, IBM, and CDC, began to produce products used for making computer graphics. Meanwhile, many small companies, started with venture capital, sprang up to produce graphics hardware and software. This expansion began with some skepticism about applications, and with a concern that designs produced with the aid of a computer would cause the artistry of the designer to be lost and that designers would be replaced by machines. As it turned out, however, designers' work became easier as computer imagery and capabilities became more sophisticated. Application costs decreased through innovation and demand, and interactive graphics capabilities became necessary to manufacturers who wished to remain competitive.

In the early 1970s researchers could see the applications of their early work in practice, and there was a growth in interest in and development of computer graphics. Thomas Hagan and Robert Stotz, then at Adage, one of the earliest companies to work in computer graphics, commented on the shift at this time:

> The optimists . . . have been disappointed because they failed to recognize at the outset the level of completeness necessary in a graphics system to do a useful job. . . . The realists, on the other hand, *knew* that achievement of successful on-line graphics applications would at first be a slow and costly process, and planned accordingly. . . . They have achieved some significant successes. Based on those achievements, we believe that computer graphics is entering a new phase which will be characterized by rapid growth.[97]

The emphasis in graphics on human-machine communication continued with the addition of new applications; and researchers in the graphics industry, as well as artists and designers, began to devote more attention to the aesthetics of computer images. It was no longer simply a matter of placing an image on a screen: the visual quality of the image became important and was part of the search for realism described above. For example, whereas color in interactive graphics had initially been seen as a means of making simple distinctions in information, now it was seen as important for visual representation and the quality of the image. The focus on aesthetics resulted from both the continued attempt to increase the amount of information communicated through an image by improving its realism, and an increasing emphasis on communication between humans. As noted above, previously the emphasis of most interactive computer graphics researchers had been on communication between humans and computers.

The new emphasis resulted, in part, from a growing attention to the user and the consumer (rather than just the engineer), as manufacturing applications and other applications gained favor. People outside computing began to see computer graphics as an important medium for human-human communication. In new applications, images were to be three-dimensionally accurate and even attractive, and they were no longer intended only to represent information. Computer graphics could even be used for appreciating the relation between form and function through the use of simulation. As Hagan and Stotz stated, "simulation will be used for what we think of as styling as well as for what we think of as engineering design. The distinction between the two [e.g., the engineering and the styling of an automobile], already blurred, may gradually disappear. In the future, dynamic aspects will be 'styled,' as static aspects now are."[98] The growing interest in the aesthetic aspects of computer imagery led to further applications in product design. Beginning in the early 1970s, complex graphics could be produced with the aid of networked computers. Design workstations began to appear in several application settings, and in the late 1970s microcomputers became available, eventually making inexpensive and sophisticated graphics possible in a variety of work settings.

Some of the researchers who promoted investigations into the aesthetics of computer images were actually artists who began to create computer-assisted art. The first computer image competition in which images were judged on their aesthetic quality was announced in the trade periodical *Computers and Animation* in 1963.[99] The first- and second-place winners were plotter drawings from the U.S. Army Ballistic Missile Research Laboratories in Aberdeen, Maryland. Some of the earliest gallery displays of computer graphics were held in 1965, most notably in Stuttgart, West Germany, and at the Howard Wise Gallery in New York City.[100] The number of computer artists and the interest in computer graphics as an art form grew. Most of the early computer graphics artists did not have a background in art. However, one of these pioneers who did have training as an artist was Charles Csuri. He worked with a programmer named James Schaffer to produce several series showing the transformation of one image into another through the use of computer averaging calculations.[101] Others applied image-scanning and picture-processing techniques to fine art and to other areas, such as advertising and product design.

One of the difficulties with creating computer graphics as art and design at this time was that the technology was only available at a few military, university, and business R&D locations. To promote collaboration between artists and computer scientists, researchers formed partnerships. For example, Billy Kulver, a physicist in laser research at Bell Laboratories, and the artist Robert Rauschenberg started an organization called Experiments in Art and Technology in 1967;[102] and in the same year Gyorgy Kepes began the Center for Advanced Visual Studies at MIT. The Advanced Computing Center for Art and Design at Ohio State University, initiated by Csuri and funded by NSF, was created to contribute to artistic as well as scientific research in computer graphics. The genesis of the interaction between art and science in the form of computer graphics enriched both fields, presenting new opportunities for expression and understanding.

Almost from the beginning, interactive computer graphics became increasingly important in the arts and entertainment community. By the early 1970s several books had been published highlighting the use of computer graphics as an art form.[103] In the mid-1970s it was still easy to tell even a high-quality digitized image from a photograph; but a decade later, distinguishing the difference was difficult. By the late 1970s, commercial television and feature films began to include computer graphics. Several companies emerged to produce high-resolution computer graphics for television and film.[104]

One of the first films produced using interactive graphics and a sophisticated level of aesthetics was *Sunstone,* produced in 1979 at the New York Institute of Technology by Ed Emschwiller and Alvy Ray Smith.[105] The film included a scanned face that went through various transformations that had been created, in part, with a computer program called a paint system. The NYIT system was the first paint system developed, and was produced with the aid of an Evans and Sutherland frame buffer. When presented in 1978, the system included almost all of the functions that paint systems have today. Paint systems were developed to simulate the effects of oil paint and watercolor. They used "brushes," stored in a shape table, that were called on to produce various line widths and textures; and they offered their users drawing functions (similar to those of CAD programs), stored patterns, and various typographic fonts for producing text in pictures. With the NYIT paint system, the developers at NYIT attempted to make graphics more accessible and interactive. It was first marketed commercially by Digital Effects, Inc., in the early 1980s, under the name Video Palette.[106] The use of paint systems has since become common, and they were introduced for microcomputers in the 1980s, as were CAD programs. Although IPTO did not play a direct role in funding such applications, many of the graphics companies and educational programs were influenced directly by graduates of the program at the University of Utah, and indirectly by the results of many IPTO-supported projects.

Starting in the late 1960s, there was an increased use of graphics for on-screen visual organization in time-shared and networked systems.[107] A growing variety of graphics-related projects funded by IPTO helped to integrate graphics functions and capabilities into other computer uses. For example, in the late 1970s IPTO supported the development of a network graphics protocol for the Advanced Command and Control Architectural Testbed, described in the next chapter.[108] Later, networked microcomputers also used a variety of graphics functions; and the graphics software developed for use on microcomputers became increasingly powerful, included more sophisticated imagery and functions, and became less expensive. In the early 1980s, for example, IPTO funded a project at Carnegie-Mellon University to develop a graphics software system for a personal computer environment called SPICE. It was developed for use with Perq personal computers. Its "Canvas" graphics package included capabilities such as scrolling display windows and multitasking controls.[109]

IPTO funded investigations to develop automated systems for image under-

standing. These investigations included projects in low-level image processing, such as pattern recognition; and projects in high-level computer vision
and robot information sensing and response which required an understanding of the visual world. They also included studies in graphical representation
in computers, image processing, and the interpretation of visual imagery.[110]
IPTO funded work on image understanding at several university sites (Stanford, CMU, MIT, and the University of Rochester). The University of Massachusetts received funding for work in the processing of dynamic images,
which involved the development of algorithms for animating realistic three-
dimensional outdoor scenes.

Researchers became concerned about the ways in which visual information
could be created and interpreted in computers, and the ways in which that information could be applied in the world. For example, the University of Utah's
1977 "Sensory Information Processing Research Proposal" which was sent to
IPTO stated that the researchers' primary goal was to influence reconnaissance
and weapons development, and that they thought this could be accomplished
through the development of better graphics and a unified theory of image
understanding. As a result, links developed between computer graphics and a
variety of other areas of research, such as reconnaissance and artificial intelligence.[111] Graphics systems were also used to solve textual and other problems.
For example, David L. Parnas at CMU was given IPTO support for the development of a graphics-based line editor with which he could show several
windows of text on-screen at the same time. Some of this development will be
considered below as part of our discussion on research in artificial intelligence.

Program visualization was another area of research that was funded by IPTO
in the 1980s. Since computer systems had become so large and complicated
that no single person could know all about any one system, and since maintenance operations such as debugging required understanding and manipulation of the systems, there was a need for efficient and effective ways to access
and work on system parts. Each system had to include ways for users to access information about it, make changes to it, and update the information; but
documentation was ineffective for this purpose because changes were made
so often in the systems. An interactive way to visualize a system was needed.
In conjunction with DARPA's Defense Sciences Office, IPTO supported several
sites, including Stanford Research Institute, Bolt Beranek and Newman, and
the Computer Corporation of America, to work on program visualization. The
resulting graphics were reminiscent of the software development flow charts
seen on-screen in earlier graphics experimental programs.

An important new set of graphics applications resulted from the incorporation of very large scale integrated circuits. IPTO supported several developments in this area, and in the 1980s the office's funding for the University of Utah graphics program began to focus on VLSI circuits. CMU also received IPTO funding for this graphics application. In 1983 IPTO funded a project that Henry Fuchs organized at the University of North Carolina to work on the display of high-resolution graphics. He and his group built a VLSI-based machine capable of supporting interactive graphics with three-dimensional shaded colored surfaces at a rate of thirty thousand polygons per second.[112] To make VLSI production more efficient, the capacity for doing difficult and time-consuming graphics calculations that had hitherto been done by software — such as calculations for clipping, scaling, and shading — was built into hardware. Fuchs and his co-workers developed "a highly parallel, distributed method of performing these calculations called 'pixel-planes.'"[113] This software system became a graphics system called Pixel-Planes, and then Pixel-Powers. Through an extension of the logic-enhanced memory architecture, and with the help of continued IPTO funding, these systems were developed further for rendering three-dimensional objects quickly and with a high degree of realism.[114] An analogous chip resulted from IPTO sponsorship at Stanford; T. H. Clark developed a "geometry engine" on a chip capable of handling this type of calculation.

An indication of the rapid growth of the field was the emergence of specialty professional groups. The first meeting of the Association for Computing Machinery's) Special Interest Group on Computer Graphics (SIGGRAPH) was held in 1974. In 1976 the group opened its conference to exhibitors, and ten corporations displayed their hardware and software;[115] and beginning in 1981 the group sponsored computer art shows in connection with the conference. The National Computer Graphics Association (NCGA) was formed in 1977 to promote industrial applications of computer graphics and had its first conference in 1980. Computer graphics professionals developed standards for graphics which were not specific to a particular computer. These standards were first initiated at the 1974 SIGGRAPH meeting. A description of a standard graphics language, called CORE, was established, and work continued on the project through the American National Standards Institute.[116]

The field of interactive graphics grew to include greater numbers of people and systems, and to have a greater effect on a range of other fields. A professional community of computer graphics artists and scientists emerged, and educational programs specifically concerning computer graphics were established. As work progressed, researchers became interested not only in producing more effective computer images in a more efficient way but also in producing images that were impossible to make with other media. These researchers wanted nothing less than to create a new visual world. Students of these researchers went on to staff other universities and corporations. For example, the institutions of higher education which became influential in the 1970s and 1980s because of this diaspora included Stanford, Cornell, New York Institute of Technology, Carnegie-Mellon University, and Ohio State University. People from these schools were often involved when IBM, General Motors, Lockheed, McDonnell-Douglas, Bell Laboratories, and General Electric organized graphics projects, and when, in the late 1960s and 1970s several smaller graphics hardware and software companies appeared on the scene.

The results obtained by research programs over the years at MIT, Utah, NYIT, Ohio State University, Tektronix, and DEC became the cornerstone of the rapid diffusion of computer graphics techniques in the 1980s. In the 1980s, a rising graphics industry became the locus of new developments. Digital Effects (which created television commercials), Walt Disney Productions and Lucasfilm (which created film animations), Calcomp (which specialized in CAD/CAM; architecture), Xybion Corporation (which specialized in business graphics), Apple (which created the Macintosh system), Cubicomp (which specialized in professional graphics), and Silicon Graphics (which specialized in parallel graphics processors) are just a few examples of corporate development among dozens. Many of these firms were founded by or hired people trained in IPTO-supported graphics projects.

These developments in interactive computer graphics illustrate that computer graphics emerged experimentally, facing some skepticism, and was in large part the creation of a few teams of scientists, designers, and funders. Even into the 1970s, graphics were considered by many to be a special area of computer science; by the 1980s, however, graphics had been integrated into computer science—indeed, also in areas well beyond computer science.[117] The field of interactive computer graphics developed and expanded, in part, because of an increasing interest in graphics applications on the part of a growing number of scientists, artists, educators, and people in business and industry. Thus, people now see animated computer graphics every day on television, and even young children can create their own interactive computer images at home and in school.

Interactive graphics enabled better communication between humans and machines, made it possible for humans to visualize what had previously been unseeable, and improved human-human communication through the illustration of complex information structures. Investigations in the field of interactive computer graphics began with an interest in improving communication between humans and computers. Early researchers focused upon getting an image on the screen and making interaction easy and fast. Improved input and output devices made interactive graphics clearer, easier to produce, and more comfortable to use.

As graphics advanced, researchers tried for a more realistic image quality. At the same time, others focused on understanding human perception of the visual world and developing a theory of image understanding. Developing a theory meant considering the representation of perceptual information in a computer and the visual representation of ideas, rather than just objects; and this led to more involvement with other areas of computing.

Graphics became a computer function with a seductive quality unknown to other areas of computer science, with the possible exception of artificial intelligence research. The images had such realism that people could no longer tell when they were seeing images produced with a computer. The rich visual effects of color, texture, and shading now seen every day on television, in design studios of all types, and in art galleries are no longer greeted with skepticism. Even the microcomputer graphics hardware and software available in the mid-1980s which included elaborations on other early techniques, such as picture scanning, were capable of high resolution, three-dimensional animation, thousands of colors, and the visually subtle simulation of a variety of media and techniques, such as Japanese brush-and-ink calligraphic strokes. New applications of computer graphics include scientific visualization, layout for advertising and publishing, and fine art and entertainment.

The history of IPTO's involvement with computer graphics illustrates the growth of a vital professional community and the importance of support for the exchange of ideas for research and development. The fundamental concepts behind the remarkable computer graphic images we encounter every day emerged primarily from research projects funded by IPTO. Work at Lincoln Laboratory, MIT, and Harvard (where Sutherland pursued more graphics research after leaving IPTO) defined the problems regarding input/output devices and the visual characteristics of computer graphic images which needed solution; and it resulted in a number of techniques that have been fundamental to graphics ever since. Researchers at the University of Utah focused specifically on the visual characteristics of images, and over a ten-year period beginning in 1968 contributed the theory and techniques needed to produce aesthetically pleasing and seductive computer images. All these developments can be ascribed to the foresight of the leaders of IPTO in the 1960s and 1970s. Through the commercial exploitation of these developments, graphic systems became available to everyone with access to a computer. But from IPTO's perspective what was most significant was that the commercial availability of sophisticated graphics systems in the 1980s led to the military use of such systems, enhancing the command and control functions of military units. Again, in this area of computing too, IPTO both contributed to the command and control function, hence carrying out its DOD mission, and contributed generic technology to the national and world economies.

CHAPTER FOUR

Improving Connections among Researchers:
The Development of Packet-Switching
Computer Networks

J.C.R. Licklider strongly believed that his purposes at IPTO, and those of the DOD, would be better served if members of the computing research community were able to interact with each other more frequently and easily through computers. The interaction he had in mind was the sharing of programs and applications that crossed research project boundaries. In 1963, users who had access to identical hardware and software could share programs by trading tapes. This trading had started in the 1950s, when little or no software was available. But it was almost an impossible task to cross computer system boundaries. Licklider started his tenure at IPTO with the assumption that especially in a field with a shortage of researchers, any new tools or improved programming languages and environments would be most useful if they could be quickly and effectively shared. His first step toward facilitating sharing in the research environment was, as we have seen, to encourage a group of time-sharing developments that made it possible for people at one site to share programs in a more convenient and more flexible manner. IPTO's program for increasing the sharing of resources in the community did not stop with time sharing; resource sharing became a continuing focus of IPTO. During his tenure at IPTO, Licklider talked about networking—a scheme for the sharing of resources among researchers.

Through time sharing, Licklider, and later Ivan Sutherland, achieved the sharing of hardware at a single site, or at most at a few sites that were close to each other. However, like other users, IPTO contractors had acquired incompatible systems, making sharing across systems exceedingly difficult and therefore impractical. Users, especially researchers, wanted the ability to share data, programs, techniques, and knowledge about computing across computer systems, and recognized the consequent need to overcome hardware and software incompatibilities. In a memorandum to the "Members and Affiliates of the Intergalactic Computer Network," Licklider vaguely explored "the possibilities for mutual advantage," asserting that to "make progress, each of the active research[ers] needs a software base and a hardware facility more complex and more extensive than he, himself, can create in reasonable time."[1] This was not a plan for a computer network like that which was eventually realized by the ARPANET, but it shows that from IPTO's beginning there was an interest in forming a research community that would build upon ideas, share existing systems, and develop new advances from the existing base of computing tools and techniques. Licklider's proximate aim was to overcome computer system incompatibilities, and his ultimate aim was to make sharing through all computers possible, indeed imperative, for all IPTO contractors.

In the early 1960s, a "computer network" was defined as "a single computer center having a multiplicity of remote terminals."[2] This definition applied to time-sharing systems as well as other kinds of remote computing, such as those used in airline reservation, banking, stock quotation, and retail business systems. With the exception of time-sharing systems, these other existing "networks" were not designed to share computer systems for interactive general-purpose use. Time-sharing systems were designed for such use and could also be used from remote, sometimes distant locations. However, remote access to existing time-sharing systems was limited. One problem was obtaining enough information about a remote system. Distant users of a time-sharing system had problems receiving help and information about the system.[3] Another problem was the high long-distance telephone charges incurred by those using remote time-sharing systems. To connect to a nonlocal system, a user dialed a long-distance telephone number using a modem, and the connection had to be maintained for the duration of the interaction. The telephone system charged its users on the basis of the length of time a connection was held, not on the basis of how much data was sent over the line.

Time-sharing systems also had functional limits. While the user was connected to a single system, there was no ability to transfer files between systems, so programs had to be run and data files used on the single system that the user dialed. Data transmission rates were limited, and there was often a noticeable delay in response time. Noisy telephone lines caused errors in transmission, which affected the display that the user saw on the terminal screen. While time-sharing systems shared a system among a limited community of users who were physically co-located, these systems were not sufficient to achieve resource sharing over long distances.

Early IPTO contracts investigated ways of sharing across systems and eliminating these problems. But it was left to Robert Taylor in 1966 to expand the concept into a full-scale networking program in IPTO. He brought Lawrence Roberts to IPTO to supervise and implement the program. Roberts, with substantial advice from members of the research community, organized the technical design and implementation of a system to interconnect time-sharing systems. Such a system would provide quick response to the users and allow the same kind of file and data sharing that was available in individual time-sharing systems. The idea at the center of this program was packet-switching wide-area computer networks. Through this program, IPTO was primarily responsible for developing packet-switching networks and developing methods of interconnecting networks. Whereas its other R&D support in the 1960s went to efforts that were built on the earlier work of other groups, IPTO itself spearheaded the development of the first functioning packet-switched network, the ARPANET, and presided over its expansion and the growth of networking that led to the development of the Internet to interconnect networks. If IPTO had accomplished nothing else in the last thirty years, these networking accomplishments would be enough to ensure its place in the history of computing.

Time sharing had "considerable influence" on the design of the ARPANET. In fact, IPTO personnel and IPTO-sponsored researchers viewed the network as a "natural extension" of earlier time-sharing ideas.[4] Roberts made the connection between time sharing and networking explicit when he noted that "just as time-shared computer systems have permitted groups of hundreds of individual users to share hardware and software resources with one another, networks connecting dozens of such systems will permit resource sharing between thousands of users."[5] The ARPANET began as "a fundamental attack on the problem of hardware and software incompatibility."[6] It ended as an effective mechanism for interactive computing. Through the development of packet-switching wide-area computer networks, a large step was taken toward effective sharing of computer resources everywhere.

IPTO's First Steps toward Networking

Licklider gave the first IPTO network contract to the University of California at Los Angeles. This contract was for studies leading to the "acceleration of development of computer network systems and techniques" and included provisions for time-sharing systems for on-line direct use at remote locations both on campus and at participating institutions in the Western Data Processing Center.[7] UCLA wanted to build a "local" network on its campus in an attempt to unite the computation center, the health sciences computing center, and the Western Data Processing Center (a consortium of some thirty participating institutions).[8] Six of the participating institutions had modem connections with the Western Data Processing Center's computer. IPTO selected UCLA for this work because of its "unique experience" in the operation of a data center using remote computing and its "outstanding capability in the design of executive and monitoring programs to manage the interactions of the several computers in a network." The three computing centers to be connected offered a favorable context for the development of computer networking techniques within the campus. The IPTO contract called for research on computer network problems and computer time-sharing problems, with an emphasis on programming. During the contract, control and monitoring languages would be developed "to facilitate effective interaction between the users and the network of computers." UCLA researchers would design other software to govern the way in which the time of the computers in the network was shared among several or many remote users, and to facilitate the programming, editing, correcting, and operation of computer programs.[9] Another reason for choosing UCLA was its proximity to Santa Monica and the SDC Q-32 computer. Interconnections would be made with the Q-32 computer at SDC and the other computers connected with the Q-32. To Sutherland's regret, the "network" was not developed during this contract because of strong local disinterest: The directors of the three centers appeared not to want to work together. Consequently, this effort did not produce any significant progress toward IPTO's objective of a network of computers. Sutherland later refocused this contract on the production of a load-sharing system involving IBM 7090s located at the three centers.

In March 1964, however, the contract with UCLA appeared to be the first step toward a network. DARPA's director, Robert L. Sproull, confidently told Congress,

> One of the next steps [after time sharing] will be the linking of an individual at MIT through the MIT computer to a computer at Carnegie Tech, Stanford, UCLA, or the Systems Development Corporation to permit the researcher at MIT to call forth programs or information stored in these remote computers for use in solving his research program.

Not to raise expectations too high, he went on to "emphasize that this experimentation is in its infancy."[10] The experimentation reflected IPTO's early concern for the issues underlying its later research area.

In congressional hearings a year later, Sproull referred again to the UCLA contract and stated that a computer network is now being defined which will distribute computation equipment for passive defense much as modern military communications systems distribute circuits and equipment for reliability in the face of enemy action. This network will serve as a test bed for various computer networking concepts. We can foresee the day when such computer networks will automatically distribute computation and information to diverse users.[11]

An effective connection would allow IPTO to fund a large system in one location and have it used from multiple locations; the beneficial effects would be multiplied without duplication of the resource. In addition, researchers could learn from the successes and failures of others around the country. This sharing was equally important for the further development of military systems. While Sproull's comments to Congress and IPTO's own thinking now focused on the development of concepts, the potential uses of such a network soon became clear to IPTO. Primarily, a network would make possible shared use of specialized resources such as hardware, applications software, system software, and data. Secondarily, and ultimately more important, sharing had the perceived advantage of fostering "a 'community' use of computers," making it "possible to achieve a 'critical mass' of talent."[12] While the initial concept of a network evolved in the early years of IPTO, the implementation of a network did not begin until 1968.

Simultaneously with the UCLA contract, IPTO issued a contract for a "network" project at the Berkeley campus of the University of California, four hundred miles north, but the justification for this contract was slightly different than that for the UCLA contract. IPTO added Berkeley to its "network" of contractors because of the presence at Berkeley of specialists in programming languages and heuristic programming, the existence of a computer center with critical facilities on the campus "at approximately the correct distance from Los Angeles to make manifest several critical problems in computer intercommunication," and the presence of several groups active in research in programming areas relevant to IPTO concerns but not supported by IPTO. The work plan was essentially identical to UCLA's plan for the development of programs to enhance interconnections among computers.[13]

By 1966, IPTO wished to connect the seventeen existing research centers it had under contract, and by so doing, to make available to all IPTO centers the results of work at any other center. Many of the centers were engaged in duplicative software efforts, the overlap of which might be lessened if software could be shared. Interconnecting the centers by leased lines seemed prohibitively expensive. A quarterly report from SDC dated June 1966 notes that a series of informal meetings involving BBN, SDC, the University of California at Berkeley (but not UCLA), and IPTO were held in which the formation of a network was discussed. The group planned at the time to use three SDS 940 computers to form the basis of this system. This homogeneous environment of computers would facilitate the transfer of data and files among the computers' users. IPTO contracted with Carnegie-Mellon University to test a homogeneous environment by connecting IBM 360/67s running TSS (the IBM time-sharing operating system) at CMU, Princeton, and IBM. The network was first operational in 1968 and seems not to have been instrumental in later IPTO decisions about a network.

The Computer Corporation of America (CCA), a small Cambridge, Massachusetts, company founded in 1965 by Thomas Marill, a former Lincoln Laboratory staff member, submitted a proposal to IPTO in 1965 for a networking study. IPTO refused to provide support to CCA because the company was not considered large enough to accomplish the task. Instead, IPTO arranged for a subcontract through Lincoln Laboratory for a preliminary study.[14] The study, "A Cooperative Network of Time-Sharing Computers," was to investigate options for connecting different kinds of computers over large distances. It focused on one of the most important problem areas for wide-area networks: geographically separated systems with different software and hardware.

Marill summarized the results of the initial study in a June 1966 report: the "principal motivation for considering the implementation of a network" was that "a user of any cooperating installation would have access to the programs in the libraries of all of the other cooperating installations."[15] He studied the available communication options, and made a comparison of tariffs for available communication services. Summarizing the results of the preliminary study, Marill could "foresee no obstacle that a reasonable amount of diligence cannot be expected to overcome." He recommended implementing a small computer network involving the Lincoln Laboratory TX-2, the SDC Q-32, and the Project MAC IBM 7094 in order to experiment, give demonstrations, and gather experience. (When approached, the MAC people were not interested in participating in the test.) Marill cooperated with Lawrence Roberts of Lincoln Laboratory in conducting the experiment. Although they connected the machines successfully, the response time and the reliability of the connection were not satisfactory. In a public presentation of their results at the 1966 Fall Joint Computer Conference, Marill and Roberts described a procedure for exchanging short messages and one for simple error recovery. Short, discrete messages that included characters to mark the header and the end of the message, were sent back and forth across telephone lines.[16]

Network research did not begin in earnest until Sutherland's successor as IPTO director, Robert Taylor, arrived in mid-1966. Taylor brought Roberts to IPTO to provide the technical leadership for this project. Roberts's experience with the network experiment conducted at Lincoln Laboratory was the starting point for his initial approach to the IPTO network project. Roberts oriented the IPTO program toward the connection of heterogeneous computers, which reflected the prevailing situation among the IPTO-supported centers. Under Roberts, IPTO's networking concept evolved into the famous ARPANET, which was a highly successful packet-switched wide-area network. But before we examine the use of the packet-switching technique and the development of the ARPANET, it is useful to review the communications options available for "networking" in 1966 and the origins of the packet-switching idea.

The Roots of Packet Switching

IPTO needed a new approach to connecting computers because the existing communication systems and terminal-oriented computer networks were too limited for the requirements of time-sharing systems. By the mid-1960s, two approaches to communication networks existed: message-switching and circuit-switching. The telegraph system is an example of message-switching and the telephone system is an example of circuit-switching. Rather than making direct connections the way the telephone system does, the telegraph system is a store-and-forward message-switching network. Store-and-forward systems accept a complete message, store it, and eventually send it on toward its destination. The telegraph system began using store-and-forward techniques to increase line utilization. Storing the messages on paper tape at switching centers and later forwarding them allowed multiple telegraph operators to work in parallel on messages for the same location. Because messages were sent in batches, an electrical connection was established only once for each batch sent to a given location, improving the utilization of the connection and reducing the number of times a connection had to be established.[17] Message-switching systems traditionally kept copies of the messages at each intermediate node and had large and complicated switching stations.[18] Automated versions of this technique required sufficient message storage space at every intermediate node. Since a message might be stored for a while at each of a number of intermediate nodes, message delivery to the receiver was always delayed. This made store-and-forward inappropriate for interactive communication.

By the early 1960s, the military used digital store-and-forward message switching systems.[19] The AUTODIN system, a digital store-and-forward message switching network composed of nineteen internationally dispersed switching centers, serviced all branches of the military beginning in the early 1960s and remained in use for several years. It maximized circuit use, and the military used it for noninteractive communications.[20]

The telephone system is a circuit-switched system. Switching stations increase the flexibility in connection set-up, determining and then establishing a direct, dedicated connection from the initiating telephone to the destination telephone, through the local switching stations. Thus, in the 1960s, telephone system customers were connected by a dedicated circuit from source to destination. The telephone network, using circuit-switching, was used to connect existing computer networks. But this was an undesirable solution for IPTO, because connecting each computer system it supported to all the others was too expensive. For example, to fully interconnect eighteen computers would involve more than 150 leased lines. The sporadic nature of interactive computer use, coupled with the way telephone service was charged, made dedicated circuits uneconomical. A new scheme was needed.

Another thread in our story emerged from a wholly different concern. Concern about the possibility of a nuclear attack on the United States was so high in the late 1950s and early 1960s that the military gave a high priority to the survivability of communications, in order to know instantly and reliably what was occurring.[21] In the early 1960s the air force turned to the Rand Corporation for research assistance in the survivability of military command and control systems. In August 1964, Rand published a series of thirteen reports, under the title "On Distributed Communications," describing a distributed, store-and-forward network using "message blocks."[22] In these reports, Paul Baran proposed a variation of store-and-forward message switching as a solution to the military problem of devising a communications system that could survive an enemy attack.

The Rand study described a "Distributed Adaptive Message Block Network," a name that reflected the three primary characteristics of the proposed system: it was distributed and adaptive (in its routing), and used message blocks (later known as packets). Baran chose a distributed network because it was more reliable than a centralized network. Centralized networks had a single point of control from which connections extended.[23] If the central node failed, the entire network became unable to function. A distributed network circumvented this problem of a hierarchical setup by having the nodes in a "peer" relationship. The destruction of a single node would not cause the entire network to stop functioning, although the network might be impaired. Connecting each node to more than one link would make the network more reliable. Information was to be exchanged in standardized message blocks, where a "standardized message block would be composed of perhaps 1024 bits. Most of the message block would be reserved for whatever type of data [was] to be transmitted, while the remainder would contain housekeeping information."[24] The message blocks were a way to keep the switching mechanisms simple. Dynamic routing, which Baran referred to as "adaptive" or "heuristic" routing, would allow each node to dynamically learn about the connections it had.

In August 1965, Rand made a formal recommendation to the air force to proceed with research on the network described in "On Distributed Communications."[25] The Air Force System Command created a review committee to evaluate various parts of the proposal. The committee recommended proceeding, but the communication network was not implemented.[26] While Baran's work was largely unrecognized for the next few years, it later influenced Roberts and the conceptual design of the ARPANET.

Meanwhile, Donald Davies at the National Physical Laboratory (NPL) in Teddington, England, considered how to achieve better computer communication. The scheme he devised was a communication network similar to that described in Baran's earlier proposal, although Davies initially developed his ideas without any knowledge of Baran's work. The mismatch in speed between people and computers, and the problems with the use of telephone systems for data communication, led Davies to his scheme. After visiting several time-sharing projects in the United States in April 1965, Davies organized a time-sharing conference to introduce United Kingdom researchers to different types of time sharing.[27] As mentioned earlier, Roberts, who was then at Lincoln Laboratory, attended the British time-sharing conference.[28] Davies, Roberts, and other participants discussed the "inadequacy of data communication facilities."[29] Davies later recalled, "Certainly the mismatch between time-sharing and the telephone network was mentioned. It was that which sort of triggered off my thoughts, and it was in the evenings during that meeting that I first began to think about packet-switching."[30]

Four months later, in March 1966, Davies presented his scheme for a new computer communication network at a seminar at NPL. At this conference he learned of Baran's work from a representative of the U.K. Ministry of Defense.[31] Davies circulated a document proposing a digital communication network in June 1966. It introduced the term *packet* to denote the collection of data and control information sent as a unit across the network. Davies explained the difference between the message the user understands and the parts of the message that the system uses. "The user regards the message as the unit of information he wishes the system to carry, but for its own purposes . . . the network may break up the users' messages into smaller units."[32] Davies distributed the paper in the United Kingdom and the United States. It identified the problems with interactive computer use and the telephone communication system.[33] Davies's group at NPL developed the idea of packet switching for computer communication by using simulations and building a prototype network that linked terminals to a computer using packet switching.[34]

While Davies was developing his ideas in England, there was a growing interest in the United States in better computer communication. For example, the Interuniversity Communications Council (EDUCOM) conducted a Summer Study on Information Networks in 1966.[35] Computer manufacturers were also interested in computer communication, but they wanted to interconnect their own computers rather than different manufacturers' machines. IBM experimented with a network to connect computers running its Time-Sharing System operating system, and it was conducting a separate experiment called Network/440.[36] Control Data Corporation connected thirty international data centers together in the CYBERNET network, which was operational by 1968.[37] Improved computer communication also concerned individual organizations, such as the Lawrence Radiation Laboratory of the University of California, Berkeley, which connected the various computers within its organization. Through the late 1960s, several different and independent groups developed computer networks. Unquestionably, the most important development in computer networking during this time was the design and implementation of the ARPANET wide-area network using the technique of packet switching.

Designing the ARPANET

While IPTO had funded time-sharing developments with encouraging results and was a vocal proponent of interactive computing, by the end of 1966 it had made little progress in connecting time-shared systems into a network. Taylor believed that an effective connection would allow IPTO to fund a large system in one location and have it used in an economical fashion without regard to the physical location of the machine or the researcher. The benefits of using the system could then be duplicated without duplication of the resource. The large-scale systems that were available or planned heightened IPTO's interest in networks; it wanted these systems to be available to and used by multiple contractors.[38] Any new tools or improved programming languages and environments would be most useful if quickly and effectively shared. Researchers could then learn from the successes and failures of others around the country.

Although the benefit of sharing resources was clear, significant obstacles existed. In early 1966, IPTO had seventeen contractors around the country whose computer resources represented a variety of computers and operating environments. IPTO had to overcome the problems of connecting many computers that were incompatible and geographically separated, with a communication channel that had enough capacity to support file sharing and a fast enough response time to maintain interactive sessions. Although connecting computers into a network was part of the office's research mission and fit into its agenda of improving interactive computing, it was the office's perceived need to share resources that drove its interest in computer networking. In late 1966 IPTO, under Taylor's direction, began preliminary planning for a wide-area networking project. During the next year and a half, the office managed both the conceptual development and the writing of general specifications for the construction contract.

At the end of 1966, Taylor brought Roberts to IPTO from Lincoln Laboratory as the technical officer to supervise the project.[39] In spite of his reluctance to leave Lincoln Laboratory, Roberts assumed his new task with enthusiasm. The objective of the new IPTO program was twofold: (1) to develop techniques for and obtain experience in interconnecting computers in such a way that a very broad class of interactions was possible; and (2) to improve and increase computer research productivity through resource sharing.

One of Roberts's first tasks was to develop a view of what the network ought to be and sketch out a design. At a principal investigators' meeting at the University of Michigan in early 1967, only a few months after Roberts arrived at IPTO, he laid out his initial plan. The plan discussed at the PI meeting proposed connecting the computers located at each site with each other over dial-up telephone lines, in the same way as Marill and Roberts had done in their network experiment at Lincoln. A program in each computer would perform a message-switching and transmission function when it received messages that were destined for some other computer.[40] Taylor remembers that most of the people he talked to at the meeting were not "enamored" of the idea. "I think some of the people saw it initially as an opportunity for someone else to come in and use their [computer] cycles."[41] Roberts also later reported that PIs lacked enthusiasm for the network owing to concerns about sharing their limited resources.[42] Nevertheless, some of the contractors in attendance formed a "communication group" to work on problems such as deciding on the conventions for exchanging messages between any pair of computers in the proposed network and determining the kinds of communication lines to use.[43]

While listening at the PI meeting to the discussion of the network plan, it occurred to one principal investigator, Wesley Clark, an IPTO contractor at Washington University in Saint Louis, that the problems of designing a network able to communicate between different kinds of computers could be simplified by adding small computers, each connected to a larger computer. Clark, who had worked on Project Whirlwind and SAGE at Lincoln Laboratory, had been attempting to develop computers meant for individual use, rather than time-sharing computers. He discussed his idea with Roberts on the way to the airport after the meeting. Roberts described the small computer as an interface message processor (IMP) in the "Message Switching Network Proposal" that he wrote to acquaint the other researchers who were at the meeting with Clark's IMP proposal.[44] Having a standard interface from the primary (host) computer to the interface processor allowed the network designers to specify most of the network without regard to the details of the various host computers. Because the host computer was connected only to the IMP, and saw the network through that interface, each contractor site would be able to concentrate on preparing only what was needed to connect to and communicate with the IMP, without concern for the specifics of the network implementation. Changes to any part of the network (except the host itself or the IMP-host connection) could be made without affecting the host systems.

Following the PI meeting, groups of researchers met in smaller, more specialized meetings. The development work was parceled out to a group that included Leonard Kleinrock of UCLA, Frank Westervelt of Michigan, Richard Mills of MIT, and Elmer Shapiro of Stanford Research Institute. Roberts directed the group, which met during 1967. A group calling itself the "ARPA network circle" met in May and adopted the interface message processor concept.[45] A subgroup, which included Licklider, formed to prepare a specification language to describe network messages. Additionally, they considered the software protocols to use when communicating between computers.[46] These protocols for communication are analogous to those used for other kinds of communication (the protocol for speaking on the telephone, for example, includes conventions on who speaks first, how the parties identify themselves, and how conversations terminate). A draft of the network communication procedures was also considered at the May meeting. Another subgroup prepared a second draft and circulated it during the summer of 1967. The group met again in October 1967 to discuss the second draft of the protocol.

Packet switching, as applied to the problem of connecting interactive computers, was an innovative application of existing communication techniques to a new problem. Roberts described it this way: "packet-switching technology was not really an invention, but a reapplication of the basic dynamic-allocation techniques used for over a century by the mail, telegraph and torn paper tape switching systems." [47] While it did use basic techniques from message switching, the new scheme required significant adaptations.

Packet-switching networks broke messages into discrete parts instead of sending the entire message intact through the store-and-forward system, as was done in message-switching systems. Each discrete part (or packet) contained header and control information along with the text. The header information allowed routing of the packets by specifying such items as the source and destination of the packet. Control information (e.g., checksums) was used for error checking. Each packet was put into the correct format and sent out into the network. A packet was temporarily stored at the next location, and then was sent out from that location, and continued to be stored and then forwarded until it reached its final destination. In a packet-switching system the data traveled over lines that were shared intermittently by others; packet switching did not require a dedicated end-to-end circuit for the duration of the message. The packets from many different messages could be sent on a single line, and thus more than one user could send information to the same location at approximately the same time on the same line. The network handled the decomposition of messages into packets at the source and their reassembly into messages at the destination, including all checking, retransmission, and error control. Because packets were routed through the system to locations that were not directly connected by a circuit, fewer nodes needed to be directly connected than would be required in a totally connected network. [48] For example, if node A was connected to nodes B and C, nodes B and C could communicate through A without being directly connected to each other.

A network may send a packet over many different paths when routing it through the network. One way of organizing the network was to choose the path in advance, on the basis of the destination of the packet. Alternatively, a packet could be dynamically routed through the network, so that at each node a routing decision was made on the basis of the destination, and the status of the network. Since in dynamic routing the path was not preassigned, each leg in the path was chosen only when the packet needed to be sent on to the next leg. This arrangement avoided unavailable nodes.

While progressing in some areas of the network plan, Roberts in October 1967 attended the ACM Operating Systems Symposium in Gatlinburg, Tennessee, and gave a paper on the proposed network. His paper focused on the reasons for the network but not on the technical details of how the network would work. He reviewed the problem of using the telephone system for achieving interactive response, only hinting at a solution to the problem by mentioning "message switching" in the last paragraph of the paper in a manner that suggested that the details of the network were still in doubt and the use of packet switching uncertain.

> It has proven necessary to hold a line which is being used intermittently to obtain the one-tenth to one second response time required for interactive work. This is very wasteful of the line and unless faster dial up times become available, message switching and concentration will be very important to network participants.[49]

In addition to being the first public presentation of IPTO's proposed network, this symposium also featured a presentation by Roger Scantlebury, representing Davies's network group at NPL.[50] Scantlebury discussed packet switching with Roberts at the symposium and pointed out that "packet switching offered a solution to his [Roberts's] problem."[51] It is not clear just how much of an impact Scantlebury had on Roberts, but Roberts's ideas for the network were modified by his discussions with Scantlebury. Among the specific changes that followed the Gatlinburg meeting were an increase in proposed line speeds and the adoption of the term *packet* for the proposed IPTO network.[52] During their discussions, Roberts also learned of the applicability of Baran's work on distributed communications to the computer network he was planning. According to his later description, upon returning to Washington from the Gatlinburg meeting he was influenced by Baran's reports: "I got this huge collection of reports back at the office, which were sitting around the ARPA office, and suddenly I learned how to route packets. So we talked to Paul and used all of his concepts and put together the [ARPANET] proposal."[53] Baran met with the participants in the IPTO network project before the end of 1967 and attended the March 1968 PI meeting.[54] In June 1968 Roberts described the ARPANET as a demonstration of the distributed network recommended in the Rand study.[55]

The building of a functioning packet-switching network presented problems for the ARPANET designers. Although Baran and Davies each had discussed basic principles of packet switching, no one had yet made a useful and functional packet-switching network of the kind being considered by IPTO. IPTO faced several problems in the network design during this early phase, including how the nodes should be connected (its topology), what were appropriate bandwidths, how packets should be routed through the network, and how much storage was required at a node to allow the packets to be stored before being forwarded.

Maintaining internal control, Roberts organized the informal committees of contractors already involved in network planning into a network working group. It met first in November 1967. The group was composed of representatives from some of the existing IPTO contractor sites: Rand, the University of California at Santa Barbara, Stanford Research Institute, the University of Utah, and the University of California at Los Angeles. The group's objective was to produce a plan for the network which would include criteria and procedures for testing the acceptability of the network and its configuration, and to recommend a schedule for developing the network.[56] The group participated in a series of meetings. By December, Roberts was ready for IPTO to issue networking contracts and awarded SRI a four-month contract to study the design of and specifications for a computer network.[57] During late 1967 and 1968 Elmer Shapiro of SRI circulated a series of working notes from the group, along with many drafts of the specifications, among the working group.[58] By the summer of 1968 this group had completed its main task.

After eighteen months of work, Roberts organized all the information and specifications and submitted the program plan for the ARPANET to the director of DARPA in June 1968.[59] It was approved by the end of July, and IPTO sent the details of the desired network to 140 potential bidders in a Request for Quotation (RFQ). A bidder's conference was attended by one hundred people from fifty-one potential bidders.[60] The RFQ was for the communication subnetwork, including the interface message processor hardware and software. A computer at each site was to be connected to an IMP, and the IMPs were to be connected over fifty-kilobit-per-second leased lines from AT&T. The request specified the functions of the IMPs, which included breaking messages into packets and reassembling packets into messages; routing; coordinating activities with other IMPs and with its host; error detection; and measurement of network functions.[61]

The most important network performance characteristic, according to the RFQ, was the average message delay time, a characteristic that was crucial to maintaining interactive response. The RFQ required an average delay of less than one-half second for a fully loaded network. Reliability was the second priority. If the network was to be effective, it was important that the users have confidence in its overall reliability. Network capacity (the rate at which data can be input at each node and still keep delay at less than a half second) was the third criterion. The RFQ required the potential bidders to design the communication part of the network, consisting of the IMPs and the communication software. The store-and-forward design was to include six critical functions: (1) dynamic routing, (2) message acknowledgment, (3) fault detection and recovery, (4) IMP-to-IMP communication, (5) message decomposition into packets and reassembly into messages, and (6) priority handling of shorter messages. A second set of requirements dealt with the writing, testing, debugging, and demonstration of the communication programs and interfaces for the prototype IMP.[62] Other requirements covered the installation of the four IMPs and the provision of documentation. The RFQ contained fifteen mandatory "Elements of System Design" along with three optional ones.

Twelve companies bid for the contract, and a DARPA-appointed committee evaluated the twelve proposals received in response to the RFQ.[63] The companies submitting bids included BBN, Raytheon, Jacobi Systems, and Bunker-Ramo.[64] Many of the proposals met, and some of them exceeded, the requirements of the RFQ. Four bidders were "within the zone of contention," and IPTO carried out final negotiations with two of them. By November 1968 the office requested additional funds to cover the prices quoted by the two bidders still under consideration for the contract.[65] IPTO awarded the contract to BBN in January 1969. BBN was well positioned to start on the network implementation; it had been involved with earlier IPTO network plans at SDC, had been recommended to Roberts by Clark, and had, like many others, been explicitly alerted to the upcoming RFQ by Roberts before it was issued.[66]

Implementing the ARPANET

In the summer of 1968, at Shapiro's suggestion, the network group was transformed into the Network Working Group, which comprised representatives from the host sites who gathered regularly to work on issues important to the sites that were to be connected. Shapiro specified that the major objective of the group was to solve whatever problems host computers would encounter in becoming part of the network. Because the group discussed and solved problems other network users would later encounter, Shapiro specified that "the work of the Group should be fully documented and available to these other users to use or ignore, as they see fit." [67] Their task would be to define the host portion of the network interface, including how to connect IMPs and hosts, how the hosts were to communicate with each other, and what network services each host should provide.[68]

This new Network Working Group, primarily composed of graduate students from the early ARPANET sites, met in October 1968, a year before the network was available, to discuss how to use the ARPANET.[69] Over the following year the group explicitly defined the specifics of some of the services needed by users of the network, such as remote interactive login and how to transfer files from one system to another. Vinton Cerf, who would continue to play a major role in computer network developments and later became an IPTO program manager, was at the time a graduate student at UCLA who belonged to the group. He explained how the group viewed itself: "We were just rank amateurs, and we were expecting that some authority would finally come along and say, 'Here's how we are going to do it.' And nobody ever came along, so we were sort of tentatively feeling our way into how we could go about getting the software up and running."[70] The group began to distribute its working notes and proposals in a series of requests for comments (RFCs). Stephen Crocker, another graduate student at UCLA and, like Cerf, later to become an IPTO program manager, had the idea for the RFCs; and another graduate student, Jon Postel, became the editor of the series of comments. One of the first requests, written in April 1969, contained documentation conventions and explained the purpose of comments. In soliciting comments the authors wanted to ease participants' inhibitions regarding sharing ideas before they were published, and to encourage informal communication.[71] The informal mechanism of the requests for comments was very effective. It is still in place today, and well over a thousand comments have been issued.

When Roberts reviewed the work of the Network Working Group, he was concerned enough about the progress the group was making to initiate a twelve-month study by Raytheon Corporation in September 1969 to determine the protocols to be used between hosts. In the contract request, he noted that the Network Working Group was deciding upon the initial protocol and other conventions. He went on to note that "a committee cannot be expected to investigate and solve the more difficult, longer range problems, particularly when the best solution may require considerable effort for some of the members."[72] Roberts changed his mind, however, and continued with the working group despite the difficulties. It is unclear what became of the Raytheon study, but the working group continued, developing the standard host protocol, known as the Network Control Program (NCP), in early 1971.[73] The inclusion of individuals from each site in the working group gave the sites a sense of ownership. A shared involvement and a role in the technical shaping of the network increased the sense of community developing among this group of IPTO contractors. By November 1969, the Network Working Group represented ten different sites.[74]

From the beginning of the project, Roberts recognized the need for a service to provide documentation on the programs and files that were available at the various sites in the network. He awarded SRI a contract for the Network Information Center (NIC) in January 1969, just after BBN received its contract for the rest of the network. The Network Information Center, located at SRI, was to serve as a clearing house for information provided by those on the network about what was available on the network and how to use it.[75] The center maintained lists of network participants and of various subgroups, and kept an archive of all the documentation available.

In January 1969, BBN, the winner of the competitive bid, went to work on implementing the physical network. Frank Heart managed the BBN team, which consisted of Robert Kahn, Severo Ornstein, David Walden, and William Crowther.[76] Ornstein and Kahn did the technical work in preparing the proposal, although they consulted others at BBN on various aspects of the proposal. Kahn had been interested in the network project before he began to work on BBN's proposal. He had written to Roberts about his interest in computer networking and had contributed some ideas to the ongoing network specification before the RFQ.[77] Most of the BBN group had worked at Lincoln Laboratory on real-time systems; some of its members had worked during the 1950s on SAGE and Whirlwind.[78] For example, Heart had been a member of the Digital Computer Laboratory at MIT in 1951, had worked at Lincoln Laboratory from 1952 through 1966, and had been the manager of a group at the laboratory working on real-time computer systems for air force projects. The BBN engineers applied this experience in real-time computing to the network project.[79] For example, both real-time computing and the early network required close attention to issues of timeliness and responsiveness.

The small project staff assembled at BBN worked together closely and was in regular contact with Roberts.[80] In addition to the implementation of the network and correction of day-to-day problems that occurred, a few changes were needed to the network design. One important change allowed more than one host computer to be attached to a single IMP. Several IPTO contractors had more than one computer at a single location to add to the network, so multiple computer connections to the IMP had quickly become an issue. A site with multiple computers connected to an IMP could use the network to communicate between its local computers.

Another early concern was the distance between the IMP and the hosts. The multiple hosts at a site were often in different locations across campus, but the original specification limited the cable length to thirty feet. BBN first developed a distant-host interface that allowed a host to be two thousand feet from the IMP. By the end of 1971, it had improved the IMP further by developing a Very Distant Host option that allowed the host to be an arbitrary distance from the IMP, and had added error detection and correction between the host and the IMP.[81] Thus, connecting local computers to the ARPANET became easier, which added to the overall growth of the network.

In addition to working with BBN as its personnel continued to design and build the physical network, Roberts continued to attend to other network activities taking place in the Network Information Center and the Network Working Group. With this initial work under way, he awarded new contracts to start research in network modeling, performance measurement, and the topological design of the network.[82]

IPTO needed a consistent way to measure and analyze the network's performance. Roberts directed BBN to install the first IMP at UCLA so that the Network Measurement Center, run by Leonard Kleinrock, would start functioning as soon as possible. A computer at UCLA, one that had been used on a previous project for IPTO, was attached to the IMP and used to provide input for measurement center tasks.[83] Kleinrock was involved in network design from its early stages and played an important and ongoing role in the ARPANET. His interest in the performance of networks was evident in his 1962 Ph.D. dissertation at MIT, in which he studied message flow in connected networks of communication centers.[84] While at Lincoln Laboratory as a staff associate from 1957 to 1963, Kleinrock had been a colleague of Roberts.[85] When Roberts was assembling computer and networking experts to help develop the ARPANET, it is not surprising that Kleinrock was one of the people to whom he turned.

Kleinrock wanted to model the likely behavior of the network. Since no packet-switching networks existed at the time, the performance of such a network was unknown. No precedent existed for measuring this kind of network, and very little reliable data about them existed. The Network Measurement Center developed specifications for the measurement software for the IMPs, which BBN then implemented. As the network began to function, the Network Measurement Center provided analysis and simulation of network performance, along with direct measurement of the network, on the basis of statistics gathered by the IMPs. This data provided invaluable feedback for tuning the network.

BBN installed the first node in the network at UCLA in September 1969, just nine months after it had received the IPTO contract. The network was functioning with four nodes by the end of that year: SRI, which was running an SDS 940; the University of California at Santa Barbara, which was using an IBM 360/75; UCLA, which was running an SDS Sigma-7; and the University of Utah, which was running a DEC PDP-10. All of the systems were interactive time-sharing systems in a nonhierarchical arrangement; no single machine was in control of the network.[86] By February 1970 the four-node packet-switching network was undergoing testing and being developed for remote interactive use of the various machines.

IPTO initially developed the ARPANET as an experimental system that would enable the research community to share computer resources, but recognized its potential use in military settings. The increasing use of computers throughout the 1960s had resulted in increasingly severe cost and software development problems, which created difficulties for everyone, but especially so for the complex and extended DOD. The purchase of different computers at different DOD centers led to the necessity of developing different software for special needs at each of those centers. The ARPANET could allow more effective use of the hardware at those centers by allowing real-time remote use of another center's facility and the sharing of software, eliminating the need for duplicate programming efforts within the network.[87] In 1970 this goal seemed achievable to DARPA. In fact, in 1969 Herzfeld already saw the value to the DOD. He reported to Congress:

> The ARPA computer network is addressing this question by electrically connecting a set of computers, permitting different users to use each other's machines and software whenever the latter's capabilities match the former's needs. . . . The network is constructed to make it possible for conversations between computers to take place at an extremely high data rate. If this concept can be demonstrated as feasible, it could make a factor of 10 to 100 difference in effective computer capacity per dollar among the users. The ARPA network will connect universities engaged in computer sciences research, the final DOD application would connect Defense installations. Initial success has been achieved.[88]

A bit premature perhaps, but within the next year the success of the nodes at UCLA, the University of California at Santa Barbara, SRI, and the University of Utah seemed to assure the value of the network to the DOD. The initial nodes demonstrated that the expectations for the ARPANET were well founded. It was not long before IPTO planned additional nodes for its network.

While it was reasonable for Roberts to examine all the possible ways to connect a four-node network, he needed a better way to determine which nodes should be connected to which other nodes in order to maximize the effectiveness of the growing network. The network was not fully connected—although one IMP might have several paths to and from it, it would not be connected to all other IMPs in the network. In 1969 a new company, the Network Analysis

Corporation, received a contract to develop a topological design of the network and to analyze its cost, performance, and reliability.[89] Howard Frank, the president of this company and principal investigator on the contract, had a background in network design and had applied network theory to reduce the costs of oil flow through pipelines for the Office of Emergency Preparedness in 1968. After that experience, he had organized the Network Analysis Corporation to continue the computer modeling of all types of networks.[90] It was Kleinrock who introduced Frank to Roberts.

At the time of the contract issued to Network Analysis Corporation, the ARPANET still consisted only of the original four nodes, three in California and one in Utah. Roberts had ordered additional cross-country lines, so the company's first assignment was to study the new configuration. This required both the analysis of the suggested configuration and an attempt to improve on it. Frank and his co-workers found a configuration that was 30 percent cheaper than the configuration initially ordered and provided significantly improved network capacity.[91] Following this success, the Network Analysis Corporation remained active in the IPTO network project.

As the early network began to expand beyond the initial four-IMP prototype, Roberts and the others who developed the network discovered several problems. For example, BBN's first implementation of the network software had a serious flaw. If any one node received too much traffic, the entire network would break down. Once the network was implemented, BBN understood the inadequacy of the control of packet flow, yet it was difficult to change it. A change to the protocol specified in the documents given to the host sites required changes by all the sites that had already implemented the host-to-host protocol. While initially such changes would involve only four sites, within two years the network connected more than twenty sites. This type of problem might have proved serious in another setting but was manageable in the collegial atmosphere of the ARPANET: ARPANET users simply agreed not to send too much data to a single node. As Walden later described the situation:

> [There were] informal rules in the network that you didn't pump stuff into it from the hosts so fast it swamped the net. Because we all knew the algorithms didn't work, . . . they broke when you pumped stuff at them too hard. And so this interim solution worked in that network . . . because everyone agreed not to break it. You know how to break it; let's not break it. Meanwhile, let's go figure out how to fix it.[92]

Roberts and BBN set up the Network Control Center at BBN to monitor and help maintain the network. The center's task was to identify the source of day-to-day problems that users encountered. It was a difficult task because the cause of any given problem could be anywhere in the network: the local host, the distant host, any IMP in the path, or a telephone line. The existence of a central location to deal with problems made the network more "user friendly." Users had a single location to contact, and the center would do the troubleshooting and debugging of the problem. Problem identification, however, required more time and energy than the network designers envisioned.[93]

IPTO wanted its network to connect existing computer systems; consequently, a person had to be a user of one of those computer systems to have access to the network. This proved to be a problem. Roberts wanted to extend ARPANET access to people who did not have access to one of the host computers already connected to the network; and to accommodate such users BBN designed the terminal IMP (known as TIP). Terminals could connect directly to the modified IMP, which would route the terminal traffic and serve as a host to the terminal users. The first terminal IMPs were installed in mid-1971 at NASA/Ames and MITRE.[94] The TIPs had to be flexible enough to allow the attachment of many different kinds of terminals. As the ARPANET became able to serve more and more users through direct terminal access with the TIPs, providing enough computing resources for the TIP users became a problem. In 1972, IPTO provided funding for the Information Sciences Institute. Founded at the University of Southern California by Keith Uncapher, it was to be a single location where TIP users could access computing resources, thereby gaining for IPTO the scale benefits of administering multiple computers from a single site.[95]

The network operated with a wide variety of hardware and software systems, thus achieving IPTO's primary objective for the network project.[96] Although the use of IMPs minimized the amount of special programming required, effort was still required to connect each new type of host system to the network. Each site needed to provide the hardware interface to the IMP, in accordance with specifications laid out in BBN Report 1822, "Specifications for the Interconnection of a Host and an IMP." This process could take from six to twelve months.[97] In addition, each new software system attached to the network required the software implementation of the network protocol for that system. It was the responsibility of each organization to write and install this complicated software.

The ARPANET expanded steadily (see fig. 3). From the four IMPs that comprised the network in January 1970, the network expanded to thirteen IMPs by January 1971 and twenty-three by April 1972.[98] The number of computers connected by the network was larger than the number of IMPs, because each IMP could be attached to more than one computer. In mid-1971, when the network

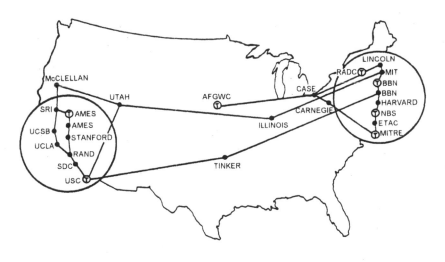

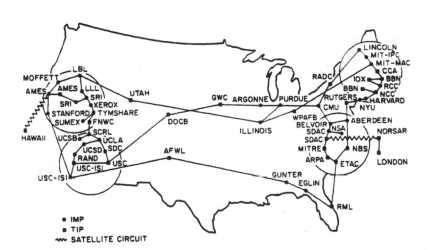

Fig. 3. The ARPANET among IPTO contractors expanded rapidly. In 1969 it linked only UCSB, UCLA, SRI, and the University of Utah. *Top,* Networked installations had grown considerably by August 1972; *bottom,* in July 1975, when ARPANET shifted to the control of the Defense Communications Agency, IMP (interface message processor) and TIP (terminal interface [message] processor) facilities numbered fifty-seven. Adapted from *ARPANET Completion Report Draft,* 9 Sept. 1977.

consisted of fifteen IMPs, Roberts judged it operational because "true user activity could usefully begin."[99] As the contractors began to use the ARPANET more frequently and fully, people saw the network as a tool in their research rather than as a network experiment. The network performed remarkably well as a communication system for interconnecting computers,[100] and its development led to better understanding of the packet-switching technique. The role played by IPTO personnel in orchestrating the resources of the community to develop the network concept was one of the most important contributions of the IPTO program in this period.

The Network Information Center (at SRI), the Network Operation Center (at BBN), and the Network Measurement Center (at UCLA) all operated to support the functioning of the network. ARPANET became a tool that IPTO used to help manage and monitor its programs, and DARPA itself used the network to help manage its internal affairs. After 1970, communication between contractors and DARPA frequently occurred on the ARPANET.

Disseminating Packet Switching

The ARPANET was a complete service to those who had access to it. New services and applications regularly became available to users. The Network Information Center offered information about the network's services to its users, and the Network Control Center helped ARPANET users with network problems. The Network Measurement Center and the Network Analysis Corporation improved the network by analyzing its cost, performance, and throughput. BBN continued to upgrade the network, developing methods for maintaining the same or compatible software levels across the network, creating new designs for the interface message processor, and continuing to install new host sites. In just under six years the IPTO network project went from a set of desired capabilities to a functioning tool both for contractors' use and for network research.

The implementation of packet switching was successful, but it was only part of IPTO's goal: as part of DARPA, IPTO had as an explicit goal the direct and indirect transfer of results from IPTO-supported research to the military. Roberts included this goal in his original plan for the IPTO network project. Packet switching reached the broader computing community through a variety of mechanisms: published papers, demonstrations, and the close ties of IPTO with civilian contractors.

IPTO permitted only certain users to have access to ARPANET, but it did not restrict network information. It planned to transfer the packet-switching technology, at least in part, through publication in the scientific and technical literature.[101] This dissemination of information was successful. Substantial publicity in the computer field surrounded the ARPANET. Roberts organized an ARPANET session at the 1970 American Federation of Information Processing Societies (AFIPS) Spring Joint Computer Conference, a major conference for computer professionals. The session consisted of five papers describing different aspects of the network, one from Roberts and Barry Wessler at IPTO, one from representatives of the Network Working Group, and one each by individuals from BBN, the Network Measurement Center, and the Network Analysis Corporation.[102] Two years later, at the 1972 AFIPS conference, another series of papers brought the audience up to date on the status of the network. IPTO deliberately encouraged publication and distribution. For example, IPTO bound and distributed copies of the papers about networking from the 1970 and 1972 Spring Joint Computer Conference proceedings.

Roberts (who by this time had succeeded Taylor as director of IPTO) and Robert Kahn (soon to leave BBN and join IPTO as a program manager) planned a public demonstration of the ARPANET for the First International Conference of Computer Communications (ICCC) in the fall of 1972. The demonstration needed to go beyond showing that communication was possible through packet switching. It needed to show the usefulness of what the new network could provide: not only physical connection between machines but also the logical extension of an individual's working environment. Kahn took responsibility for organizing the demonstration of the ARPANET. He believed that until 1971 there had been an apparent lack of incentive to spend the time and effort to start useful networking applications. Kahn described the state of the network as being analogous to a highway system without on-ramps and off-ramps.

> The reality was that the machines that were connected to the net couldn't use it. I mean, you could move packets from one end to the other. You could run all the test programs you want[ed]. The network nodes could even send test traffic. I could sit at the teletype at one place, connect it to an IMP and talk to somebody at the other end, but none of the host machines that were plugged in were yet configured to actually use the net.[103]

The incentive to configure the host machines for actual use was provided by the requirement to prepare a demonstration of the network. Kahn later recalled that the demonstration "would put some pressure on the community to make the connections useful."[104]

The three-day demonstration showed the successful functioning of the ARPANET. Visitors used several programs running on machines located across the country. The available programs demonstrated cross-country interactive graphics, aids to collaborative research across geographically distributed groups, an interactive formal mathematics system, and a chess-playing program.[105] The programs ran on time-sharing systems such as the MULTICS system at MIT and an IBM TSO system at UCLA.[106] The demonstration of programs and computer systems available on the network, and the response time that the network achieved, provided a powerful proof-of-principle for packet switching and the ARPANET.

Those who had developed the ARPANET considered the demonstration crucial to displaying the feasibility of packet-switching networks to those outside of the group of researchers supported by IPTO. They saw it as causing a major change in attitude about the feasibility of packet switching.[107] Alexander McKenzie, who was the non-IPTO liaison to the outside world as part of the project team at BBN, recalled that "after 1972 when the ARPANET had its public demonstration . . . the whole world suddenly began to believe that packet-switching was real, and organizations all over the world wanted to learn a little bit more about it."[108]

In addition to disseminating results, Roberts wanted to remove IPTO from continued close involvement in the functioning network. He had mentioned the desire to transfer the network experiment to a communications common carrier in his original plan for the network.[109] The ARPANET, however, stayed under the control of IPTO until its transfer to another DOD agency, the Defense Communications Agency (DCA), in 1975.

The use of the ARPANET began to have an unexpected consequence. Although the possibility of using a network for electronic mail had been mentioned in 1967, and mail systems were already in use in time-sharing systems, electronic mail had not been a motivating factor for the development of the network.[110] However, it later became an important part of the ARPANET's usefulness. By 1975, the ARPANET directory listed well over one thousand people who had network electronic mail addresses. Many of the people involved in the early design and implementation of the ARPANET listed the usefulness and widespread use of e-mail as an unexpected consequence of the network.[111]

Use of the ARPANET was restricted to DOD contractors or others connected to the government;[112] and because the DOD owned and operated it, the network did not address certain issues of concern to commercial computer networks. A commercial network would have needed to involve the Federal Communications Commission, which regulated communication carriers. The

FCC first approved a handful of "value-added" data communication carriers in 1973. This recognized a new class of common carrier, a value-added carrier, which provided communication services to the public by leasing communication channels from transmission carriers and then "adding value" through services such as switching, error control, and code conversion.[113]

Telenet Communications Corporation, which had one of the first functioning U.S. commercial packet-switching networks, was one of these value-added carriers. BBN had formed Telenet as a subsidiary in 1972 in order to use the experience and knowledge gained during the development and management of the ARPANET.[114] Roberts became the president of Telenet when he left IPTO in 1973. In 1975, the Telenet network began servicing seven cities, and the number grew rapidly, to twenty-two in 1976 and sixty-eight in 1977. The Telenet network is still in existence today. Telenet is perhaps the most obvious example of the several networks directly influenced by the ARPANET. The ARPANET influenced several other new computer networks that incorporated packet switching, including DATAPAC in Canada, CYCLADES in France, and TIDAS in Sweden.[115]

The Expansion of Networking

Network use advanced faster than anyone anticipated. In the late 1970s a number of important issues associated with network use became a concern in the IPTO program, including questions of security and accuracy in systems. Questions of reliability and cost arose within the DOD. As the number of computer systems in the DOD increased, the department's computer experts wondered whether networks might alleviate the increasing cost of adding more systems.

In late 1972, after organizing the ARPANET demonstration at the ICCC, Robert Kahn joined IPTO and became program manager for the packet communication programs, including packet radio and packet satellite. He promoted a number of projects such as research on moving speech through packet-switching networks and providing end-to-end security over the network.

In the 1970s, IPTO began a new program in distributed systems. The high cost and poor performance characteristics of the DOD communications facilities, together with the desire to eliminate duplication of hardware, software, and personnel efforts wherever possible and to share special-purpose resources whenever feasible, motivated the office to develop distributed networks. Research and experience with networking contributed to this new program. IPTO coordinated this program of research and development with the DOD and other government agencies that planned to acquire new computing technology. The partners inside DOD included the air force logistics and intelligence agencies and the air weather service, while the partners outside the DOD were NASA, NSF, and the Federal Reserve Board.

Another area of networking was the IPTO-supported research at the University of Hawaii, under the direction of Norman Abramson and Franklin F. Kuo, during Roberts's tenure as IPTO director.[116] The University of Hawaii could not use existing forms of remote access to computers because the cost and quality of local telephone system service were unsuitable: the noise in Hawaiian telephone lines interfered with the transmission of data.[117] In 1968 Abramson and Kuo began research to learn if transmission via radio waves would work and be less noisy. Their experiment successfully showed the feasibility of packet-switching radio transmission to and from a central computer. The ALOHA system permitted the linking of terminals throughout the Hawaiian islands to a central computer at the University of Hawaii. Specifically, researchers at the University of Hawaii investigated the sending of radio communications from small interface computers at user sites to a central computer. The Hawaii group then experimented with packet-switched satellites using a low-capacity VHF satellite and small earth stations. Because they designed this experiment to demonstrate proof of principle, they did not worry about sophisticated techniques for designing and managing their network. The results were promising, and IPTO decided to go further with the idea.

Kahn set out to expand the use of packet switching in two areas: he wanted to make it possible to use satellites to do packet switching, and to build radio networks for mobile computer communication on the ground. The packet radio program built on the prior research at the University of Hawaii to investigate approaches other than conventional wire communications for computer-to-computer connection.[118] At one point in the packet radio program, approximately twenty contractors worked on different parts of the technology, theory, and implementation of the radios and stations. The packet radio network program was a direct application of packet-switching techniques to military needs, since the network would allow mobility and operation over a wide area. DARPA approved a new program for the development of packet radio systems in May 1972 as an ongoing program to concentrate on the analysis, design, and evaluation of a packet-switching telecommunications network for military requirements. The military wanted to convert to all-digital communication systems, a process that they described as "long and arduous," and that IPTO hoped to facilitate through packet switching.[119]

In addition to the packet radio project, Kahn directed IPTO's packet satellite program. The two options available to him for satellite connections were the International Telecommunications Satellite Organization (known as Intelsat) and the military satellite system. Kahn reported what was involved in establishing this system:

> We chose to go with Intelsat, although the military satellite system was an option that we were seriously exploring. But Intelsat had no mechanism for supporting this kind of technology at the time. Their tariffs did not even permit it. We had to work through the Board of Governors at Intelsat to get new tariffs; we had to get all the countries to approve it; we had to get countries to agree to participate; plus we had to get the systems actually installed and working. So there was—I do not know what to call it—a political/administrative challenge to actually make that all happen.[120]

IPTO decided to go with the Intelsat system. Comsat, Linkabit, and BBN built the technology for a quasi-operational packet satellite system using Comsat facilities in the United States, and using Post Telephone & Telegraph facilities abroad through the Intelsat IV satellite. At the time, no domestic satellite capability existed in the United States. The communications satellite industry was itself in the development stage. Intelsat sent aloft its first satellite, called *Early Bird*, in April 1965. Telecommunications services using it commenced in June 1965. During 1967, Intelsat placed three Intelsat II satellites, each with a 240-circuit (or one-television-channel) capacity, in orbit, and between 1968 and 1970 five Intelsat III satellites were added to the system, each with a circuit capacity equal to that of all of the earlier satellites combined. In the first half of the 1970s, seven Intelsat IV satellites joined the system, and the combined channel capacity of the system substantially increased. Multilateral agreements that had been under discussion since the beginning of the 1960s became effective in February 1973. At this point Intelsat existed as a global commercial communications satellite system with defined operational procedures and limits.[121] It was a propitious time for Kahn to begin discussions with that organization about data transmission via satellite.

IPTO's packet satellite program grew out of the needs of the DARPA seismic verification program.[122] IPTO initially planned to extend the ARPANET to countries overseas so that seismic data could be gathered from European locations.[123] Cost considerations quickly led to a satellite solution in which, by the use of packet switching, a single channel would be shared among multiple ground stations. By September 1976, packet satellite communication experiments, using available satellites, had been used to install and debug the satellite interface message processor (SIMP), initiate channel access schemes, and design and install measurement software in the SIMPs.[124] The SIMP was initially very similar to the ARPANET IMP, a Honeywell model 316 computer with 16K additional words of memory and special support for satellite operations. In the same way that packet switching got around the requirement that dedicated telephone lines connect pairs of computers that needed to communicate, the satellite system permitted many ground stations to share a single satellite channel so they could send and receive data without preallocating satellite channel bandwidth to any single pair at one time. All ground stations could hear the satellite broadcasts, so it was possible to address the information to many destinations simultaneously.[125]

Kahn came to understand that another packet-switching network, using satellites, would be incompatible with both the ARPANET and the packet radio network. The packet radio network already included several IPTO contractors, but the researchers connected to it could not benefit from any of the systems or programs already connected to the ARPANET.[126] As the packet radio network developed, Kahn realized that these separate networks needed to be connected to each other. Even when packet-switching techniques were used, the networks and protocols were different and incompatible with each other. A way for different packet-switching networks to communicate with each other had to be found. The problem was similar to the earlier problem of connecting incompatible computer systems, except that now incompatible networks needed to be connected.

IPTO wanted to connect both its new packet radio network and its planned packet satellite network to the ARPANET. While the ARPANET had solved the problem of interconnecting time-sharing computer systems for the IPTO research community across the United States, it had limitations. The existing ARPANET protocol would not work for unreliable transmission, yet that capacity was required by the packet radio network: the protocol had to work through a radio network that was much less dependable than a telephone line. Also, the first ARPANET protocol assumed that the underlying physical network would maintain sequencing of packets. The details of the networks were quite different. For example, ARPANET ran over fifty-kilobit-per-second lines, with packets of one thousand bits, and assumed reliable end-to-end transmission; the packet radio network worked at one hundred and four hundred kilobits per second, had two-thousand-bit packets, and assumed an unreliable transmission medium.[127] The protocols designed for the ARPANET would not work over the packet radio network.

The Interconnection of Networks

IPTO's attempt to overcome the incompatibilities of its packet-switching networks led to a major new project area. The problem that IPTO faced in interconnecting its networks was generic in that it would be experienced when any two networks were interconnected. As soon as organizations wanted to connect networks, the problem of how messages would be recognized and sent across the networks arose. For which computer in the second network was the message destined? The ARPANET had been designed without the notion that other networks would be connected to it, and the original ARPANET addressing contained insufficient information to route packets to another network. To examine the problem of interconnection IPTO embarked on the next stage of its networking program: the development of a packet-switched supernetwork through which networks could communicate.[128] With the design and implementation of the Internet, IPTO addressed and solved the problem of communicating across its networks, and did much more.

Concern with interconnection was not limited to IPTO. As other wide-area networks developed, there was a growing interest among their sponsors to find ways to interconnect them. Those involved in commercial networks and a growing number of international networks were interested in the problem of interconnecting networks. For example, at the ICCC ARPANET demonstration in 1972, the ARPANET was interconnected to the commercial TYMNET network developed by TYMSHARE.[129] As the director of telecommunications systems at TYMSHARE noted, "The experimental connection that was used during the recent ICCC '72 demonstration has created much interest in the area of network crossover technology. Since this will undoubtedly be of great value to future data services, it seems logical to continue the experiment."[130] Although this particular experiment did not proceed, it showed the growing interest in a general solution to the network connection problem.

In 1973 Kahn enlisted the help of Vinton Cerf to work with him on the problem of interconnecting different packet-switching networks. Cerf, a member of the ARPANET Network Working Group, had just received his Ph.D. degree in computer science from UCLA (1972) and joined the faculty at Stanford University. Cerf and Kahn cooperated on the design of the Internet strategy, creating an addressing scheme for multiple networks and a new protocol for end-to-end communication.[131] Then Cerf and his students at Stanford carried out the first Internet protocol implementations with IPTO funding.

Cerf's connection with the international researchers influenced the protocol that he and Kahn designed. He credits several people other than Kahn, especially Herbert Zimmerman and Louis Pouzin, with having influenced the interconnection protocol.[132] Both Zimmerman and Pouzin had been doing packet-switching experiments in France. Connecting the various European networks together presented problems similar to those posed by connecting IPTO's networks together. Other researchers involved in early discussions included Gerard Lelann, who had been working with Pouzin, and Robert Metcalfe, a former member of the ARPANET Network Working Group, who was then working at Xerox PARC and would later be a key figure in the design of the Ethernet approach to local area networks. Lelann, Metcalfe, and Cerf were all involved in the International Network Working Group (INWG) and were discussing interconnection issues as early as 1973.[133] In 1976 Cerf went to work for IPTO and assumed responsibility for running the Internet program (as well as managing the packet radio, packet satellite, and network security programs).

The protocol proposed by Cerf and Kahn eventually became two different protocols that work together, the Transmission Control Protocol (TCP) and the Internet Protocol (IP), or TCP/IP. The ideas of the original paper were defined and extended by another group, the Internet Engineering Group.[134] In the internetwork design, computers known as gateways connected the networks by being able to participate in both of the networks. It was impossible in the existing protocol to specify a sufficiently detailed address to proceed through the gateway and find the correct location within the second network. To make this possible, a packet from one network was embedded in a larger packet. The larger packet itself had to contain more addressing information in order to be delivered to the correct destination. Thus, it was designed to include the information needed to route the packet in the second network.

While Kahn and Cerf worked on the TCP/IP protocol for connecting DOD networks, the INWG was working on the problem of other networks. The International Network Working Group had been organized, and had first met, at the 1972 ICCC to explore the interconnection of computer networks. Cerf was chair of this group. In 1976, when the INWG published its proposal for international standards, there were already experimental implementations of Cerf and Kahn's TCP/IP in the United States and elsewhere.[135] In the United States, especially within DOD, TCP/IP rather than the alternate international proposal continued to be used, and with the backing of the DOD, the TCP/IP protocol became widespread. All the host computers in the ARPANET were required to support TCP/IP by January 1983.[136] On that date, the ARPANET replaced its original host protocol (NCP) with TCP/IP.[137] The international group eventually created the ISO/OSI protocols.

An IPTO contract under which BBN was to implement the new TCP/IP protocols for use with the popular UNIX operating system strengthened the hold of TCP/IP. This implementation in a widely available operating system, the Berkeley UNIX system, meant that TCP/IP would become available to 90 percent of the computer science departments in the United States.[138] This widespread availability led to increased use and understanding of the protocols. The development and use of internetwork protocols within the DOD and the university communities began to make the ability to communicate internationally via computer networks commonplace.

To complete our story of the IPTO networking program, we need to indicate the fate of ARPANET in this new world of the Internet. In 1982 the DOD formed the Defense Data Network (DDN) to expand and further deploy the ARPANET for military use.[139] The ARPANET had become both a useful tool and an experimental system. This dual nature of the ARPANET, however, was a problem. Some IPTO contractors needed to experiment with new network procedures, but other network users required a stable environment and did not want to experiment with the network or to be forced to modify their network software to adjust to changes in the network.[140] In 1984 the DOD separated the MILNET from the ARPANET in order to provide a separate, nonexperimental network for the ARPANET's military users. The MILNET provided protection from unauthorized users yet retained some connection to the experimental community, offering access to all the resources already connected to the ARPANET.[141]

The Use of the ARPANET in the DoD

The accomplishments of IPTO programs in the development of the ARPANET, the Internet, and related distributed-systems technology, along with the development of new control technology, sharply focused the IPTO command and control (C^2) mission. In the 1960s, Licklider and others expected research in computing to aid the DOD's command and control R&D; and in the 1970s IPTO paid considerable attention to integration of the elements of command and control and added communications (called C^3) to its mission. IPTO included its involvement in advanced C^3 as a sub-element in its program during FY 1973 and presented it to Congress as part of the program in the following fiscal year.

George Heilmeier, director of DARPA from 1975 to 1979, testifying before Congress in February 1977, gave an eloquent description of the need for advancing the technology base for C^3, and the way in which that advancement would be accomplished. He started by acknowledging that "for the foreseeable future, people, rather than computers, will be making command decisions.

Human limitations in formulating and communicating commands will be a central difficulty in increasingly complex command, control, and communications systems." He then noted that DARPA had already initiated research for new types of command systems, ones that involved the development of new technology for information management. In his remarks about the capabilities of the new systems, we can see the level of integration which DARPA officials expected to take place. The capabilities would include spatial rather than symbolic information storage and retrieval; techniques for allowing high-speed, consistent selection of information from large amounts of data; techniques for dealing with the problems associated with information overload; new approaches to structuring information in order to improve comprehension and memory; and rapid presentation of pictures and text to increase comprehension and memory. Heilmeier also included the "complimentary capabilities" of technologies to aid reasoning and problem solving, as well as heuristic modeling. These would enhance a commander's ability to make effective decisions, particularly in crisis situations. Finally, he turned to problems of "*group* decision making, *group* problem solving, *group* reasoning" which required "automated aids to assist heterogeneous groups in understanding communications, combining resources, and reaching decision consensus in an optimal fashion.[142]

IPTO's efforts to advance the state of the art of information technology for command and control in accordance with Heilmeier's vision focused on advanced technologies in several areas. One area was computer communications. In addition to the packet radio and packet satellite systems described above, these technologies included the integration of digital and voice communications into future defense communications networks that included speech-understanding capabilities. Other areas included software production systems, a new Advanced Command and Control Architectural Testbed (ACCAT), a flexible design tool to test operational credibility, introduction of secure interactive computer techniques into military message systems, and crisis management using computers. Although IPTO had been working with military agencies all along, the expanded command and control program in the 1970s meant more involvement with the military services to demonstrate to them the value of new IPTO technology. One way in which IPTO increased its involvement was by including military officers on its staff. These officers, including IPTO director Russell and program managers Dolan, Perry, and Ohlander, facilitated contact with the military services. IPTO also increased its contact with the military by occasionally awarding contracts for the building of specific systems for military use. In the mid-1970s, at Heilmeier's insistence, IPTO increased its participation in testbed programs with the military services. Significant increases in IPTO's budget in the 1970s, and greater increases to DARPA budgets as well, came about through funding designated for testbed activity. These projects were all in the nature of command and control initiatives. We next examine a few of these programs to illustrate how IPTO attempted to transfer this advanced technology—the ARPANET, packet switching, and other results of IPTO programs—to DOD agencies. The ARPANET concept and technology became very important to DOD. For example, a program called the National Software Works (1974), a joint effort of IPTO and the air force Data Automation Agency, sought to improve the software production process using the ARPANET. The development and maintenance of computer software was becoming increasingly critical to the air force in terms of cost, reliability, and development time, as indeed it was to almost every user. During the winter of 1973–74, IPTO and the air force's Data Automation Agency worked out an agreement to organize a program to expedite the creation of computer software within DOD.[143] The air force hoped that this program would produce "a totally new approach to software development." Through ARPANET, programmers and documentation writers were to have continuous access to a shared interactive facility that was specifically designed for software and documentation production and would be completely separate from any site's production hardware. The first phase of the program involved two air force programming centers; later phases involved a wider variety of air force users and encouraged the use of tools that companies had produced based on the earlier results of the program. Software production was a central activity in the development of weapons systems, command and control systems, communication systems, intelligence systems, and logistics systems; and software systems that would materially improve the cost-effectiveness and reliability of the production process would materially improve the nation's defense posture.[144] IPTO saw the program as a transfer mechanism for automatic programming technology.

The IPTO interest in automatic programming emerged from earlier DOD programming concerns of the 1960s. Throughout the 1960s, the ratio of software costs to hardware costs increased rapidly. Programming tasks drained financial resources and absorbed all the available programming talent, and there was no end in sight to the demand for more programmers. The view that software was a significant problem was becoming common among computer professionals.[145] Evaluations of the software production process noted that software was "the major source of difficult future problems and operational performance penalties."[146] *Datamation* reported in 1971 that the procurement of the military's World Wide Military Command and Control System (WWMCCS) had been estimated to involve expenditures of between $42 and $206 million for hardware (depending on how many systems were to be purchased) and $722 million for software.[147] The software production process received important attention in the DOD as one answer to these spiraling software costs. And to stem the rapidly growing costs of software production, especially for the DOD, IPTO in 1971 established a separate program in automatic programming. IPTO expected, with the results of this program, to be able to assist the DOD in reducing its annual expenditure of $450 million on computer programs, a sum independent of the WWMCCS software costs.

In the early 1970s, automatic programming was also thought to have the potential to overlap with the interests of the IPTO AI programs. IPTO designed the automatic programming effort to generate improvements in automatic problem solving and planning. "Currently it is possible to automatically form simple, two- or three-stage plans to solve problems like constructing a three- or four-module computer program or, in human terms, planning how to get around an unexpected roadblock."[148] MIT, CMU, SRI, Stanford, and BBN received major contracts in this research area, and some smaller contracts were given to Syracuse University, the Information Sciences Institute at the University of Southern California (USC/ISI), and the Computer Corporation of America. In the mid-1970s, the new DARPA director George Heilmeier took a critical stance with respect to IPTO programs, especially AI, and this negatively affected the automatic programming activities and the National Software Works.

The National Software Works was an attempt by IPTO to incorporate AI technology into automatic programming, which IPTO hoped would be an example of an effective technology transfer to the military. What was significant about the National Software Works agreement document is the report of the contracts that IPTO already had in force related to this work, the way in which the results of these contracts were to be used in the effort, and the number of air force sites either on the ARPANET or planning to arrange for a node. Several new ARPANET protocols needed to be defined to allow exchanges of this kind. A new filing system, under development through the CCA, was in operation using a PDP-10 computer. IPTO intended to use the TENEX operating system for security in this new enterprise. The filing system and TENEX were developments of other projects, once again illustrating the convergence of IPTO projects. The air force reported that six air force organizations had terminal interface message processors for access to ARPANET. The air force systems command linked its CDC 6600 computers to ARPANET, and the Rome Air Development Center (in Rome, New York) added its MULTICS and GCOS machines to the net. This project successfully used the ARPANET. However, the National Software Works itself was unsuccessful: because it used a system with too many layers of protocols and operating systems, it was too slow and clumsy.[149]

Another example of how the network concept and technology became important to the DOD can be seen in the use of testbeds. Several testbeds used packet radio technology. One of Russell's prime responsibilities was visiting army commands to brief them about packet radio and networking technology. The purpose of these visits was to stimulate army personnel's interest in this technology and increase their willingness to use it in exercises. For example, an army group at Fort Bragg expressed interest in working with IPTO on a mobile packet radio system. As part of the work on packet radio Russell took responsibility for establishing at Fort Bragg a packet-switched mobile radio test project involving packet radios, small processors, and computer terminals. A microprocessor was built into a small portable radio about one cubic foot in volume. The army tested the system in exercises to judge its practical utility. Russell helped to formulate the project, found the contractors to carry it out, and handled the liaison with the Fort Bragg personnel.[150] Under an IPTO-sponsored contract, packet radio technology was deployed, and the army used that technology in exercises in order to evaluate its behavior in real applications. The packet radio testbed was the start of a series of experiments by IPTO in advanced C^3 technology.

Heilmeier had more advanced management ideas in mind as well. The packet radio testbed focused on the integration of components to form a better command and control system. As Heilmeier put it: "Decisions [to acquire the technology] are based on insufficient data. The best of candidate architectures is not possible, and system issues are not resolved in 'try-before-buy' fashion." According to Heilmeier "one doesn't design aircraft without a wind tunnel or ships without a tow tank. A C^3 testbed will serve as the integrating function and 'final exam' for modern, expensive C^3 technologies."[151] For Heilmeier, integration was of utmost importance.

In collaboration with the navy, IPTO sponsored a testbed project called the Advanced Command and Control Architectural Testbed. The ACCAT facility used time-shared computers securely interconnected through the ARPANET for research on navy command and control problems. ACCAT began in mid-1976 as part of a five-year joint effort by IPTO and NAVELEX (the Navy Electronics Systems Command) to "speed up the application of new artificial intelligence, computer, and networking technologies into the military command and control area." The ACCAT testbed demonstrated and tested new technologies that incorporated AI techniques, and extended ARPANET to include special military computers and new security requirements.[152] In 1981 IPTO described ACCAT as providing "a realistic Navy Command and Control environment for the use and evaluation of information processing and computer communications technology."[153]

IPTO became involved in the project (at least in part) because Heilmeier believed that its AI efforts should be directed more toward applications.[154] Heilmeier and IPTO director Russell brought Floyd Hollister from NAVELEX to be program manager for the ACCAT program element in IPTO; and Russell and Hollister defined the elements of the program. Over the five-year period, IPTO spent $15.7 million on ACCAT, while NAVELEX put in $1.5 million.[155]

The ACCAT project had several objectives: "to test current AI and related technologies, to acquaint those in command and control R&D with their potential, and to challenge AI researchers to come up with useful applications."[156] The ACCAT system was initially set up using DEC computers with UNIX and TENEX operating systems. ACCAT was linked to ARPANET with packet network security devices developed by BBN with IPTO funding. The early facilities linked to ACCAT via ARPANET included the command and control laboratory at the U.S. Navy Postgraduate School; the Fleet Numerical Weather Center at Monterey, California; the Navy Electronics Laboratory Center (NELC; later the Naval Ocean Systems Command [NOSC]); the CINCPAC Command Center in Hawaii;[157] and the University of Southern California/Information Sciences Institute.

The navy used the ACCAT testbed for war games and technology evaluations involving new operational fleet tactics and weapons characteristics data that required a classified environment. ACCAT also served as a testbed for three IPTO-supported AI projects: the Rand Intelligent Terminal Agent (RITA), SRI's natural language system (Language Access to Distributed Data with Error Recovery [LADDER]), and NOSC's Naval Message Disambiguator. RITA and LADDER were commissioned in 1977 to help solve the typical C² problem of obtaining timely information from distributed data bases at the Fleet Command Centers in Hawaii and Norfolk. RITA and LADDER assisted operators unfamiliar with data bases in information searching. LADDER contained a new naval vocabulary and new display systems used for war games played on the Warfare Evaluation Simulator at NOSC. Both fleet and laboratory personnel used these displays for interaction and evaluation. The navy used RITA and LADDER to support a modified version of the Computer Corporation of America's SDD-1 relational distributed database management system.

ACCAT served also as a testbed for a Naval Message Disambiguator, funded in 1981 under IPTO's Intelligent Systems program. The disambiguator's purpose was to "attack the problem of computer understanding of free (unstructured) natural language messages before the message is sent, thereby taking advantage of the sender's knowledge to eliminate ambiguities." It was based on IPTO-funded work at Yale on natural language.[158]

Although ACCAT's results indicated the potential of the new AI technology, a number of technological limitations existed. The modified SDD-1 system was too slow because of ARPANET communication bandwidth limitations; RITA was found to be "slower than the standard manual procedure"; and LADDER was "too limited." Although RITA and LADDER themselves received little use, the involvement of the office of the commander of the U.S. Navy's Pacific Fleet (CINCPACFLT) in IPTO's Strategic Computing program included similar natural language capability testing.[159] The project did not lead to the navy's directly adopting any of the new technologies, although the project did test those technologies in operational settings and thus achieved part of its objective. The technologies successfully transferred to NOSC included networking techniques, techniques for communication between military computers, and ARPANET security measures.[160] IPTO's participation in the ACCAT joint project ended in 1981, and the ACCAT facility was transferred to the navy.[161]

An example of CMU's work on testbeds was an experimental testbed: the NAVLAB robotics project. This project employed both image-understanding research and knowledge-based systems technology. As mentioned above, in the Fort Bragg packet radio testbed program, later renamed the Army/DARPA Distributed Communications and Processing Equipment project, the army and IPTO explored innovative concepts in distributed automated data processing for use on the battlefield. The applications were tactical information reporting and management, communications and transportation resources management, and real-time fire control.[162] Also in the late 1980s, an expert system called FRESH (Force Reallocation Expert System) was developed in an operational environment by Texas Instruments as part of the Strategic Computing program. FRESH provided staff decision aids for use in force reallocations and capabilities assessment. Earlier IPTO-sponsored work in heuristic hypothesis formation and constraint-based allocation and scheduling formed the basis for FRESH. The testbed involved IPTO and CINCPACFLT and is now operational.[163] Although diverse, these examples illustrate IPTO's interest in moving new technology, especially networking developments, into systems that the military would find useful. All of these testbeds used the ARPANET and the Internet in some way; in fact, IPTO could not have initiated them without the ARPANET.

Officially, IPTO's network project was the implementation of a specific tool to solve resource-sharing problems that the office encountered in administering and guiding its programs. It was also a research project that put the office at the forefront of computer network development. IPTO's direct management involvement was the key that allowed the network to succeed. ARPANET was a large-scale experiment that was started to solve the problem of resource sharing between computers and among researchers. The goal was to connect computing systems, and through the systems the researchers, so that research could be accumulated and duplication of effort avoided through the sharing of resources and improved communication. Packet-switched networks, as applied to the problem of connecting interactive computers, were an innovative application of communication techniques. Through the use of packet switching to connect computers—and later, computer networks—distributed computer resources could be economically shared; and access to the resources, and to the people using them, could be improved.

Although the ARPANET solved the problem of interconnecting time-sharing computer systems across the United States, certain limitations became clear as the network grew in the early 1970s. These limitations resembled those found earlier during attempts to connect incompatible computer systems, except that now there were different and potentially incompatible networks that needed to be interconnected. Various forms of networks developed, such as local area networks and commercial networks, but it was the development of packet radio and packet satellite networks that led to IPTO's continuing involvement in computer networking. The theoretical and practical knowledge obtained through the realization of a distributed, wide-area, packet-switched network led to the understanding and growth of computer networks well beyond the ARPANET itself. While IPTO's greatest technical achievement in networking was the development of the ARPANET, its lasting contributions to networking are packet switching and the Internet.

One of the outgrowths of the development of time sharing was the need to connect computers together in a way that allowed interactive access. Packet switching, as demonstrated by the ARPANET, met this need. IPTO directors and program managers brought together a group of contractors and a collection of new ideas to design and construct a network that allowed users interactive access to computers around the world. IPTO's support of packet switching spread packet-switching techniques to different kinds of networks that were connected by different media. Just as the sharing in a time-sharing system expanded when a network was implemented, so the sharing and communication in a network expanded when networks were interconnected; and they have continued to expand to the present as increasing numbers of networks continue to be interconnected.

However, describing the accomplishment in this way does not explain why the development occurred through the intervention and activities of IPTO personnel, nor does it indicate why they were successful in the endeavor. The work on networking during Licklider's early tenure at IPTO was consistent with the overall IPTO program: that is, contracts were awarded to researchers to support investigations of possible avenues for networking. But this did not work. The UCLA contract for connecting three centers is a good example of the failure. The problem of networking transcended the needs of a given laboratory, a given group, and a particular way of addressing computing problems. The time was not right for achieving the result; specifications needed to be developed first, and they were not forthcoming from the computing community as had been the case with time sharing and graphics. Incentives for networking were lacking in the community. Such incentives, however, did exist in DOD, where there was a need to reduce the high cost of software development, improve communications among military units while increasing computer use, further develop command and control systems, and so on. Moreover, the problem of specifications was a typical military problem in R&D for new systems. Licklider's successors understood how the military accomplished its R&D through consultation with the community: consultations directed and evaluated by the military personnel themselves. Taylor also observed this style of R&D at NASA, another organization using the military R&D model and military personnel. So IPTO adopted this inside model to define specifications for a network.

Inside the DOD, there was freedom of action to pursue investigations by means other than contracting alone. And DARPA management allowed this approach to R&D on occasion. Using the inside model when specifications did not exist allowed IPTO to focus on the DOD's needs and IPTO's desires, with rapid time scales and staff unencumbered by other activities (Roberts was engaged to perform the work). IPTO was not restricted to a particular group of contractor personnel, and had no need to define milestones at the outset of the project as a contractor must do. Taking the problem inside the DOD allowed Taylor and Roberts to go anywhere for advice, define the specifications for a system that IPTO desired and that would meet military needs, and focus on the overarching objective—universal connectivity—rather than local enhancements. By doing this, Taylor and Roberts elevated the role of IPTO in the computing research community, enhanced IPTO's reputation, and contributed a new communications system to the military and a generic technology to the nation.

IPTO's Robert Kahn and Vinton Cerf made even greater contributions with the design and implementation of the Internet. The Internet grew out of the

DOD's need to interconnect its various research networks. The solution to the problem of the lack of intercommunication across networks was a new protocol and architecture, designed by Cerf and Kahn in IPTO and published in 1974. For implementation efficiency, this protocol was eventually split into two protocols that work together, the Transmission Control Protocol and the Internet Protocol. The key to this development was the notion of a gateway between dissimilar packet-switched networks, and the use of the IP protocol to provide a global, uniform addressing structure. By the end of the 1970s, IPTO's support of packet switching had spread packet-switching techniques to various kinds of networks. The development of internetwork protocols and their acceptance and use within the DOD and the university communities began to make flexible, worldwide interconnection of computer users commonplace.

The most common use of networks currently is interpersonal message exchange, called electronic mail, or e-mail, perhaps the most interesting long-term general effect of networking. A user can send a message to anyone on the network, provided he or she knows the network mail address. Whereas conventional postal systems provide for the exchange of written communications in a few days or weeks, networks cut that down to a few hours or less. Computers also provide tools to create and edit messages. Because the exchange is nearly in real-time, the communication is often more conversational and much less formal than letters sent through postal services. Computer networks connect locations throughout the world. More and more people are being affected by networks and are able to communicate globally on an informal basis. As of 1992 the Internet connected more than five thousand networks, with an estimated number of users in excess of 4 million. People who are physically distant from one another can communicate quickly and easily. The use of electronic mail across time zones can enhance communication, particularly international communication. For example, a message can be sent to Europe at the end of one day in the United States and a response can be received at the start of the next day. Both parties would have time to consider the communication and work on it during their normal working hours; there would be no need to try to coordinate telephone connections at inconvenient times. This can be particularly useful when the message is not in the receiver's native language, as it is often easier to read and decipher a message at leisure than to understand the communication in real-time. Users wanting to access a host computer across the country or the world needed only to connect to a local host, usually not more than a local telephone call away. The development and proliferation of worldwide packet-switched networks, and the ability of these networks to communicate with one another, have changed basic communication patterns in the scientific, business, and technical communities. Thus, the impact of packet-switched computer networks has reached far beyond technical achievement. Although the ARPANET was dismantled in the late 1980s and no longer exists, the research and developments that it led to, specifically the TCP/IP protocols and the resulting Internet, have had a lasting and significant influence on computer networks.

Users of computerized communication have noted many benefits. In 1977, after more than two years of experience with electronic mail as provided by the ARPANET, the U.S. Army Materiel Development and Readiness Command reported, "This is a major new medium of human communication and interaction, with a very positive impact on the way we do business." [164] They reported better communication through electronic mail for a number of reasons, including the greater degree of communication, the ease of sending multiple copies of messages, the opportunity to stay in touch while away from the office, and the ability to send and receive messages whenever and wherever appropriate. Because the messages were not received in real time, the network also improved communication between various time zones. This rapid but written form of communication was credited with saving time because the "turnaround time in getting decisions on urgent actions" was reduced from a few weeks to a few days. Better coordination was "a major advantage" because of the ability to "coordinate complex actions among many people, independent of geography." [165] Meetings were still required, but there were fewer of them, and they were more productive.

The DOD gained a direct benefit from IPTO's network research when it formed the Defense Data Network. In 1984, a set of ARPANET nodes was split off as a distinct, nonexperimental network called the MILNET, which offered the military a network that was protected from unauthorized users and provided access to all the resources on ARPANET.[166] By 1987, the MILNET consisted of more than two hundred nodes worldwide. The successful emergence of the MILNET out of the ARPANET, and the benefits that the former provided to the military, were due to the existence of a functioning packet-switched network, the ARPANET, and to the interconnection techniques and the network security developed in IPTO throughout the 1970s.

While concentrating its efforts in wide-area networks, IPTO also influenced other kinds of networking. One of the most widely used local area networks, Ethernet, was developed by a group of researchers that included Robert Metcalfe, who had been part of the IPTO Network Working Group and had participated in the early stages of the packet radio program. Ethernet was developed at Xerox's Palo Alto Research Center, in a group put together by Robert

Taylor, after he finished putting the IPTO network project together. Taylor was explicitly sought after by Xerox PARC management owing to his experience at IPTO.[167]

We are constantly aware today of "the network," exploited in the United States of the mid-1990s as the "Information Superhighway," the road to tomorrow's world. Once again government is trying to take the lead in providing a way to increase the productivity of scientists and engineers and the learning efficiency of students, to enhance international economic competitiveness, and to create a leading-edge high-technology image for the nation through networking. It is well to remember that the basis for this program to connect us all to the "Information Superhighway" is only the latest chapter in the story of the invention, development, and implementation of networking, a technology begun by IPTO.

CHAPTER FIVE

The Search for Intelligent Systems

Where would artificial intelligence be without the Information Processing Techniques Office? IPTO provided large sums for the support of artificial intelligence research in a selected set of United States institutions from 1962 onward. Indeed, it was the largest funder of AI in the world for at least a decade and a half after 1962 — providing an amount far greater than the total provided by all other groups — and was a substantial funder thereafter. In a sense, AI's history from 1962 to 1975 is IPTO's AI history, but IPTO's continued presence in support of AI suggests that we take a longer view. If we do, we note that AI's history over the last thirty years is sprinkled with significant developments, interesting intersections with other research areas outside computer science, and a record of controversy. The field of AI is extensive. However, even as the largest funder, IPTO was not interested in the entire field. Rather, the office was specifically interested in AI techniques that could contribute to the development of the most capable, flexible, intelligent, and interactive computer systems for use in research and in military systems. Therefore, in examining IPTO's interest in AI, we do not require a full history of AI, no matter how desirable such a history might be. Examining those parts of AI promoted by IPTO is important for our overall study of IPTO for three reasons. First, IPTO-supported activities made fundamental contributions to understanding and accomplishing artificial intelligence. Second, they were important to the overall IPTO program. And third, they set the stage for much of the AI work to follow. Therefore, we will limit our discussion to three areas of technical development — problem solving, speech, and robotics — that contributed to IPTO's objectives for computer systems, and to the institutional contexts from which developments in those areas emerged; we will not undertake the larger task of providing a complete history of AI research.

Interest in intelligent machines is longstanding. Over the years, researchers have explored a range of topics in efforts to produce a machine with capabilities similar to those of a human. They have constructed logic machines and robots, automatons and smart weapons. In the middle of the twentieth century, the search for intelligent machines stimulated researchers to examine questions in control theory, problem solving, psychology, natural language, learning, pattern recognition, and philosophy, among many other fields. The appearance of the electronic digital computer in the mid-1940s, with the capability to process both numbers and symbols, generated interest in the use of computers in the quest for intelligent machines.

In the 1950s, researchers interested in using the computer in research on symbol-processing problems in artificial intelligence focused primarily on two themes, which are still evident in AI today. On the one hand, some researchers view AI as the experimental and theoretical study of perceptual and intellectual processes using computers. The goal has been to understand these processes well enough to make a computer "perceive, understand, and act" in ways that formerly were possible only for humans. Put another way, this group employs programs as tools in the study of intelligent processes, tools that help in the discovery of the thinking procedures and epistemological structures employed by intelligent creatures.[1] This group focuses on whether AI can produce truly intelligent behavior in a computer and includes researchers who work in the area called cognitive science. Much of the controversy that surrounds AI is associated with the views and efforts of the researchers in this group.

On the other hand, some researchers, most likely a majority, believe that AI is the study of ideas that enable computers to be intelligent.[2] Put another way, AI is the "part of computer science concerned with designing intelligent computer systems, that is, systems that exhibit the characteristics we associate with intelligence in human behavior — understanding language, learning, reasoning, solving problems, and so on."[3] These researchers are experimental pragmatists who are willing to use any method that will result in intelligent machine behavior that helps to solve complex problems. In other words, they explore how AI as a discipline can provide techniques that are beneficial in solving very difficult problems. Most researchers in AI today would align themselves with this second group. Although this group of researchers believes that truly intelligent behavior in a machine is possible, the fact that achieving such behavior will take a very long time makes them view intelligent behavior in machines as an ideal. While the science of AI moves toward this ideal goal, workers in this second group pursue research to make steady progress in achieving more useful techniques for the solution of complex problems. IPTO focused its attention on this group, and so shall we.

What distinguished the pragmatic form of AI research from other computer science activities was a quest for intelligent machines that used a model of thinking based on behavior rather than on physical structure. Researchers in artificial intelligence understood that the behavior of humans might provide clues to structuring programs for problem solving. They could not resort to information about the physical structures of human minds, because we know so little about what mediates perception and thinking in humans. The goal of the early pragmatic AI researchers was focused on developing computer programs that would perform some aspect of human behavior that could be clearly called intelligent.

The statements defining the research of the pragmatic research group are deceptively simple, but they hide a monumental research problem, the full dimensions of which continue to unfold forty years after the research area's founding. The principal dimensions of the research problem for the pragmatists are (1) to decide what kinds and amounts of knowledge are needed to set up a problem in a computer (representation and rules of change); (2) to express explicitly the actions that have to be performed (constructing a solution via search, employing inference techniques); (3) to learn how to devise a machine (or program) to interact with the world; and (4) to develop models describing how all this activity takes place (programming, rules of thought, etc.). All four of these dimensions were investigated simultaneously after 1960, and there has been a significant amount of overlap among them. The titles of specific subelements in these dimensions have varied slightly over time, but the basic tasks in each research area have remained the same.

Some researchers interested in these dimensions concentrated on developing a capable machine assistant for use in human decision-making systems. This effort required the design and construction of a system that could understand data about a problem situation, frame hypotheses about the task to be performed in order to solve the problem, select from among the hypotheses, present a recommended or rank-ordered solution, and state the justifications for the recommendations. This was not an easy undertaking; we spend many years promoting the development of these capabilities in humans through the education process.

In the early years, AI researchers interested in developing this kind of system often focused on carefully delimited problems: problem solving (e.g., writing a program to solve an algebra problem); performing simple tasks moving blocks (e.g., constructing a program to simulate simple actions in the three-dimensional world); and robotics (e.g., how to mimic human motor capability with a mechanical arm). The search for general proof methods and simple speech and sight systems directed this focus on small specific problems. In particular, the application of heuristic search techniques to these problems allowed researchers to address a range of problems in this field, with some remarkable results. A heuristic is a strategy, simplification, or "any other kind of device which drastically limits the search for solutions in large problem spaces." A useful heuristic provides solutions that are "good enough most of the time."[4] Among the heuristics useful in the early days of AI were the technique of "working backwards" (e.g., doing an algebra problem in reverse, from known solution to problem statement); attacking AI problems with methods successful in other domains; and means-ends analysis; along with some special-purpose heuristics. Research in problem solving used these heuristic techniques with substantial success. Research into vision, speech, and automated mechanical motion increased our knowledge of the fundamentals in these areas and eventually produced machines with good vision and speech qualities.

Research in problem solving, speech, and robotics was supported continuously by IPTO. As time passed, researchers identified issues attendant to those tasks which transcended one or a few applications, until it became possible to combine research results into large systems such as the IPTO-sponsored automated land vehicle of the 1980s. Inasmuch as there is no adequate history of AI research, we begin with a brief overview, which provides a rather episodic summary of the history of AI from 1950 to 1980, describing IPTO's entry into the field, highlighting the continuing evolution of AI in areas associated with IPTO programs, and providing a background for the remainder of the chapter, which deals with research in problem solving, continuous speech recognition, and robotics. These are the three main themes in IPTO's support for the development of flexible and interactive computer systems.

An Overview of the History of Artificial Intelligence

There are a number of ways to categorize the problems investigated in AI at the opening of the 1960s. In 1963 Feigenbaum and Feldman, in *Computers and Thought*, divided the papers in their volume under the rubrics "Machines that Play Games," "Machines that Prove Mathematical Theorems," "Question-Answering Machines," and "Pattern Recognition" machines. This division was suitable for the early 1960s, when it was not so obvious where the field was headed.[5] In the 1990s, the array of topics in AI research suggests other categories with which to analyze the field of the 1960s: languages, machine designs, problem solving, modeling human knowledge and thought, hand-eye coordination, natural language processing, and human-machine interaction. When the field is examined in this way, we see much more overlap among projects over time. Moreover, it is then easier to appreciate the emergence over the years of well-defined AI problems and a consequent separation of subareas in the field of AI.

A general history of AI from 1955 to 1990 would show that support for research in the field came from a wide variety of sources, but that in the formative years (1962 to about 1975) virtually all the support for university AI research in the United States came from IPTO. While IPTO did not invent the field of AI, AI's inclusion as a major effort in IPTO-supported laboratories helped IPTO-supported researchers play the major role in organizing the field. Developments in some areas, such as speech and expert systems, attracted corporate support. However, even the basic developments in speech recognition and the fundamental concepts of expert systems had their origin in IPTO-supported laboratories.

From 1962 to 1975, the preponderance of funding for research in artificial intelligence came from IPTO, although some funds came from the Office of Naval Research, the Air Force Office of Scientific Research, and the National Institutes of Health. In 1974 Steven Lukasik, director of DARPA, described IPTO-supported research in artificial intelligence as "devoted to creating a fundamental understanding of how to represent and manipulate knowledge in a formal manner." Lukasik expected AI research results to make it possible to exploit computers in problem areas that "cannot be approached merely by increasing the speed or size of conventional numerical computations."[6] With support from IPTO, researchers in AI addressed some major problems in knowledge representation, speech understanding, and image understanding. While working on projects in these areas, they made great strides in understanding rule-based systems, the nature of problem solving, techniques for dealing with colored, curved, and textured surfaces in computer analysis of complex scenes, and understanding natural language.

At the outset, AI researchers found that the computer languages developed in the 1950s, such as FORTRAN, did not allow convenient operation in symbol manipulation tasks. They set out to design new languages for solution of AI problems. Because they needed AI languages to fit the specific characteristics of their computers, more than one AI language was developed.[7] In the mid-1950s, Herbert Simon at Carnegie-Mellon University and Allen Newell and J. Clifford Shaw at the Rand Corporation worked on computational tools for AI programming. They collaborated to produce a program they called the

Logic Theorist, for the purpose of solving mathematical problems. As part of the Logic Theorist, the three men developed a new language labeled IPL. In fact, they developed a group of languages that culminated in IPL-V.[8] In IPL the arrangement of the various groupings of symbols and data, called lists, was independent of the actual physical geometry of the memory cells, and the structures underwent continual change during computation. After their work on Logic Theorist, Simon and Newell explored the idea of a general problem solver, and Newell invested considerable time in a computer chess program.[9] This collaboration included Shaw, at Rand, as well. The AI group at CMU evolved from Simon and Newell's interest in problem-solving programs as aids in decision-making systems. Later in the decade, John McCarthy at MIT designed the LISP language, which resembled IPL, for use in a problem concept named Advice Taker, which was analogous to Logic Theorist.[10] Others at MIT found LISP useful in designing other problem-solving programs. In 1960, for example, James Slagle used LISP in his symbolic integration program SAINT. People outside MIT also used LISP. For example, Herbert Gelernter at IBM worked on a plane geometry theorem–proving program. As AI researchers designed and developed more programs, they most commonly chose LISP as the programming language.

One of the earliest problems addressed by AI researchers was how to represent knowledge. Knowledge representation is a combination of data structures and interpretive procedures which, if used in the right way in a program, leads to "knowledgeable" behavior.[11] Various representation schemes were developed in the 1960s and 1970s in IPTO-sponsored research programs. At CMU, Simon and Newell (who had moved from the Rand Corporation to CMU) developed production systems, based on the interpretation of contingency rules, for their models of human cognition. Ross Quillian and others, also at CMU, investigated semantic nets. Systems—such as PLANNER and MICRO-PLANNER—employing procedural representation emerged from the work of Carl Hewitt, Gerald Sussman, Terry Winograd, Richard Greenblatt, and Eugene Charniak at MIT. The most extensive use of PLANNER occurred in the design of Winograd's blocks-world system SHRDLU. Richard Fikes, Peter Hart, and Nils Nilsson used the predicate calculus as a representation scheme in the development of the Stanford Research Institute Problem Solver (STRIPS). Two later representation schemes developed in the 1970s were frames and scripts. In the mid-1970s at MIT, Marvin Minsky devised frames as a basis for understanding visual perception, natural language dialogues, and other complex behaviors. Shortly after this, Roger Schank and Robert Abelson at Yale developed scripts for representing sequences of events. Both of these concepts were designed for organizing knowledge to facilitate recall and infer-ence.[12]

In the area of computer designs, researchers investigated a number of ap-proaches. For example, at IBM the goal of one research project of Herbert Gelernter and Nathaniel Rochester was "the design of a machine whose behav-ior exhibits more of the characteristics of human intelligence."[13] Frank Rosen-blatt at Cornell developed a computer using the neural net concept and called it a perceptron. Perceptrons made decisions about possible patterns by assem-bling evidence obtained from many small experiments.[14] Meanwhile, Marvin Minsky at MIT was working on a similar idea. Several students in the MIT AI group investigated computers to replicate human motion.

In order to use an electron-beam machine to manufacture electronic cir-cuits, Charles Rosen, who was for years a principal researcher in AI activities at the Stanford Research Institute, became interested in pattern recognition by machine. Rosen, Alfred E. Brain, and George Forsen developed MINOS I, a trainable pattern-recognition device, in 1961. By the time Richard Duda joined the group in 1963, the team was engrossed in building MINOS II with money from the Office of Naval Research. In 1961 Nils Nilsson arrived at SRI and began work on decision rules and training procedures for use in collecting by computer sample statistics on observed patterns, which used MINOS I, a per-ceptron built by SRI, and resulted in his 1965 publication *Learning Machines*.[15]

While researchers addressed the issues in large AI system designs, others investigated the kinds of input accessories that were needed to improve the researchers' work environments. For example, out of one of these efforts came the Rand work on human-oriented environments—JOSS, the Rand Tablet, and the Rand Video Graphic System, which we described in chapter 3. Research-ers expected these devices to be useful in linear programming, network flow analysis, game theory, and the investigation of the relationship between neu-ral nets, learning, and automata.

Many AI researchers concentrated on problem solving as "the key type of task for machines to do—game-playing, theorem-proving, and puzzle-solving."[16] Researchers at MIT, Stanford, and CMU maintained a loosely con-nected set of research projects that examined the broad range of AI concerns in problem solving. The basic concern was to provide a problem statement, that is, input, to a "problem solver" that incorporated certain methods for at-tacking the problem. The problem solver transformed the problem statement into an internal representation and then attempted to solve the problem by applying the methods available to it.[17]

Researchers also examined certain information-processing tasks that could not be specified mathematically: they created problem-solving programs; visual, acoustic, or linguistic pattern-recognizers; and language manipulation programs. We mentioned above the work of Gelernter and Rochester at IBM. At RCA, Saul Amarel, who two decades later would be an IPTO director, offered an information-processing mechanism that automatically constructed computer programs to accomplish information tasks that were specified to them by example.[18] Research groups at IBM and Bell Laboratories in the 1950s investigated word recognizers, which are discussed below.

When IPTO came on the scene in late 1962 with its interest in human-oriented computing concerns and its ample funds, many of these AI research issues were folded into the overall IPTO program. IPTO funds for AI work at MIT came through Project MAC and later through other IPTO contracts. IPTO provided significant support for work at CMU and Stanford, enabling these schools to organize major AI laboratories, and mounted smaller efforts at SRI, Bolt Beranek and Newman, and IBM. Licklider, the first director of IPTO, believed that AI research was as much a part of command and control as interactive computing was.[19] He therefore supported work in problem solving, natural language processing, pattern recognition, heuristic programming, automatic theorem proving, graphics, and intelligent automata. Various problems relating to methods of interfacing between humans and machines — tablets, graphic systems, hand-eye coordination — were all pursued with IPTO support. Licklider was also director of the Behavioral Science Program in DARPA. The way in which he funded various activities makes it clear that he saw an overlap between the areas of the behavioral sciences and command and control, and thus supported AI-related studies in both areas. The goal of all this support was the improvement of interaction between humans and machines.

In his first year as director of the two programs, Licklider supported a variety of projects. From the command and control program he funded heuristic programming as part of Project MAC at MIT; as well as SYNTHEX, a program for processing natural language which answered questions by searching stored text, at Systems Development Corporation; a program investigating natural language communication with computers at BBN; and a computer speech recognition project at the University of Michigan. At the same time, out of the behavioral sciences program he funded research related to cognitive science, specifically the development of dynamic models for pattern recognition at CMU and the University of Michigan, and heuristic programming and the theory of computation at Stanford University and the University of California at Berkeley.[20] As the examples reveal, Licklider put no clear boundary between the two program funding areas, but he did justify the individual contracts within each program area according to a separate rationale for each research area.

Because AI research required extensive and expensive computing facilities, only an organization with a budget like IPTO's could support the research. The funding that IPTO made available at a few of the major educational institutions in the United States in the early years made it possible for those institutions to excel in AI. It was difficult for others to embark on AI research until much later, when funding was more generally available from IPTO and other sources. The institutions with funding had an additional edge: their reputations and funding allowed them to attract the very best people. The conjunction between the funding available and the presence of investigators such as Newell, Simon, McCarthy, and Minsky gave CMU, Stanford, and MIT a major head start in this field.

At the start, MIT had the most ambitious program; and through Project MAC, it had access to new IPTO-funded equipment that enlarged the number of activities within its already broad scope. CMU and Stanford took a little time to develop a similar broad scope, but expand they did. Within four years, Stanford's program rivaled MIT's. CMU continued to emphasize the modeling of human thought and decision processes throughout the 1960s, but in the 1970s it expanded into projects in speech, vision, robotics, and new machine architectures. Thus, it initiated efforts similar in appearance to the programs at the other two institutions. In the 1970s, additional universities — Rutgers, the University of Pittsburgh, the University of Pennsylvania, the Information Sciences Institute at the University of Southern California, and Yale — entered the growing field IPTO supported. In the 1990s, AI problems are investigated in many more organizational settings in the United States, and in many other nations as well, with funding from many sources.

We are getting ahead of our story, however. Before 1963, Marvin Minsky and John McCarthy at MIT received support from the Research Laboratory of Electronics, which provided what Minsky described later as funding for students, space, and equipment. With the inauguration under IPTO support of Project MAC at MIT, Robert Fano, director of the project, wanted to include the various MIT AI activities in Project MAC and offered support for research to the AI group headed by Minsky, specifically in the areas of heuristic programming and natural language processing. The AI group received two primary types of support from Project MAC. One was support for faculty and students, both graduate and undergraduate, and the other was the use of time-shared computing facilities. With access to these facilities as well as the development of their own time-sharing system (the Incompatible Time-Sharing System [ITS]), the MIT AI group pursued a range of loosely connected programs aimed at developing systems that exhibited humanlike intelligent behavior in their activities, without regard for the way the process might occur in humans.

The first MAC proposal in 1963 noted a pressing demand from researchers for computation time; and heuristic programming was specifically mentioned as an area that demanded a large-scale computer system. The natural language processing component was intended to develop in a computer the ability to understand English.[21] For example, in one of these programs, the machine solved many word problems presented to it in a "comfortable, though restricted, subset of English." Richard Greenblatt and Stewart Nelson continued to develop LISP as part of Project MAC. Also as part of Project MAC, Adolfo Guzman and Harold McIntosh embedded CONVERT, a language for graphics, in LISP.[22] These and other studies by the AI group were categorized in the first MAC proposal as research on recursive functions, symbolic manipulations, heuristics, and problem solving.

Minsky and the AI group developed techniques for computer interaction with the physical environment. They pursued research on vision, manipulation, hardware, theoretical work, and systems and time sharing. Greenblatt and Jack Holloway worked on computer tracking of human eye movements to find a new means to input data. Henry Beller and Arnold Stoper did work on visual perception at the laboratory while visiting from Brandeis University. David Silver developed a "bug" that rolled around the lab memorizing the scene; if the scene changed, the bug would roll to the place where the change had occurred. M. Beeler developed a hill-climber algorithm (using an error-reducing iterative method). Guzman's POLYBRICK program was written in CONVERT to do scene analysis and recognize partially hidden objects in a scene. The AI laboratory's PDP-6 computer was coupled with a mechanical hand to manipulate objects. Mechanical arms for industrial use were under development at the laboratory in the early 1960s, and one lobster-claw hanging from the ceiling was dubbed the "Minsky arm."[23]

Since IPTO personnel hoped that work in AI would "provide a technological base for developing and building logical automata and extensors for use in intelligence applications,"[24] the AI laboratory at MIT worked on an intelligent automaton. MIT's AI staff noted in a 1965 proposal that "the most advanced computer-controlled automaton constructed to date was the mechanical hand, 'MH-1,' constructed at MIT by Henry Ernst in 1961,"[25] and that "in the interval since 1961, there has been *no* important work in this area."[26] The proposal argued that more research was needed on visual and tactile sensors and that "as soon as we get beyond the earliest stages, we will encounter the kinds of problems that have recurrently appeared in the area of Heuristic Programming and Artificial Intelligence." As Terry Winograd described it years later, the AI laboratory at MIT contained a group of "people who were basically working on their own and then knowing enough about what each other were doing to join up," that is, to relate one problem to another, searching for overarching results.[27]

At CMU there was a similar range of research efforts in AI, with a weighting toward the cognitive side of AI in the early years. Simon published a major work in 1962 entitled "The Architecture of Complexity," in which he argued that building a system requires many levels before interesting behaviors begin to emerge. This became an important work in determining how to program a stable system.[28] A year later, he published a work entitled "Experiments with a Heuristic Compiler."[29] Also in 1963, Julian Feldman argued for the importance of viewing the mind as a symbol processor.[30] This became one of the key concepts for the development of AI in the years to come. In 1964, Feldman moved from CMU to Stanford and became part of the work being done there on hand-eye coordination.

Newell published a seminal article in 1962, in which he argued that learning can only be defined in relation to performance processes.[31] Newell's criteria for learning were accepted by the AI community throughout the decade. In 1963, Simon and Kenneth Kotovsky designed a model of discovery for scientific laws, which was framed as a computer program in a list-processing language; the program was capable of taking a standard psychological test.[32] Similar to Simon and Kotovsky's work was the work that Simon did with Edward Feigenbaum, then a graduate student at CMU, on the Elementary Perceiver and Memorizer (EPAM). EPAM was a model of well-known verbal-learning behavior which involved memorization of a series of nonsense syllables. Nonsense syllables avoided the problem of meaning in the learning process. Through models of this kind, knowledge of short- and long-term memory was gained.[33] These activities demonstrated Simon's continuing interest in many disciplines, although his overreaching concern remained understanding what it means to solve a problem. It was to facilitate this work in AI that the CMU group appealed to IPTO in 1963 for funds to develop a time-sharing system.

When McCarthy moved to Stanford in 1962, he created the Stanford Artificial Intelligence Laboratory (which became known as SAIL). Within four years, the staff and funding for AI work at Stanford had reached a level that was on a par with MIT. SAIL was intended to work on McCarthy's own interests in formal logic and LISP, and he also wanted to arrange the laboratory to include projects in speech understanding, hand-eye coordination, problem solving, higher mental functions, and computer music—a range similar but not identical to MIT's. Expert systems research was centered in DENDRAL, a project in the Stanford Medical School initially supported by NASA and NIH. D. Raj

Reddy had started the speech-understanding area in 1963 in connection with his Ph.D. research. (In 1965 Reddy and William McKeeman received the first two Ph.D. degrees granted in Stanford's newly formed computer science department.) After receiving his doctorate, Reddy became the program director for speech understanding in SAIL. He worked with Lee Erman until they both moved to CMU in 1969, where Reddy established the major project in speech recognition described below.

In machine designs, or hand-eye research, Stanford worked on a two-arm system for mechanical assembly tasks. A new language, HAL, was developed for hand-manipulation tasks. A group worked on stereo vision and motion parallax to locate objects and to draw contour maps, as well as a program to recognize specific objects. McCarthy continued his research into the mathematical theory of computation, and worked with Zohar Manna and David Luckhman on formal reasoning and correctness of programs. Various people at Stanford worked on game playing. For example, Arthur Samuel continued his earlier work on a checkers-playing program and also worked on the game of Go. McCarthy and Barbara Huberman Liskov worked on chess. David Wilkins also worked on chess, and Don Waterman worked on poker-playing programs. As part of this effort in game playing, a proof-checker was designed to aid in the verification of programs for AI written in the programming language PASCAL. Some additional work was done in theorem proving—a system was created to do programming in PASCAL and demonstrate the program's correctness. In the area of automatic programming, Stanford researchers designed an experimental system to accept library routines, programming methods, and program specifications and to return a program written in ALGOL. And in the area of natural language, they completed MARGIE, a natural language system that utilized Roger Schank's concept dependency theory, which became a useful tool in Schank's later work on scripts in stories at Yale in the 1970s.[34]

Kenneth Colby, a psychiatrist who was a research associate in Stanford's Computer Science Department, developed PARRY, a question-and-answer program. PARRY was a natural language processing system, which could "read" grammatically incorrect sentences. Colby's ultimate goal was to explain how a paranoid belief system operates, and he went on to publish widely about using computers to understand autism and paranoia. Schank developed conceptual grammars because of the natural language processing needs of Colby's project. He attempted to provide a representation of all actions using a small number of basic sentence constructions. This developed into his concept dependency theory.[35] PARRY, like Joseph Weizenbaum's ELIZA, which had been developed at MIT, relied upon pattern-matching techniques. Unlike ELIZA, however, PARRY did not seek out key words; rather, it used parsing rules to recognize patterns so that it could give an appropriate response. For example, the pattern "I . . . you" can be answered with "I . . . you too," or "Why do you . . . me?" Thus PARRY did not need to know the content of any given input sentence. If the program failed to recognize a pattern, it continued the conversation with a remark unrelated to the input sentence.

In the 1960s the portion of AI involved with problem solving (the cognitive part of AI) was called heuristic programming. In 1965, Sutherland justified the IPTO activity in heuristic programming simply by saying, "We are studying the ways to get computers to solve difficult problems. This [heuristic programming] project will assist our contractors in that study." The specific project to which he was referring was a contract with IBM regarding a large memory facility that was under development to be used by researchers working on heuristic programming. The specifications required the facility to employ both LISP and IPL. Sutherland planned that the facility would be used at MIT, CMU, and Stanford, "and [by] such other persons as shall be authorized from time-to-time by ARPA."[36]

In the 1960s, researchers constructed a number of approaches to examine different aspects of problem solving. M. Ross Quillian at CMU focused on semantic nets to study memory and knowledge organization. At MIT, Daniel Bobrow continued research into algebraic problem solving, but used word problems to explore linguistic issues. Also at MIT, Terry Winograd used the "blocks world" concept developed by Roberts to study natural language processing. While a student at Stanford, Cordell Green had worked with Bertram Raphael at SRI on the design of the question-and-answer programs QA2 and QA3.[37] At the time, AI researchers believed that analysis of a diversity of systems would lead to a general understanding of problem-solving behavior.

Natural communication with computers, including speech understanding, was another important objective of the early IPTO AI program. As early as 1964, Sutherland had prepared a program plan that stated the objective in the following way: "When computers can converse in the human language, people in all walks of life will be able to use computers without detailed training in computer programming."[38] Nowhere would this computer ability be more valuable than in the military. But first, IPTO understood that the computer had to be able to know the meaning of questions in context before total integration was possible. Therefore, research was conducted on natural language, understanding of context, and examination of the operation of computer programs. These were very far-reaching and demanding tasks.

By the 1970s, CMU's AI research was equivalent in scope to the investigations in process at MIT and Stanford. In addition to problem solving and cognitive modeling, CMU engaged in a number of investigations of new computer designs. These were a group of parallel machines, the C series, which started with Gordon Bell's C.ai.[39] Later, researchers at CMU developed the C.mmp as a general-purpose machine to investigate the true power of the machine. At least ten separate research projects were run on the C.mmp. Two of them involved speech-understanding systems in the mid-1970s: Hearsay-C.mmp and Dragon-C.mmp.[40] The CMU speech work is highlighted below.

By comparison to MIT, CMU, and Stanford, SRI had a more muted history in AI, although the AI work it did was equally important. IPTO supported SRI through its Applied Physics Laboratory. In the 1960s it was strong in machine design, especially robotics, and it later emphasized speech, but it never pursued AI research as extensively as did the university groups, perhaps because of its organizational structure. Cordell Green came to SRI and developed QA1, a question-answering system based on an ad hoc theorem-proving method. Green went on to develop QA2, which was used for high-level problem solving in Shakey, a mobile robot designed at SRI. John Munson developed the MINOS III pattern recognition system, which was delivered to the army Signal Corps in 1968.

We noted above the work on representation that Saul Amarel did while at the RCA Laboratories in the 1960s. In 1969 Amarel became professor in, and chair of, the Department of Computer Science at Rutgers University. He extended his research interests to explore AI in the context of complex "real world" problems and to focus on reasoning that was not as systematic and clear as in his previous research in theorem proving.[41] With Casimir Kulikowski, Amarel began a project called Research Resource on Computers in Biomedicine. They used the ARPANET to connect it with work at Stanford. In parallel, Amarel started a project on AI and automatic programming. These two projects brought Rutgers into the IPTO orbit. After a decade of this type of research, Amarel returned to his earlier interests in representation, but now in the context of problems of modeling and search in engineering design. With support from IPTO, Rutgers's research in this area concentrated on VLSI design. Subsequently, in the late 1980s, the research shifted to the problems of the engineering design of large structures (e.g., the hydrodynamic design of boats).

Even with all this IPTO-supported activity during the 1960s, there was no AI program as such in IPTO. Licklider, Sutherland, and Taylor funded AI research as part of large contracts for work in computing, or for specific projects in AI. During his directorship Roberts began to organize AI research into subprograms. In the hope that artificial intelligence research would develop techniques that would enable machines to improve decision making, IPTO continued to support a wide range of projects in basic research through its Intelligent Systems subprogram, one of the new subprograms for AI. The objective of the Intelligent Systems subprogram was "to develop the theory and capability to permit the automatic solution by computer of very complex problems involving planning, generalization, learning, and the interpretation of sensory information . . . to build automatic linguistic, visual, manipulative, and problem solving systems."[42] In this period, IPTO supported work in the automatic interpretation of visual scenes, in the machine understanding of printed English text, in improved methods for information retrieval and query analysis, and in the planning of a course of action to accomplish a given command; it also supported the theoretical work necessary to achieve these developments. IPTO did this, for example, by supporting work at MIT on vision and understanding of natural language, and at SRI on problem solving, sensory perception, question answering, and machine learning. Both MIT and SRI worked on system integration of elements. Syracuse University received a contract for work in computational logic, particularly for work on sophisticated mathematical logic areas, to develop a formal logic system using this logic, and to explore inference making in problem solving by computer. CMU received support for the development of methods in automatic programming and for more work in problem solving. IPTO funded work in robotics and an array of AI software projects, and made substantial contributions to work on medical expert systems—additional work on DENDRAL and later on MYCIN, for example.

According to IPTO, the work in intelligent systems provided the problem-solving and computerized planning capabilities that were important in making progress in automatic programming and speech understanding, two other subprograms for AI. The speech-understanding program enhanced problem solving using machines, and would increase the efficiency of use of computers in various sectors, if a system could be developed. New schemes for handling graphic data and for picture processing contributed to better use of data and more effective decision making. Work in vision helped researchers to understand effective ways to interpret sense data about a scene and to produce movement of the computer system in the scene.

Throughout the 1970s, IPTO-supported work in intelligent systems emphasized problem solving, natural language processing, speech, parallel computers for AI research, vision, sensors, and robots. Simultaneously, the results obtained from these fields gave rise to new research areas. Researchers at the

University of Utah, Lincoln Laboratory, and Stanford produced image analysis systems valuable in photograph interpretation for intelligence uses. In the 1980s, IPTO sought the integration of many of these areas into applications for the Strategic Computing program, a subject of the next chapter. In fact, AI was one of the two infrastructural foundations for SC, the other being VLSI. Making AI an infrastructural element of SC shows that IPTO was convinced of the significance and sophistication of past AI research and confident that such research would contribute to success in SC and other DOD R&D, such as the Strategic Defense Initiative. Today, AI studies are a major component of every computer science educational program around the world, and the number of interfaces with other areas of computer science and other disciplines continues to increase. From a concern for creating greater flexibility and more capability in computing machines by the use of AI techniques came a major disciplinary endeavor. IPTO's role in this development was substantial, as we will now show in three AI research areas.

The three areas — problem solving, speech understanding, and vision, besides being prime areas in which IPTO provided support for the development of more flexible and intelligent computer systems, are three different examples of the approach of IPTO directors to support of research areas. Our discussion of these areas highlights research that illustrates the nature of the questions investigated, the research progress made, and some of the influences that laboratories had on each other. We can note at once that one of the principal sources of influence was the fact that many researchers moved from one IPTO-supported laboratory to another and worked on similar problems in each. Problem solving was part of overall AI research. In the 1960s, while the research area underwent definition, IPTO staff members hardly entered the discussion about the goals and objectives of problem solving research. One reason for this may be the incremental nature of progress in research on problem solving. This area has none of the characteristics of speech and vision research, in which either the machines perform or they don't, and the only evaluation is how well they perform. The evaluation of problem-solving developments was much more complex until the development of expert systems, which is one development that can be directly tested. After the Mansfield Amendment and increasing concern inside the DOD about congressional attitudes, demands for more relevance to the DOD mission led to increasing emphasis in the expert systems area of problem solving. At this time also, IPTO demanded that more attention be given to the areas of overlap with other research fields such as vision and mobility, as we will see. Hence, the following description of the history of problem-solving research in AI is almost devoid of any mention of

IPTO — the office was integral to its progress but was not as directly involved in setting requirements as it had been in networking.

This is clearly not the case for research in speech understanding, where IPTO played a commanding role in setting objectives and evaluating results. IPTO directors Roberts and Russell, along with a string of program managers, were directly involved, and DARPA director Heilmeier was indirectly involved, in all decisions about the Speech Understanding Research Program, from its conceptualization to its demise.

Research in vision shows both aspects of IPTO's approach to the management of programs. At least in the 1960s, vision research was part of the search for better sensing techniques in hand-eye coordination systems and part of robotic studies, both of which were done as part of overarching contracts such as Project MAC. As the 1970s progressed, an interest in the use of vision systems in image processing and integrated AI systems resulted in more directed projects, in which IPTO involvement took the form of PI meetings and site visits. As vision techniques and problem-solving systems became more sophisticated, research results became part of the design of a new program — Strategic Computing — to design and test fully integrated computer systems undergirded by automatic AI functions.

Problem Solving

The Search for Generality

Humans confronted with a new problem often reason by analogy during the solution process. Therefore, when solving problems they include in the solution process many factors that have not been explicitly presented to them. Often, they do not realize what they need and use in problem solution; yet they solve problems. In computerized problem solving, until learning processes are sufficiently developed, all of the information needed for the analysis and solution of a problem must be brought to the attention of the machine through programming. To construct a machine to participate in problem solving, therefore, requires much understanding of how to represent knowledge of situations and processes for solving problems. The power of a computer program to perform "intelligently" depends primarily on the quantity and quality of the knowledge it contains about the task to be performed.

By 1965, the search for a general system for developing machine intelligence had become one of the most pressing issues in AI.[43] Researchers sought "a formal system so general that all problems, however represented internally, could be translated into the form of proving a theorem in this system; yet so specific that general proof methods could be developed for it."[44] In spite of a decade of effort in this area, Feigenbaum could comment in 1968 that "we lack a good understanding yet of this problem of generality and representation." He noted that there seemed to be two roads for research on generality: "the high road and the low road." The best example of approaches following the "high road" was Newell and Simon's General Problem Solver. Those researchers on the "low road" did not seek a general problem solving system, but instead explored specific complex domains and/or tasks in order to (1) test whether ideas and methods developed so far in AI could be used in significant problems, and (2) identify new issues for AI basic work.[45]

The AI researchers at MIT who worked on Feigenbaum's "low road" made a distinction between their approach and that of other laboratories. Minsky and Papert reported on these distinctions between laboratories, and their words on the subject are worth quoting:

> [When] faced with the *apparent diversity* of kinds of knowledge, the common approach at [non-MIT] centers is to seek ways to render it more uniform so that very general, 'logically clear' methods can be used; [the MIT] approach is to accept the diversity of knowledge as real and inevitable, and find ways to manage diversity rather than eliminate it.[46]

The laboratories referred to, of course, were CMU, which tried only to use "logically clear" methods, and Stanford, which employed both styles. An interesting array of problem-solver programs was developed at MIT in accordance with the dictum of accepting "diversity of knowledge." One important characteristic of these programs developed at MIT (and at other places as well) is that, according to Feigenbaum, they grappled "in various ways with the problem of the meaning of ordinary natural language utterances and the computer understanding of the meaning of these utterances as evidenced by its subsequent linguistic, problem-solving, or question-answering behavior."[47] In Daniel Bobrow's program STUDENT, developed at MIT, high-school-level algebra word problems constituted the domain of discourse. According to Minsky, Bobrow's work "is a demonstration *par excellence* of the power of using meaning to solve linguistic problems." STUDENT reflected several levels of organized knowledge. On the lowest level, knowledge of algebra was carefully packaged in a function called SOLVE. On the next higher level, knowledge of how to convert the pseudo-English input into equations was partly algebraic in character. Some syntactic ambiguities in the input could be decided on the overall basis of algebraic consistency. At the next level, the program contained some knowledge-classifying ability. This did not mean that the program could acquire knowledge in operation. In fact, at this time Bobrow set aside the problem of acquiring knowledge until he better understood how to represent and use knowledge.[48]

Another approach to natural language systems at MIT was Bertram Raphael's question-answering program SIR. This system involved sophisticated information retrieval that required logical deductions to be made from information stored in a database. Raphael concentrated on the deductive and associative mechanisms needed to accomplish this.[49] As a result, SIR was a theorem-proving program, an attempt to build a memory structure that converted information received into a systematic, efficient representation. "The model takes the form of a network in which objects are interconnected by a variety of different relations; the system answers various kinds of questions by examining the network."[50] The program worked by "understanding" statements when they were made, and consolidating this understanding by adding to or modifying the network. The program worked very smoothly within the limitations of the relations the model was built to handle, but a significant amount of knowledge had to be put into the model in advance to obtain this smoothness.

At CMU, M. Ross Quillian, a graduate student who was studying with Newell and Simon, investigated both human memory and knowledge organization through the use of a network model, called a semantic network. The chief concern of his work was to show how a more sophisticated plan for understanding the relation between meaning and grammar might come from developing network models. Quillian distinguished between recognition memory and recall memory for the meaning of words, and his program concerned only the former. Using contemporary linguistic theory, he subordinated meaning to syntax in search of rules to produce all possible grammatical sentences, and only grammatical sentences.[51] Partly as a result of Quillian's work, semantic nets became a major topic of AI research.

Another link to modeling and representation of knowledge came through work done by McCarthy and his colleagues at Stanford. In an article written with Patrick Hayes in 1969, McCarthy wrote that he and Hayes wanted "a computer program that decided what to do by inferring in a formal language that a certain strategy will achieve its assigned goal. This requires formalizing concepts of causality, ability, and knowledge." This proposal used a type of symbolic logic called the predicate calculus, which can deal with the logical relations between sentences, as a basis for introducing a "situation calculus," and other explications of *cause, can,* and so on. The authors drew upon work in philosophical logic to develop adequate representations of the world and then to provide an explanation of the action verbs *can, causes,* and *knows* and how they are to be employed in that representation of the world using a system of interacting automata.[52] This concept influenced development of the QA3 and STRIPS programs at SRI. Both of these programs are also examples of modeling and representation research.

The special concern of AI groups with programming and simulation meant that AI researchers had a special interest in languages and programming aids. Various languages (e.g., IPL and LISP) were developed to handle the heuristic nature of problem solving in AI systems. A new language, PLANNER, was developed at MIT in the late 1960s by Carl Hewitt in order to learn how to build courses of action into programs. Hewitt designed PLANNER to implement both representational and control information through the use of assertions (LISP lists) and procedural data in the form of theorems. PLANNER used the theorems to evaluate the use and implications of the data. PLANNER's properties include varied data structures, associative memory, pattern matching, automatic and intelligent back-up, and procedural embedding of knowledge. As Chandrasekaran pointed out in his 1975 review of AI, "the power of PLANNER lies in the fact that while the language itself is independent of any subject matter, very detailed and arbitrary procedural heuristics can be readily supplied by means of the language."[53] Reporting to IPTO, Minsky called PLANNER "a new style of programming and thinking about programs for" AI, which "has almost explosively spread through our own laboratory and has even already begun to affect the work at other centers of research."[54] Languages developed elsewhere with similar properties were POP-2, QA4, and SAIL (Stanford Artificial Intelligence Language). PLANNER was used in a number of new investigations at MIT; for example, Terry Winograd's natural language–processing system SHRDLU; Gerald Sussman's HACKER, a model of the process of acquiring programming skills; and Eugene Charniak's STORIES, a model for comprehension of children's stories.[55]

An important step forward was Winograd's study in natural language processing. Winograd was interested in applying symbol-processing ideas to syntactic analysis.

The driving problem was the question of how can you use extra information [to describe the situation], part of which is textual—what's mentioned recently—and part of which is world knowledge—like pizzas and vehicles, and so on—to disambiguate natural language utterances into clearly defined [knowledge], [get] *that* block and put it in *that* place.[56]

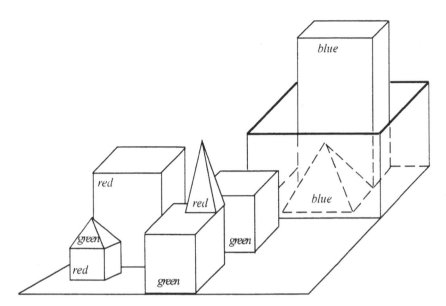

Fig. 4.   Typical scene in the blocks world. By 1972 robots (or processors representing them) could understand and interpret simple commands like "PICK UP A BIG RED BLOCK" and more complicated questions like "HOW MANY BLOCKS ARE NOT IN THE BOX?" After Terry Winograd, *Understanding Natural Language* (New York: Academic Press, 1972), 9.

SHRDLU was written in LISP and MICRO-PLANNER, a LISP-based programming language. The program simulated the operation of a robot arm that manipulated blocks on a table. (Fig. 4 provides a visual representation of the blocks world.)

The system consisted of four basic elements: a parser, a recognition grammar for English, programs for semantic analysis (to change a sentence into a sequence of commands to the robot or into a query of the database), and a problem solver (which knew about how to accomplish tasks in the blocks world).[57]

SHRDLU used a heterarchical organization of these various components, an organizational structure in which responsibility for control is equally distributed throughout the system, and internal communication in the program is facilitated.[58] SHRDLU's knowledge base included a detailed model of the blocks world it manipulated and a simple model of its own reasoning process, so that it could explain its actions. This program went beyond the other early question-answer systems because of its superior flexibility and its use of common sense.[59] Even though the approach in SHRDLU was difficult to generalize, the program was a milestone because it showed that for a restricted and well-defined domain, the building of natural language interfaces was feasible.[60]

Despite these advances, at the end of the 1960s there were severe limits to problem-solving techniques. It was still not possible to prepare problem-solving programs that would accept problems in a rather general representation at the start but then alter the representation as more problem-specific information became available during the problem solving.[61] Various types of knowledge needed to be represented: objects in the world and facts about these objects; events and actions in the world; performance, or knowledge of how to do things; and a level of meta-knowledge, or knowledge about what we know.[62] In the crafting of a problem-solving program in AI, the major effort most often involved gaining an understanding of how humans learn and therefore of how to represent knowledge, so as to set up the conditions for learning in machines. Building on the work described above, various researchers developed representation schemes in the late 1960s and the 1970s. Among these schemes were state-space search, as is found in game programs such as chess-playing programs; formal logic, for example using the predicate calculus; procedural representation, for example in some of the task-oriented languages; semantic nets, found in Quillian's operational model for word meaning; production systems, as illustrated best in early expert systems; and some special-purpose representations, such as Raphael's program SIR (created at MIT), Robert Lindsay's program SAD-SAM (created at CMU), and Feigenbaum's program EPAM (created at CMU).

One of the most influential experimental approaches to resolving Feigenbaum's concerns about the problem of generality and representation of knowledge was the concept of representational shifts introduced by Saul Amarel, then at RCA Laboratories. In 1968 Amarel published an article in *Machine Intelligence* in which he examined researchers' choices of basic concepts for problem solving, and the successful evolution of algebraic systems that describe language constructions in computers, called grammars, on the basis of assimilation of new knowledge about the problems in a particular class.[63] He noted that the way a researcher represented a problem affected the choice of a solution. Amarel developed a mechanism that would shift the representation as more was learned about the problem during the solution process. He helped to clarify the nature of representational changes, showing the universality of representation problems and their significance in problem solving, and identifying a few important classes of representational shifts, as well as mechanisms for implementing them. Amarel explored several types of representational shifts in a number of problem classes, and studied the impact of these shifts on problem solving performance. Earlier investigators had explored only a few examples.[64] Even with all these developments, Nilsson wrote in a 1974 survey of the field that "we are still quite a way, it seems, from having a sound theoretical basis for knowledge representation."[65]

In a survey delivered at the International Federation for Information Processing (IFIP) meeting in 1974, Nilsson presented a two-level structure: basic methods and techniques (or core topics), and major (first-level) application areas.[66] The core areas were in application areas other than theorem proving, game playing, and puzzle solving. The major application areas described by Nilsson were machine vision, robotics, game playing, automatic theorem proving, automatic programming, natural language systems, mathematics, science and engineering aids, and information-processing psychology. Problem solving was divided into three core areas: heuristic search; modeling and representation of knowledge; and common-sense reasoning, deduction, and problem solving. Researchers had become so interested in what could be learned in search and representation that these topics had themselves become the subject of research. Earlier, search and representation had been seen as means to ends, but the problems were so extensive that it was necessary to engage in research on search and representation methods to increase their effectiveness.

At the beginning of the 1970s problem-solving systems began to converge. But the very act of convergence led to a recognition of shortcomings in the approach to problem solving by computer systems. We described above some of the early work in problem solving—Logic Theorist, SIR, and the General Problem Solver (GPS). Newell, Shaw, and Simon's Logic Theorist had a direct influence on GPS. Raphael's SIR developed separately at MIT. A few years after SIR, Cordell Green and Raphael extended it into the question-and-answer programs QA2 and QA3, making use of McCarthy's Advice Taker concept. QA2, QA3, and GPS came together in STRIPS, developed at Stanford by Fikes and Nilsson in 1971. Through these developments, question-and-answer programs, commonsense reasoning, and puzzle solving had come together by the early 1970s.[67] Researchers realized that to deal with a large domain a system would need a large network of basic knowledge. This realization led some AI researchers to shift their focus from an inference-based paradigm to a knowledge-based paradigm, thereby abandoning the search for generality.[68] The result was the development of expert systems, the path to which was already becoming clear in the early 1970s.

Knowledge-Based Systems

Computer-based expert systems began in the mid-1960s with the initiation of the DENDRAL project in organic chemical analysis at Stanford and the MACSYMA project in symbolic integration and formula simplification at MIT, although in their early stages these systems were viewed as experimental systems for research in AI. The DENDRAL project at Stanford University began with a NASA contract awarded to Joshua Lederberg, a professor in the medical school and a Nobel laureate (1958), for data interpretation in trying to find evidence of life on Mars. Lederberg believed that mass spectrometry would be powerful enough to provide such evidence if the data and an interpretation mechanism could be sent to the planet.[69] The prevailing method of interpretation was catalog look-up. Lederberg believed that to use an earth-based catalog in a machine on Mars would certainly be limiting. He developed a notational algorithm that used a unique naming convention for organic molecules. In effect, this algorithm was a systematic hypothesis generator for simple translation of the notation, which was capable of generating all of the topologically possible isomers of a chemical formula.

Members of the DENDRAL project's staff, including Lederberg, Feigenbaum, Georgia Sutherland, and Bruce Buchanan, wrote a program to infer structural hypotheses from mass spectral data. DENDRAL contained a generator that was essentially a topologist that knew nothing about chemistry except for the valences of atoms. The generating algorithm served as the guarantor of the completeness of the hypothesis space, in a fashion analogous to the legal-move generator in a chess program. Since the generating process used a combinatorial procedure, it produced for all but the simplest molecules a very large set of chemical structures, almost all of which were chemically implausible although topologically possible. Implicit in its activity was a tree of possible hypothesis candidates. Various heuristic rules and chemical models were employed to control the generation of paths through this space.[70] These rules and models had been determined in consultation with chemists at Stanford and included a model of the chemical stability of organic molecules, a "very crude but efficient" theory of the behavior of molecules in a mass spectrometer (the zero-order theory of mass spectrometry), and a set of pattern-recognition heuristics (the Preliminary Inference Maker) that allowed a preliminary interpretation of the presence or absence of key functional groups.[71] The output from this process and the application of a Predictor program was a set of mass spectra for candidate molecular structures. This was compared to the input data. The Predictor program was an elaborate procedure for computing the mass spectrum of a candidate molecular structure, based in a detailed theory of mass spectroscopy. The results were evaluated, and candidates that seemed to match the input data were presented in rank order. NASA made use of some of this work in the missions to Mars, but there were larger implications of the work for AI, namely the new area of expert systems.

The MIT researchers Joel Moses, William Martin, and Carl Engleman developed MACSYMA to assist researchers in the solution of complex mathematical problems. An extremely large system, MACSYMA can perform some six hundred different mathematical operations, including differentiation, integration, solution of equations and systems of equations, Taylor series expansions, matrix operations, vector algebra, and order analysis. The system helps mathematicians, scientists and engineers to simplify large, unwieldy mathematical expressions that often result from symbolic algorithms.[72] MACSYMA includes a number of specialized, heuristic problem-solving procedures.

The program contains a variety of explicit expression-transformation commands (expansion, factorization, and partial fraction decomposition) and a simplifier that automatically applies a set of mathematical rules to every new expression as it is constructed. In applying simplification rules, MACSYMA employs a semantic pattern matcher to find patterns in the problem under investigation similar to those in its program; and it uses a search-oriented simplifier to find the simplest result.[73] MACSYMA was created in an AI environment, and as more and more capabilities have been added to it, the system has become a boon to researchers everywhere.

By the early 1970s, the use of DENDRAL and MACSYMA had helped AI research to shift its focus from the search for general principles for problem solving to a knowledge-based paradigm in which the three major activities are the representation, utilization, and acquisition of knowledge. In combination, these activities constitute what is called today "knowledge engineering," that is, the process of designing and building expert systems. As Buchanan and Feigenbaum characterized these issues, representation was the same as we have described above; utilization was the attempt to identify what designs were available for the inference procedure to be used by the intelligent artifact; and acquisition concerned the systematic gathering of knowledge in a field for computer use.[74] The early expert systems, including DENDRAL and MACSYMA, were custom-crafted. Buchanan and Duda noted that "the major difficulty [in designing an expert system] was acquiring the requisite knowledge from experts and reworking it in a form fit for machine consumption"—the problems were those of acquisition and representation.[75] In the late 1970s and early 1980s, an important development was the design of several "knowledge engineering frameworks"—including EMYCIN, ROSIE, KAS, EXPERT, and OPS—as aids in building, debugging, interpreting, and explaining expert systems.[76] As Buchanan and Duda pointed out, with these tools it is possible to construct a new expert system with "one or two dozen rules" in a few days, after which, even though considerable effort was required "to refine" the knowledge base, the effort was "focused on the knowledge and not on the code."[77]

The design of these frameworks facilitated the move of expert systems into commercial use. The frameworks were the result of "a dozen or so" experiments in the 1970s by university researchers and the pioneering industrial efforts of Schlumberger Ltd. and Digital Equipment Corporation.[78] Feigenbaum, Pamela McCorduck, and Penny Nii asserted in 1987 that there were some fifteen hundred expert systems available in the United States.[79] Here we offer as illustrations discussions of three commercial systems, and some military systems:

Because of a slow growth rate in power consumption, public utilities have used older equipment for longer periods of time rather than purchasing new generators. For a company such as Westinghouse, this meant an essential shift from selling new equipment to maintaining old equipment. A shortage of diagnostic engineers made this shift difficult. Westinghouse personnel and knowledge engineers from CMU developed an expert system for the company's Orlando (Florida) Diagnostic Center to aid in the maintenance of older equipment. The prototype was finished in 1979, and the expert system itself was on-line by 1980.[80]

A second example comes from the FMC Corporation plant in Pocatello, Idaho, that processes phosphorus-bearing ore. As the engineers in charge of the processing furnaces retired, their expert knowledge was being lost to the company. FMC, together with Teknowledge, a company founded by Feigenbaum, developed an expert system to help monitor the slag furnaces. The expert system prototype was developed in 1987, and the system was operational in 1989.[81]

Our third example comes from the banking community. The Intelligent Banking System was designed to use natural language processing techniques and rule-based expert system techniques to allow a computer to scan and understand a natural-language text message. The system was developed by Consultants for Management Decisions, of Cambridge, Massachusetts.[82] This system is now in use by Citibank.

A military problem for IPTO consideration was the transportation of large amounts of supplies and materiel to distant places for military units. This is ordinarily a routine problem, not unlike the problem faced by any transportation company that wishes to use its facilities efficiently. An additional problem of military transportation was the shipping of materiel to a service site that demands that the materiel be unloaded in a specific order. Thus it was necessary to determine both in what sequence the material should be loaded into the transport planes (e.g., C130s and C5s) to facilitate unloading, and how it should be positioned to create the load distribution appropriate for the plane being used. The loading plan development was not a straightforward process. Sometimes the planes supplied were not the ones ordered or scheduled to be sent, which caused delays as the master sergeant in charge hurriedly replanned the loading procedure. IPTO proposed an automated system using AI principles to reduce these delays.

The loading procedure became the Automated Air Load Planning System, now in use by the army. The system was developed for use at Fort Bragg by SRI under a contract sponsored by IPTO and the Army Logistics Center. The packet radio testbed established by IPTO at Fort Bragg was also used to address this problem. AALPS was very similar to knowledge-based systems and used several results from other artificial intelligence research sponsored by IPTO over the years. According to Robert Simpson, IPTO program manager responsible for AI research in the 1980s, AALPS was undertaken

> to look at how we could use networking and advanced computing to help in command and control. . . . SRI was contracted to develop the prototype. It went through the usual rapid prototyping, evolving methodology, until the final AALPS system in 1983 was basically declared the right functional system. From 1983 to about 1985, it went through some test and evaluation.

In 1985, the successful AALPS was officially turned over to the army.[83]

In 1990, there were no fewer than eighty specific applications of expert systems in the DOD. Some were already in operation. Others were operational prototypes, testbed prototypes, or conceptual prototypes. Some specific systems, similar to AALPS, developed by DARPA and their users are

— CAT (Combat Action Team; used by the navy) — the first major real-time expert system operationally deployed aboard a military platform;
— AFTI-VIS (Voice Interactive System; used by the air force) — a testbed prototype speech recognition technology;
— Pilot's Associate (used by the air force) — a conceptual prototype combination of expert systems and noisy speech recognition;

— ALV (used by the army) and NAVLAB (used by the navy) — respectively, a conceptual prototype Autonomous Land Vehicle, and a testbed for robotics at CMU;

— FRESH (Force Reallocation Expert System; used by the navy) — an operational prototype; and

— SCORPIUS (Strategic Computing Object Directed Reconnaissance Parallel-Processing Image Understanding System; used by the DOD) — a testbed prototype that detects and classifies objects at known enemy facilities.[84]

DOD agencies contemplate the introduction of many more expert systems during the 1990s.

If machines were ever to be "intelligent," it would be necessary for researchers to build in schemes for problem solving. At first, researchers sought to formulate overarching principles of program development which would aid a machine in problem solving. Investigations into problem solving revealed, however, that there were significant difficulties in representing knowledge, constructing procedures for using the knowledge, and integrating problem-solving aspects into machine designs. Increased understanding of the details of problem-solving programs was acquired in task domain studies, and this led to an emphasis on specific systems, called "expert systems," rather than a continuation of the search for universal principles.

By the early 1980s, as a result of this and other work in problem solving, the field of AI was in a position to contribute applications for general use. Expert-systems work seems to have the most visible profile of all the work in AI, beginning with the DENDRAL system at Stanford and the MACSYMA system at MIT. These systems pointed the way to all successive expert systems development. Expert systems design evolved into an important field in AI, and is one area of problem solving that contributed significantly outside AI. Meanwhile, adequate computing-machine-based problem-solving systems are difficult to develop, and research into and use of problem-solving programs continues unabated.

Speech Recognition

Speech is a highly complicated process. While speech research has a long history, twentieth-century researchers have advanced the state of analysis by decomposing speech into simple wave-forms by a frequency spectrum analysis using electronic filters. Words and their component sounds contain many frequencies because of the construction of the human vocal system, and variations in human systems offer a range of possible speech examples. To understand speech, however, humans employ more than the information found in an acoustic signal:

> [Speech is] supplemented by what we know about English [or any language], about the subject discussed, about the person we are talking to — by what we know about the world. Listening is not a passive process. We are constantly trying to find meaning in the stream of sounds that pours through the air. And in the process we tend to hear what we expect to hear.[85]

AI researchers have contributed to the development of automatic speech recognition by seeking to integrate knowledge about semantics into the computer's interpretation process. The work in this area has been based on a significant literature on speech: studies of language, sound, physiology, psychology, and automata. Our specific interests in this study of IPTO are related to automatic systems for speech recognition and understanding, especially the use of the electronic digital computer after its public introduction in 1946.

Research on speech recognition in the 1950s using digital computers was carried out in the United States primarily at Bell Laboratories, IBM, Bolt Beranek and Newman, and Lincoln Laboratory. A history of Bell Laboratories notes that the work on automatic speech recognition was to make computers more "user friendly." [86] There was also work in behavioral sciences at Bell Labs in language comprehension and word recognition, but this was focused on human memory processes.[87] During the 1950s, Bell Laboratories started work in automatic speech recognition in order to develop a telephone that could be dialed through spoken commands. A number of isolated word recognizers were developed, starting with the first complete one by K. H. Davis, R. Biddulph, and Stephen Balashek at Bell Laboratories (1952), which divided the frequency spectrum into two bands that produced patterns that could be compared with stored patterns. They developed a speaker-dependent system that could with a high degree of accuracy recognize digits spoken singly. It was developed on an analog, vacuum tube computer. While it performed well for certain speakers, it did poorly for arbitrarily selected speakers.[88]

Around this same time Licklider, while at Harvard, developed what came to be called the "watermelon box," which was designed to recognize the vowels in the word *watermelon*.[89] In 1958, Homer Dudley and Stephen Balashek designed "Audrey," which used ten frequency bands for resolution and was a more sophisticated version of the earlier Bell system. Audrey recognized the digits 0 to 9 as spoken by seven speakers in isolation. Digit recognition was accomplished by reference to a preprogrammed time frequency spectrum pattern. The system would adjust the signal to normal time before the pattern would be compared. The IBM 704 computer selected the pattern that best matched the spoken digit. This technique reduced the error rates associated with speakers.[90]

In speech recognition research in the 1950s, analysis was based primarily on phonological, or voice-related, input. In 1953 Thornton Fry and Peter Denes at Bell Laboratories used a device they called the "diagram frequency," the probability of one sound following another in a word or sentence, to help disambiguate phonemic sequences. As another example, James W. Forgie and Carma D. Forgie at Lincoln Laboratory developed an isolated-word recognition system. In fact, an entire array of limited devices of this sort were built in the 1950s. All involved analysis of words and digits in isolation rather than connected speech.[91]

In 1965 John Hemdal, of the University of Michigan, and George Hughes, of Purdue University, developed a computer system to serve as a model for human speech recognition. The system recognized ten cardinal vowels and nine diphthongs, as well as most consonants in simple random form. Recognition was on the basis of the acoustic signal alone, without considering semantics or syntax. A total of 227 isolated, nonsense words were fed into the system by an isolated speaker. The computer achieved a 92-percent recognition rate. The Hemdal and Hughes system analyzed the "features derived from fundamental acoustic parameters like format frequencies, the presence of turbulent noise, noise spectral shapes, discontinuities in sound level and format positions, and silence." This analysis was followed by pattern matching at the feature level, rather than at the acoustic level, as had been done in earlier recognizers.[92]

At this same time, IBM developed a system for recognition of isolated words. Researchers at IBM taped twenty-five male and twenty-five female speakers. They manually edited and digitized the speech signals, and ran them through a bank of forty filters. The system segmented the words into vowels and consonants and then compared each word to an internal dictionary. If a match was found, the system considered the word identified. The system identified 493 words with 97-percent efficiency.[93] Within a few years, IBM researchers developed another system for the recognition of isolated words. The system achieved 92 percent correct recognition for a single speaker for 278 isolated words.[94]

IBM researchers continued their work, and in 1970 Norman Dixon and C. C. Tappert took a step toward automatic recognition of continuous speech. Boundary events, or the boundaries around meaningful units of speech, were proposed which would function to let the machine know the proper time segment for a given phoneme, the smallest unit of speech. The system utilized a bank of twenty filters and an automatic segmentation procedure that displayed information on a spectrogram. The experimenter listened to a recording of the utterance, used a light pen to mark the machine-located boundary of a given utterance, and entered that information into the system.[95] The system stored this data as patterns for later comparison with new input data.

Although researchers had much success with these early speech recognition systems, all of the systems were quite limited. Virtually all contained limited vocabularies, were talker-dependent, and required extensive training to use. They could not handle even the simplest connected speech of a single speaker. Error rates varied, but increased markedly with untrained, new talkers.

The AI community's interest in speech recognition research grew out of concerns in natural language processing rather than these speech recognition systems. Natural language processing is a computer's analysis of, and attempt to understand, a natural language (e.g., English). During the 1960s, interest in natural language processing at Stanford was concentrated in the Higher Mental Functions Project headed by Kenneth Colby.[96] At MIT, research on natural language processing was part of Project MAC. And BBN was engaged in a program to "enable persons unskilled in 'computer programming' to make much better use of their machines."[97] All three projects were supported by IPTO. If it could be achieved, natural language processing by machine would be of great value to DOD.

While the various speech recognizers were under development, researchers in AI became interested in the problems associated with understanding connected speech. The important difference between programs for the recognition of isolated words or numbers and programs for the understanding of connected sounds and words is that in speech understanding the speech signal does not contain all the information necessary to decode the message uniquely. Like humans, the system needs to employ linguistic and context-dependent knowledge sources to infer or deduce the intent of the message.[98] This was where AI techniques, some of which came from the problem-solving research discussed in the previous section, became useful in speech-understanding work. The knowledge sources for speech-understanding research are the physical characteristics of sound (phonology), the rules describing variations in pronunciation (phonemics), the rules that describe how units of meaning combine (morphemics), vocabulary (the lexicon), the rules describing fluctuations in stress and intonation (prosody), the grammar and rules of sentence formation (syntax), the meaning of words and sentences (semantics), and the (pragmatic) rules of conversation.[99] Researchers used combinations of these knowledge sources in speech-understanding research. Speech-understanding systems can be thought of schematically as shown in figure 5. In the figure we observe that researchers added the AI concerns of syntactic analysis, semantic processes, and pragmatic analysis to understand connected speech. To appreciate how this was incorporated into a speech-understanding system, we begin with a report on the early work at Stanford University, supported under the general IPTO contract with Stanford.

D. Raj Reddy began his work in speech recognition while taking a course at Stanford from John McCarthy in the early 1960s. The course required the completion of independent student projects, which the student selected from a list circulated by McCarthy. Reddy chose speech recognition "because I was . . . interested in languages . . . having had to learn three or four languages, and I thought it might be interesting to understand and study what it is that we could learn from [the] use of computers."[100] He distinguished his work from what came before in the following way:

> People in many fields have been carrying out research in speech for some time with a variety of goals. The motivating factor in most instances has not been Man-Machine communication in itself as is the case of the present investigation. In the field of medicine the motivating factor has been the detection and correction of speech defects. In the field of psychology it has been the better understanding of the articulatory and perceptual process of the human being. In the field of telecommunication it has been the efficiency of communications and development of band-saving techniques. In the field of linguistics it has been the desire to discover the underlying structure of a given language. As a result, much of the literature on this subject is unrelated to speech recognition by machine.[101]

The most significant aspect of Reddy's project, which grew into his Ph.D. research, was the way in which he employed the computer. Up to this time, spectral analysis of the speech to be analyzed accounted for a significant amount of the computer time involved. In Reddy's work, spectral analysis had a relatively minor role. Instead, he did ad hoc measurements on the speech waveform, which took place during on-line computer sessions.[102] The system took a microphone signal and directed it to the computer memory through an analog-to-digital converter. The main difference between Reddy's work and the recognizers that preceded it was that almost all of the earlier work had used the output from a bank of filters to preprocess the sound signal through

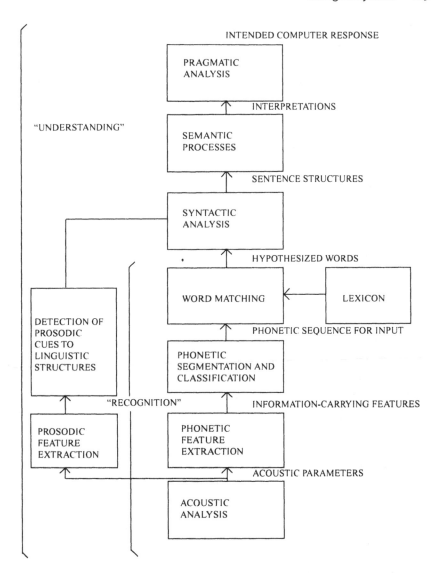

INTENDED COMPUTER RESPONSE

PRAGMATIC
ANALYSIS

INTERPRETATIONS

"UNDERSTANDING"

SEMANTIC
PROCESSES

SENTENCE STRUCTURES

SYNTACTIC
ANALYSIS

HYPOTHESIZED WORDS

WORD MATCHING   LEXICON

PHONETIC SEQUENCE FOR INPUT

DETECTION OF
PROSODIC
CUES TO
LINGUISTIC
STRUCTURES

PHONETIC
SEGMENTATION AND
CLASSIFICATION

"RECOGNITION"   INFORMATION-CARRYING FEATURES

PROSODIC
FEATURE
EXTRACTION

PHONETIC
FEATURE
EXTRACTION

ACOUSTIC PARAMETERS

ACOUSTIC
ANALYSIS

Fig. 5.   Processes involved in computer "recognition" and "understanding."
After Wayne A. Lea, ed., *Trends in Speech Recognition* (Englewood Cliffs, N.J.:
Prentice-Hall, 1980), 58.

a series of filters.[103] The faster processing of the computers available to Reddy allowed him to eliminate the need for preprocessing of the sound signal.

In Reddy's first system he used a light pen; the operator acted on a CRT-displayed waveform to select portions for analysis. The system joined segments of stored speech that the operator could hear by way of a loudspeaker. Reddy developed a procedure to segment the speech wave into what he called sustained parts, in which the characteristics of the sound were regular; and transitional parts, in which the characteristics varied with time.[104] A one- to two-second utterance was first divided into sustained and transitional segments. For each voiced segment, a typical pitch period was then identified near the center of the segment, and mathematically divided into the first one hundred harmonics of the fundamental frequency. The output from these computer routines was a list of twenty-one parameter values related to duration, intensity, and frequency measurements. On the basis of these values, the segments were then classified as vowel-like, fricative-like, and so on, and tests were made to reject those segments that were due to transitions between speech elements. The remaining segments were then associated with English phonemes.[105]

The result of this investigation was the classification of thirty-two continuous speech utterances into the sound, the intended phonemes, and the computed phonemes. The time required for analysis of the utterances on an IBM 7090 computer varied from forty-five to seventy seconds for sequences of one to two seconds duration. Recognition of sounds reached 81 percent. In spite of the long recognition time, this was a marked improvement on earlier recognition systems.[106]

Shortly after Reddy completed his dissertation in 1966 at Stanford, another graduate student, Pierre Vicens, embarked on another speech recognition project under Reddy's direction. In this case, however, preprocessing of the speech signal was accomplished using bandpass filters, breaking the speech system into three bands. The bands were then classified by examining a segment of speech in the band and selecting characteristics that could be averaged over the entire segment. These descriptions of sounds were stored for later recognition comparisons of new sounds. When processing a new segment of sound, the program employed a heuristic procedure to search the lexicon, building a list of possible recognitions by considering rough features of the utterances. Following this, the program made detailed comparisons and evaluated similarity scores on the basis of the closeness of parameters for the corresponding segments.[107]

In addition to using heuristic procedures to match the segments of speech, Vicens and others in the Stanford laboratory pursuing similar research devel-

oped artificial languages to simplify the problem of determining word boundaries and to resolve phonetic ambiguities in the analysis of long connected-
speech utterances.[108] The results for a 54-word vocabulary were quite good. The
system achieved 98-percent correct recognition in two to three seconds per
word with a trained single speaker. Extending the group of speakers lengthened the recognition time with only a slight lowering in correctness scores (to
85–90%). Seventy French words were also used, with similar results. Vicens
then extended the list of words to 561, used in short sentences. The notable
difference was that the recognition time grew to between sixteen and seventeen seconds. Vicens's system was used in the robotics hand-eye coordination
project going on at the same time at Stanford, which we describe in the next
section, on vision research.

Vicens concluded from this research that the type of preprocessing was not
as important as the power of the algorithms in the system. He also believed
that techniques of artificial intelligence, such as simplifying the problem of
searching for a solution by means of heuristics, appeared to hold great promise for speech recognition. Building more powerful syntax-directed sentence
analyzers would be more fruitful than a great amount of effort spent in devising preprocessing techniques.[109] Other researchers in this area were drawing
similar conclusions from a variety of work in progress. Future work at Stanford would be directed toward speaker normalization, so the machine could
recognize random speakers; a language for human-machine voice communication somewhere between LISP and English; more accurate phoneme recognition; and the building of a machine that could carry on conversations.[110]
IPTO supported the work of Reddy and Vicens as part of its overall contract
with Stanford for AI research.

At the time Reddy and Vicens were engaged on their projects, other people
were also exploring ways to use the computer for speech recognition. For
example, Bernard Gold, of MIT's Lincoln Laboratory, who had been a colleague of Roberts while the latter was at Lincoln, focused his work on different speakers. He extracted fifteen features by segmenting the sound, detecting
the stressed vowel, and making measurements on the stressed vowel and its
neighboring segments. The system stored measurements for 540 words spoken by ten speakers. Gold developed a decision algorithm that analyzed and
stored the data, and comparisons were made during a second pass. His system
achieved satisfactory recognition results.[111] A major obstacle in the use of filtering systems for speech synthesis had been the lack of any practical system
to estimate the pitch period for the human voice. (The pitch period is a measurement of the relative rapidity of the vibrations by which a tone or sound is

produced.) In order to process pitch contours of speech on a digital computer, there was a need for pitch-period estimation systems that operated in real time. Gold developed a pitch-period estimator based on parallel-processing techniques. To aid in this work, Gold developed two simplifications of his original parallel-processing algorithm.

At BBN, Daniel Bobrow and Dennis Klatt designed a system called LISPER for research in speech and pattern recognition. Signals entered a spectrum analyzer directly from a microphone or tape recorder. The spectral representation of a word consisted of two hundred spectral samples, corresponding to two seconds of speech material. LISPER compared the distinctive features of the signal extracted directly from the outputs of nineteen bandpass filters. Rather than use continuous speech, in which segmentation is a problem, they worked with messages with "easily delimited beginning and termination points."[112] The vocabulary contained one hundred items: words and short phrases. Bobrow and Klatt developed a set of algorithms for extracting those features in the signal that characterize speech utterances. They coupled this set with a recognition algorithm capable of high-quality message identification. The system achieved 97-percent correct recognition scores on a fifty-four-word vocabulary for a trained single speaker.

This combination of research efforts at IPTO-supported laboratories — Reddy and Vicens at Stanford, and Bobrow and Klatt at BBN — convinced the researchers that systems to analyze and understand connected speech were within their grasp. One of the things that became obvious during these research projects was that a substantial amount of additional equipment with greater capability was needed to pursue this work. With it, the researchers believed they could achieve recognition of connected speech. At the IPTO PI meeting in 1970, the researchers expressed this view rather strongly, and apparently pressed IPTO director Lawrence Roberts to support more research in speech. According to Allen Newell, informal requests were made of Roberts to have IPTO fund a major research undertaking in this area.[113]

Although intimately familiar with speech research at Lincoln Laboratory, Roberts organized a study group, headed by Allen Newell of CMU, that met several times between April and July 1970 and produced a report advocating a five-year program on speech-understanding systems.[114] The study group assembled a list of specifications for a total system. According to the study group, any system developed should be able to

accept continuous speech
from many cooperative speakers of the general American dialect

in a quiet room
over a good quality microphone
allowing slight tuning of the system per speaker
but requiring only natural adaptation by the user
permitting a slightly selected vocabulary of 1,000 words
with a highly artificial syntax
and a task like the data management or computer status tasks
with a simple psychological model of the user
providing graceful interaction
tolerating less than 10% semantic error
in a few times real time
on a 100 MIPS [million-instructions-per-second] machine
and be demonstrable in 1976 with a moderate chance of success.[115]

The group outlined what they believed to be the technical requirements and
the type of program to pursue to achieve these goals. It advocated expansion
of the research community to include other technical communities not usually
associated with computing, and suggested that the program be a two-stage,
ten-year program. It also proposed organizing a summer institute, as Project
MAC had done, at the beginning. The group suggested creating a steering
committee composed of its own members and the researchers who would be
supported under the proposed initiative. The group also stressed that as much
work as possible be done through the ARPANET. In the evaluation report five
years later, Dennis Klatt, one of the members, described the study group's
emphasis on the concept of speech understanding, as opposed to speech recognition:

> [We] believed that the hope for the program lay in analyzing speech within
> the context of specific tasks that employed strong grammatical constraints,
> as well as strong semantic and dialogue constraints, so that many sources
> of knowledge could be brought to bear to attain successful understanding
> of what was said or intended by the speaker.[116]

Originally the study group noted that "on balance we believe that the program
outlined has a high enough chance of success, and of payoff, if achieved, so
that we can enthusiastically endorse its pursuit."[117]

Roberts took the advice and inaugurated a five-year Speech Understanding Research (SUR) program. The objective was to obtain a breakthrough in
speech-understanding capability—that is, to develop a system that would recognize speech in order to perform some task, and could then be used in the
development of practical verbal communication between humans and ma-

chines. There were two differences between this program and other research in speech recognition. First, whereas previous projects had sought to improve the ability of machines to recognize speech, the advisory group suggested the use of AI principles, that is, the application of syntactic and semantic constraints, in order to have the system judge words by context as well as by pattern recognition of words or parts of words. The system would not need to recognize every word to understand the meaning of connected speech. Second, accuracy would be judged on the basis of the correctness of the response and not on the basis of whether all of the words were correctly recognized.

IPTO awarded to CMU, Lincoln Laboratory, BBN, SRI, SDC, the University of Michigan, and the UNIVAC division of Sperry Rand contracts specifying the goals the study group had proposed. The office established supporting efforts for the analysis of speech waveforms; the recognition of phonemes (the smallest units of speech); dictionary development; the organization of grammars oriented toward spoken language; the use of emphasis, pitch, and semantic context to resolve input ambiguities; and system integration.[118] Three of the contractors produced systems in the ensuing period. Two of the systems came from research at CMU—HARPY and HEARSAY-II. SRI and SDC working together produced a third system. Later, BBN submitted a fourth system, Hear What I Mean. IPTO tested three systems in 1976—HARPY, HEARSAY, and the SRI-SDC system. Of the three systems tested, HARPY performed best.[119]

At SRI, Don Walker became the leader of the speech-understanding project; at SDC, Beatrice Oshika led the speech-understanding effort.[120] SRI worked on high-level syntax, semantics, and pragmatics. SDC did work in acoustic processing and phoneme recognition.[121] The collaborators designed the system to work on a document retrieval task. The work on speech understanding involved a large team of people at SRI. Bill Paxon and Ann Robinson designed algorithms such as the "best first parser." Barbara Grosz focused on discourse and the effects of referring one expression to another in a computer program. Jane Robinson was the chief grammarian. Gary Hendrix worked with semantic nets. While SRI and SDC's work in speech understanding did not yield a complete, fully functioning system, it did serve as the basis of other natural language processing research sponsored by IPTO at both organizations.

CMU researchers developed the HARPY and HEARSAY systems. Like SRI and SDC, CMU designed its systems as document-retrieval systems. HARPY contained several knowledge sources, a network compiler, and a recognition process. The knowledge sources were a new Mini-Query Language (MQL) to include the rules of grammar; a pronunciation dictionary for the MQL, in which alternative pronunciation of each word was shown in a representation of a pronunciation graph;[122] a set of "juncture rules" to handle word boundaries, which contained examples of the insertion, deletion, and change of phonemes that occur at word junctures; and a source of phonemic knowledge.

The diversity of the knowledge sources required the introduction of a "knowledge compiler" to develop a network knowledge structure to synthesize the knowledge from different sources into a unified framework.[123] A mechanism like a blackboard (described below) served as controller. The workings of the knowledge compiler are not so important to the present discussion as are the compiler's essential elements, and its output. The knowledge compiler generated representations of all possible sentences using HARPY's syntax and lexicon. The compiler substituted words from the lexicon into the grammar to generate all possible sentences, and substituted phonetic representations of words for the words themselves. This process, which set up the template network for HARPY's 1,011-word vocabulary, consumed a large amount of computer time.[124] With the output of the knowledge compiler, HARPY began the recognition process using an approximate heuristic search method called "beam search," which selects for further evaluation what appear to be the "best paths" to explore, by comparing sounds with stored sounds in the recognition network of fifteen thousand states. HARPY kept about 1 percent of the states at each step in the evaluation of the network.[125]

Table 6 provides a comparison of the performance of the four systems developed under the IPTO program. What is most significant is that, through these projects, important advances were made in the speech-understanding field, advances that went beyond empirical demonstration and addressed fundamental questions in speech understanding. Some of these advances had an impact outside the speech area. For example, researchers learned about the architecture and control of large AI systems and developed new system organizations.[126] HEARSAY-II offered a new organizational concept, the "blackboard," which controlled a set of parallel asynchronous processes that communicate with each other through the blackboard. The blackboard model for problem solving consists of two basic components: the knowledge sources and a global database. All communication between knowledge sources takes place through the blackboard. Knowledge sources produce incremental changes in the blackboard, which leads to a solution to the problem. The advantage of the blackboard model is that there is no need for separate parts of the knowledge base to be represented in the same manner or for the separate parts to use the same inference engine. Each can use the optimal representation and inference engine. The blackboard concept may have been one of the most far-reaching outcomes of this program because of its later use in many other systems outside of the speech area.

Another advance from these speech understanding systems involved gram-

Table 6 Goals and Final (1976) System Results for the IPTO Speech-Understanding Research (SUR) Program

Goals	Results of SUR Systems				
	HARPY	HEARSAY II	HWIM	SRI-SDC	
To accept continuous speech	184 sentences	22 sentences	124 sentences	54 sentences	
from many cooperative speakers	3 male, 2 female	1 male	3 male	1 male	
in a quiet room	computer-terminal room	computer-terminal room	computer-terminal room	quiet room	
with a good microphone	inexpensive close-talking mike	inexpensive close-talking mike	inexpensive close-talking mike	good mike	
with slight adjustments for each speaker	20 training sentences	60 training sentences	no training	no training	
accepting 1,000 words, using an artificial syntax	1,011	1,011	1,097	1,000	
yielding less than 10% semantic error	5%	9% or 26%	56%	76%	
in a few times real time (on a very fast computer)	yes	yes, slower than HARPY	not quite	yes, slower than HARPY	

Source: Adapted from Wayne A. Lea, "Speech Recognition: Past, Present, and Future," in Lea, ed., *Trends in Speech Recognition,* 69.

mar design. According to the fifth-year report on the speech-understanding program presented to IPTO by the steering committee, an improvement in grammar design was an important factor in CMU's success. This improvement was the ability to manipulate grammatical complexity and observe changes to the system's performance as the task was simplified.[127] Other advances were contributions to control strategy, the handling of semantics and syntax, word identification improvements, acoustic-phonetic processing and analysis, and the computer generation of possible responses to indicate understanding of spoken language.[128]

In evaluating the overall outcome in 1976, the steering committee reported to IPTO that the speech-understanding technology was only in its early stages. More development work was needed to produce systems that could work in applied contexts. The committee went on to say,

> In 1971, when the program started, perhaps the majority of informed technical opinion put general speech recognition by computers as not possible in the foreseeable future and perhaps not possible at all. Throughout the five years, discussions of applications have been dominated by trying to imagine those special favorable circumstances where an expensive limited connected speech understanding system could have practical benefits. For the immediate future that concern may perhaps still hold. But the results we have reported changed the terms of the problem. Informed technical opinion can now be that general cost-effective speech input to computers is an attainable goal.[129]

Regardless of the progress attested to by the steering committee, there was no equipment available that could understand speech in anything near real-time. The systems needed to know a great deal about the subject area that they were supposed to understand, and there was no way to get the information into the system except on a case-by-case basis. Moreover, the systems were too inefficient to be practical, and had little promise of overcoming their limitations. George Heilmeier, DARPA director, considered this achievement insufficient to justify continuing the program with any expectation of major improvements. No immediate follow-on program occurred, and work on the problem of connected speech understanding slowed considerably.

Research did continue at IBM under the company's own auspices, based on the results obtained at CMU. In 1978, IBM announced a system that achieved high-percentage recognition of one hundred sentences with a trained speaker using a 250-word vocabulary and a 1,010-word vocabulary in a task domain similar to the one used in HARPY.[130] Around this same time, Bell Laboratories and Texas Instruments reported on systems developed at their facilities. These systems are no easier to compare with each other than are those produced in the SUR program; they certainly do not go beyond the IPTO-supported systems.

Writing in 1983, D. Raj Reddy, of CMU, and Victor Zue, of MIT, rued the fact that research on connected speech recognition had been "relatively dormant" since the middle 1970s, with exceptions such as IBM's work to use speech recognition to activate a typewriter.[131] Not long after this assessment by Reddy and Zue, speech-understanding research funding by IPTO had a resurgence under the Strategic Computing program. Speaking in 1986, Craig Fields described the speech-understanding program in SC.[132] The goal of the revived speech program was still to get a computer to understand spoken input. For a noisy environment the goal would be isolated word recognition in a cockpit. This part of the program would be developed in conjunction with the "Pilot's Associate" testbed, one of three testbeds under SC, and would have a vocabulary of two hundred words. For a quieter environment, continuous speech recognition would be part of a "battle management" program, one of the other SC testbeds, with a vocabulary of two thousand words and 95-percent accuracy. The main research project for isolated word recognition was at Texas Instruments; for continuous speech the main research projects were at CMU and MIT. Other contractors included were BBN, Dragon Systems, SRI, Schlumberger, Columbia University, and MIT's Lincoln Laboratory.[133]

One example of the developments in the SC program was the SPHINX system, created at CMU in the 1980s by Kai-Fu Lee with IPTO funding. The SPHINX system is a large-vocabulary, speaker-independent, continuous speech recognition system. The system builds on the earlier work of James K. Baker at CMU (the Dragon system, predecessor of HEARSAY-II) and on the work of Baker's later group at IBM. In the later development at CMU, Lee sought to overcome the four principal problems in the development of speech-understanding systems: First, researchers needed to find a representation of speech; Lee used "hidden Markov models" (HMM) to represent all knowledge sources, from individual sounds to complete sentences. Second, to overcome the computer's lack of human knowledge, Lee used knowledge-engineering techniques to improve SPHINX within an HMM framework. These techniques added to the large body of knowledge about phonology and phonologic variations of English acquired over the decades, and helped the group to improve the set of speech sounds, or phones; word pronunciation by computer; and phonological rules. Third, since a good unit of speech was lacking, Lee developed two units of analysis: generalized triphone models and function-word-dependent phone models. Fourth, to overcome the lack of learning and adaptation, Lee developed two algorithms to allow the machine to adapt to individual speakers. By utilizing these four new developments, SPHINX achieved speaker-independent word accuracies of between 71 percent and 96 percent on the 997-word DARPA resource management task.[134] The SPHINX system is one of the best examples of a practical system with a very large vocabulary, but it still has accuracy limitations that make it necessary to speak one word at a time, with short pauses between words.[135]

Computer Vision and Robotics

In the early 1960s, Stanford University, SRI, and MIT (particularly its Lincoln Laboratory) pursued a number of AI projects in or related to robotics. These projects came under various labels such as hand-eye coordination, pattern recognition, sensors, vision, and robots. Two of the most important aspects of robotics were (1) the sensing of the surrounding scene, which could be further divided into a sensing problem and an "understanding" of the sense data for inclusion in the machine's knowledge database, and (2) the problem-solving system that made use of this new sense knowledge to decide on a course of action. These two aspects were referred to as sensing, or vision; and acting, that is, moving in some fashion. Researchers explored a range of robotics projects ranging from simulations of action on the basis of supplied databases, to integrated robotic systems with sensing and motion capabilities.

Like knowledge representation and speech research, vision research required the deconstruction of vision into its component parts: sensing, analyzing, and reacting. In the early years, IPTO supported research on the development of hand-eye coordination systems, robots, and sensors. This research provided substantial information on the interplay among the component parts of vision and evolved into more elaborate application systems, especially for photograph analysis in intelligence gathering.

We emphasize again the work of Sutherland and Roberts which was under way in the early 1960s at MIT's Lincoln Laboratory. Several projects contributed research results that had far-reaching consequences for future research in vision. These projects included work on pattern recognition with "adaptive networks" by Roberts,[136] before he assumed a position at IPTO; and the development of systems for machine processing of line drawings, a vision problem studied by Leo Hodes.[137] Oliver Selfridge, who distinguished himself in many aspects of computing, and Ulric Neisser, a cognitive scientist with training in psychology, developed computation routines to extract useful information from the input/sense data and to select appropriate weighting factors in the output for moving the robot.[138] The accomplishments by Hodes, Roberts, Selfridge, and Neisser played an important role in the subsequent work by Roberts at Lincoln Laboratory on pattern recognition and the development of techniques for manipulating lines and three-dimensional objects, work we described in chapter 3.

Roberts's work in pattern recognition became foundational in AI, and the pattern recognition techniques he devised were broadly useful in problem solving, graphics, and vision. Much of the pattern recognition work up to 1960 had concerned the recognition of two-dimensional objects such as letters, because it was believed that achievements with two-dimensional objects would lay the groundwork for explorations of solid objects. This proved not to be the case. Roberts came to believe during his dissertation research that the reason for this was that the techniques for two-dimensional objects did not converge with the techniques needed for solid objects. According to Roberts, it was necessary to base perception of solid objects on "the properties of three-dimensional transformations and the laws of nature."[139] By remodeling pattern recognition into a research question in AI, Roberts's work opened the way for cross-fertilization of research methods and results of research in several areas of AI, both inside and outside the field of robotics. It is in this sense that Roberts's work on pattern recognition is foundational to AI.

Roberts's contribution was to show that recognition of a scene was essentially description of the scene. Description of a scene is symbolic, and was therefore outside of the techniques of pattern recognition of the time because the symbolic structure has to be constructed. The construction requires some knowledge obtained by adopting a model of the situation, a strongly inferential process that was and still is a hallmark of AI.[140] In his major publication on this subject in 1965, Roberts listed some of the assumptions useful to humans in depth perception: the use of perspective, the fact that objects occupy a definite region of space, and the conception that observed objects can be constructed out of familiar parts. The process involved breaking down the scene into a series of points and lines, analyzing them into a set of simple, linear, two-dimensional geometric figures, and reconstructing the image. The program found the polygons described by the lines, matched the polygonal structures to the models topologically, and reconstructed an image out of the appropriate polygons. With this reconstruction, the three-dimensional model could be viewed from any angle. Roberts also built in a formula for eliminating hidden lines. This work was going on at Lincoln Laboratory at the same time as the work of Ivan Sutherland on Sketchpad, also described in chapter 3. Roberts used Sutherland's concept of a ring structure, in which a block of memory registers is used for each item and the block contains pairs of ties to other blocks, although the exact block form used by Roberts was different owing to the different data requirements for three-dimensional displays.[141]

Roberts's "blocks world" became an important approach to solving AI problems; it was used as a context for examining questions in natural-language processing described in an earlier section, and as a system for studying the representation of rules for motion of robots. Two particularly important examples of this influence are Winograd's use of the blocks world in SHRDLU, and the employment of the hidden-line technique in the graphics simulator developed at the University of Utah by David Evans and Ivan Sutherland, after the latter completed his tenure as IPTO director.

In 1965, MIT began building an "intelligent" robot arm based on the work of Henry Ernst, taking robotics in a direction that paralleled and built on the work of Roberts. Ernst had developed a mechanical hand, MH-1, as part of his Ph.D. dissertation work in 1960–61. MH-1 had tactile sensors and photoelectric sensors that enabled it to sense objects on a tabletop. The photoelectric sensors were only useful at distances of a fraction of an inch, but they did enable the program to estimate the shape of the test objects Ernst used. The hand collected blocks, and put them in a box or made a tower out of them. The program contained an interpreter with a search function that could deal with unexpected conditions. This ability was the beginning of the use of heuristic programming in physical manipulation.[142] While tactile sensation controlled Ernst's arm, researchers intended the new MIT arm to have visual perception. The work started at a simple level of pattern recognition to allow block stacking and ball catching, and worked its way up to "real world pattern recognition."[143] The researchers based their work on the techniques used by Roberts in his three-dimensional work at Lincoln Laboratory, and also turned to AI techniques similar to those used in chess and theorem-proving programs being developed in their own laboratory.

MIT's first status report about this work in August 1966 noted substantial progress in the vision and object-manipulation programs.[144] Researchers had already demonstrated ball catching and cube handling, and were incorporating these accomplishments into cube building. In addition, they used the techniques for abstract scene analysis which classified the plane into distinct regions and gave generalized descriptions of region boundaries. The identification of overlapping and partially hidden objects employed ideas from Guzman's POLYBRICK program. This work included designing a general-purpose model of the machine's working space. By this time, AI researchers at MIT had stopped using the term *pattern recognition* because it was too narrow to describe what they were doing.[145]

One year later, in September 1967, MIT researchers further distinguished the vision part of the robotics problem from the pattern recognition part, and explained their reasons for believing that any significant advance in vision would have to come from AI techniques. They claimed that the solution to the vision problem lay in combining all methods in a large system:

> There are many simple techniques that are very useful in a vision system. . . . The point is that each of the simple methods works on some problems, but not on all. One could take each and try to extend it to be more universal. But our experience is that the cost and complexity shoot up rapidly when a method is pushed even a little past its "natural" limits. Our counter approach is to absorb all the methods we can, filing them with statements about where they can help, and in what situations they are liable to fail. Then we try to put the major efforts into organizing higher-level programs to use these growing banks of knowledge.[146]

The many techniques included localization of objects in space, determination of contours and regions, discovery of objects in a scene, and recognition of objects in three dimensions. These techniques were put into a "vertically organized" system to allow them to interact in a flexible manner.

These task-oriented programs developed in the 1960s were essentially hierarchical — they communicated in only one direction. The program tried to recognize geometric objects by passing through a number of successive stages from raw data to the final identification, but made no adjustment if the data was incomplete. In the early 1970s, MIT researchers set out to develop further the "vertical" organization of their systems under the new name "heterarchical system." In a heterarchical system, such as Winograd's SHRDLU, the program went back and forth if necessary, as a feedback system would do, to build a better identification of the figure.[147] To maximize flexibility in the system, MIT abandoned the lowest-to-highest form of organization — the hierarchical, which was highly successful in some problem-solving programs — and substituted the heterarchical alternative: "We are now starting to remove completely all boundaries between programs — or rather, between the heuristic and cognitive ideas embodied in programs — and embed them all in the form of 'theorems' in a PLANNER-based language."[148] By this time, PLANNER was in use in several laboratories around the country.

Patrick Winston's LEARNING program, which learned through example, was an attempt to use the PLANNER language platform to achieve integration. Examples were given to the machine in the form of picture analysis and scene analysis. When attempting to duplicate a picture, say of an arch, it employed the concept of the "near miss." Near misses were used by the machine to evolve concepts by learning what aspects of the problem were important. Winston tried to build in discriminators, or theorems, so that the system could attempt to explain, for example, why the hole in the arch is important but the color of the arch is not.[149]

Each of these programs in the blocks world and for the remote arms was an approach to a piece of the robotics problem. In 1969 Jerome A. Feldman at Stanford defined the larger problem and characterized the robotics problem in the following way:

> The robot[ics] problem, in some sense, encompasses the entire field of artificial intelligence — there is nothing in artificial intelligence work which would not be useful in the ultimate robot. The precise degree to which various other efforts should be coordinated with a robot[ics] project is unclear. Traditionally (for the past three years[!]), the MIT group has kept quite strictly to hand-eye problems while the SRI group has concentrated on combining as much of its work as possible. The Stanford group is somewhere in between — there are a large number of artificial intelligence projects at various distances from the hand-eye efforts.[150]

Feldman, who had received his Ph.D. degree in CMU's AI program in 1964, arrived at Stanford in 1966, following a year at the Lincoln Laboratory, and assumed management of the project on hand-eye coordination. In addition to his work in AI, he had a strong interest in three-dimensional graphics and the representation of three-dimensional shapes.[151] Illustrating further his attempt to combine techniques from various parts of the AI laboratory, he combined his interest with McCarthy's interest in computer-controlled robot manipulating arms, and developed a system for processing television signals in a robot.

The Stanford hand-eye project also involved Bernard Roth and Victor Scheinman, of Stanford's Department of Mechanical Engineering. When Scheinman arrived at Stanford, he continued work on a robot arm design begun by Donald L. Pieper. Scheinman's career is interesting as an example of the far-reaching influence of IPTO programs. While at Stanford, he developed a hydraulic six-axis rotary actuator manipulator, which ran under the control of a DEC PDP-6 computer. In 1972, Scheinman moved to the mini-robot laboratory at MIT and designed the MIT manipulator — a small, lightweight, electrically powered, six-axis arm. A year later, he left MIT and began

Vicarm, a company to market the MIT manipulator. He also developed the first computer-controlled articulated robot with an all-electric drive; it later became known in industry as the PUMA.[152] Also in Stanford's hand-eye project, Karl Pingle and Manfred Huekel worked on vision problems such as edge detection. They developed a device that for many years was used for edge detection and line following in numerous vision systems.[153]

The Stanford project closest to MIT's hand-eye effort was vision research involving the recognition of facets, embodied in a children's-block-stacking program. According to McCarthy, the heart of the Stanford block-stacking system was a PDP-6 computer. The system was designed to visually locate cubical blocks scattered on a contrasting background. It could not locate partially hidden blocks until the hand removed the intervening blocks. The commands to move or stack blocks could be given orally, an activity that employed results from the speech research at Stanford by Reddy and Vicens. A command might be, "Pick up the small block standing on the right side," or "Rescan the scene." A command to pick up a block had to contain the location and orientation of the block; and a command to stack blocks had to include the location of the stack. There were remote and local teletypes, CRTs, a printer, a plotter, and tape units, as well as two cameras and two audio inputs. Two arms could be attached to the system. The first was designed to be attached to a paralyzed human arm. Despite a considerable amount of effort, this first arm was inaccurate: its position could vary by as much as a centimeter from the computed positional value. The second arm was more precise, but it broke down frequently and tended to leak hydraulic fluid.[154]

The vision work for this hand-eye system focused upon two main problems that Feldman called "levels of detail" and "strategies for attention." The lowest level of detail contained data about the intensity and color of the light at a particular point in the visual field. The second level held information about the outlines of the various objects in the field. At the third level, the system could distinguish interrelationships and relative motion among objects in the field. And at the highest level, the system tried to comprehend the total situation to produce action.[155] Since the available data had the potential to overwhelm the system, the system had to contain some "strategies for attention" in order to decide what data was worthy of its attention and what level of detail was best suited to the current perceptual goal.[156] The Stanford approach involved routines at various levels, cooperating in an attempt to understand a scene.

Researchers at MIT also investigated vision in conjunction with the design and development of mini-robots. Minsky organized the mini-robot laboratory to develop a miniature hand-eye system and a remote network for robotics, so that simple jobs could be handled locally and complex jobs could be done on remote, larger machines.[157] He had in mind research on the practical automation of production techniques. The laboratory's R&D program comprised the design of a new, heterarchical vision system; the generalization of labeling theories; grouping and tactile scene analysis; analysis of curved-line drawings; color vision; touch and tactile programming; low-level vision programs; electronic assembly; analyzing complicated objects; and group descriptions.[158]

We should point out that there was already in the literature a large body of psychological evidence indicating the dependence of perception upon global information and preconceived ideas. The interaction between "levels of detail" and "strategies for attention" in the Stanford view would lead to a "grand design,"[159] which was another form of MIT's heterarchical system: it was a vision system that had selective attention and could integrate information from various levels of perception. The system had a PDP-10 computer that monitored all operations of the camera and the mechanical arm. The language employed in the system was an extension of Stanford's ALGOL compiler along the lines of LISP. The data structure, LEAP, provided for associations between a given item and a local or global structure. As part of the "grand design," the system had a set of flexible basic-vision routines, which included the ability to read raw data, change the camera position and parameters, find edges, fit corners, find regions, analyze into distinct bodies, identify particular objects, and do scene analysis.[160] By the early 1970s, Lou Paul and Hans Morovec at Stanford had combined all of these ideas into a robot arm that could assemble a water pump for a Model T Ford, and a computer-controlled cart that could circle the lab. By the mid-1970s, the "grand design" was sufficiently developed to structure the research into description, visual guidance, and picture analysis.

Contrast the robotics work at Stanford and MIT with that pursued at SRI. SRI had been working on a robot design conceived in 1964 by Charles Rosen.[161] This robot received inputs from television cameras and other sensors, and used the data to drive effector motors (a specialized name for motors in robots) to carry the machine purposefully through the environment.[162] Rosen and his colleagues took a proposal for such a machine to Ruth Davis of the Office of the Defense Director for Research and Engineering, and to Ivan Sutherland of IPTO. In November 1964, SRI received a contract from IPTO to develop an automaton, and it started work on what became Shakey. IPTO's rationale for supporting the R&D on this intelligent automaton was "to provide a technological base for developing and building logical automata and extensors for use in intelligence applications."[163] In the 1960s this program was sponsored by IPTO and other parts of the Office of the Secretary of Defense. DARPA was

Fig. 6. Built at SRI in 1972, Shakey included a video camera, obstacle detectors, and sensors to measure distance. Photograph courtesy of SRI International, Menlo Park, California.

interested in developing systems that generally exhibited more autonomy from human control, for use in circumstances where "human access to information is difficult, hazardous, or otherwise undesirable or impractical." IPTO's interest in intelligent automata was not only in the construction of such systems but also in the achievement of "an understanding of their inherent capabilities and limitations under a range of conditions and circumstances."[164]

The first Shakey was controlled by a Scientific Data Systems SDS 940 computer with radio and television links. (Shakey can be seen in fig. 6.) Peter Hart and Richard Duda designed the vision subsystem. SRI also developed "R.E.," a sophisticated robot simulator using LISP. Nilsson described the vision system of a mobile robot at SRI which was capable of pushing objects around on the floor: "The robot lives in a rather antiseptic but nevertheless real world of simple objects—boxes, wedges, walls, doorways, etc. Its visual system extracts information about that world from a conventional TV picture."[165] Two different programs accomplished this extraction of information. The first was a line-drawing program that the robot used to identify empty space on the floor, to determine where it could move. The second was an object-identification program that allowed the robot to locate and identify nonoverlapping objects.

Before continuing the robotics story, however, we need to pick up another thread in the vision portion of the IPTO program: picture processing and image understanding, part of the artificial vision and sensing research supported by IPTO. The results of these two subprograms had important effects on later IPTO-supported work in robotics. At the beginning of the 1970s, IPTO directors began to focus their attention more sharply on AI. Roberts brought in new program managers to oversee the newly defined subprograms—intelligent systems, automatic programming, and speech understanding. As noted earlier, the major goal of the activity was "to build automatic linguistic, visual, manipulative, and problem solving systems." The new programs contained an important new applications dimension, as we shall see below. The various projects in vision and robotics came under these subprograms, and the effort was expanded through increased funding.

IPTO's picture-processing and image-understanding subprograms set objectives for the research community which would aid in the development of image analysis. The Cold War and the Vietnam War created an incessant need for more and more intelligence photographs. Since the end of World War II, military agencies and the Central Intelligence Agency had been collecting photographs of everything from Cuba to the Soviet countryside. Over the two decades following the war, satellite after satellite sent photographs back to earth at transmission rates that exceeded the speed at which humans could continuously analyze the photographs. At the end of the 1960s, photo interpreters turned to computers for help in the analysis of images. This group wanted the computers to be taught how to analyze photographs, using various types of recognition criteria that had accumulated in data banks over the years as a result of human analysis of photographs.[166]

DARPA's director, Stephen Lukasik, told Congress in 1972 that IPTO's picture-processing effort was a prerequisite to achieving systems that could perform a number of tasks in military and intelligence contexts. The tasks he enumerated were interpretation and correlation of electronic sensor-gathered intelligence data; tactical information analysis; development of strategic defense systems; planning of very complex operations and projects; photograph interpretation; target finding and target tracking; sophisticated information retrieval, query analysis, and question answering; and intelligence analysis—all of which were IPTO AI activities of the 1960s.[167]

Lukasik pointed out that the primary objective of the picture-processing and image-understanding activities was to develop the knowledge necessary to transmit and store pictures economically, and to learn how to recover or enhance obscured details in a picture. He reported to Congress,

> By computer processing, we can now achieve a bandwidth reduction by a factor of 6 to 7, with equivalent reductions in transmission and storage costs, and yet there is no loss in picture content regardless of how the results are measured. We are making significant progress in restoring pictures blurred by poor focus or motion. We expect to improve these methods, largely by optimizing them in terms of human vision characteristics. We are developing means to generate realistic pictures of solid objects in motion by means of a computer.[168]

Lukasik pointed out to Congress that practically nothing was being done in this area outside of DARPA, and that this lack of research was the reason IPTO had become involved and placed considerable emphasis on the work. The IPTO subprograms involved basic research in graphics and applied research in the compression of picture data for storage and transmission. Their objectives included hardware to display computer-generated images, new techniques for the digital representation of color images, interconnected facilities for transmission through the ARPANET, picture processing on IPTO's "supercomputer," the ILLIAC IV, and new hardware for real-time generation of artificial images by computer.[169]

Pursuant to the requirements of other DOD agencies, IPTO wanted to develop the ability to interpret imagery automatically. The successes with microprocessors throughout the computer industry in the early 1970s indicated to the office that the computing power required to analyze images in small, cost-effective systems at real-time rates was at hand. Research on human vision had produced many new insights for the automatic analysis of imagery, and computer science research had successfully demonstrated many scene analysis techniques that automatically converted image data into symbolic information. Thus, program managers in the IPTO Image Understanding subprogram designed an effort to build by 1982 a concept demonstration system that could automatically find objects of interest in images.[170] IPTO gave contracts for basic research to the universities that had been most active in this area for the preceding decade—MIT, CMU, and Stanford.

The direction in which IPTO and DARPA funding was trending was quickly obvious to the AI research community. As more and more success was achieved in the design and construction of vision and robotic systems and further development occurred in sensing systems that could locate themselves in an environment or read intelligence information, there was a shift in vision research, similar to that in problem solving, to include more application-oriented projects. MIT researchers expressed this view in their 1976 proposal to DARPA: "The Artificial Intelligence Laboratory proposes to work on applications of computer vision with the emphasis shifting from tool building toward solving real problems."[171] Patrick Winston, by now the director of the MIT AI laboratory, remembered later that the mid-1970s were days in which "you had to find a way of explaining the work in applications terms." Winston accomplished this by seeking common areas between the interests of the AI researchers and IPTO:

> I was seeking to find intersections between what the laboratory was doing and what [IPTO] either was interested in or could be persuaded to be interested in. So in some cases, it was a matter of pointing out the potential application of a piece of work. And in some cases, there were attempts to interest [IPTO] in new areas.[172]

New basic research areas were sometimes selected for funding, but increasingly the emphasis was on applications. While the trend took a few years to gain momentum, IPTO and its research community rolled into the 1980s on the wheels of their accomplishments in applications. We look now into some of these research activities and the systems designed to employ them.

Like Winston and his colleagues, Thomas Binford and Barry Soroka at Stanford took up the challenge of undertaking AI projects that were more applications oriented. Stanford concentrated its applications work in image processing on photo interpretation, guidance, and cartography. This work constituted an interpretation system dubbed ACRONYM, which contained "subsystems for geometric modeling, geometric reasoning, prediction, planning, description, and interpretation."[173] In its first demonstration, ACRONYM recognized a Lockheed L1011 aircraft in photographs of San Francisco airport. The Stanford team also developed algorithms for a high-performance stereo vision system. In addition, they made progress toward a new algorithm for finding edges in images.[174]

Just as Roberts was focusing IPTO's AI activity in 1970, Reddy joined the faculty at CMU and established programs in robotics, vision, and image understanding; and CMU became an equal partner with MIT and Stanford in this area of the IPTO program. CMU's research in vision has, from the beginning, concentrated on studying naturally occurring images. This distinguishes the work done at CMU from work done at Stanford, SRI, and MIT, which have had as their origin the needs arising from research in robotics — the hand-eye projects at Stanford and MIT, and the Shakey project at SRI. A major thrust of research in image understanding at CMU in the 1970s was the work of Ronald Ohlander and Reddy in natural-color scenes such as images of houses, buildings, and cityscapes.[175] Using color and texture features, and a technique known as "region splitting," they were able to segment images previously unsegmentable. The same images were studied in human perception protocol experiments by Omar Akin, Ohlander, and Reddy, in which human subjects attempted to figure out what was in a scene by posing questions to an experimenter through a computer link. The CMU researchers hoped that this research could provide a paradigm for acquisition of knowledge through image analysis.

Other research in image understanding at CMU during the early 1970s included work on biological image processing, with a focus on constructing three-dimensional models of invertebrate nervous systems, by Ohlander and Reddy; work on detection of objects in aerial images by Keith Price and Reddy; and work on model-based vision for interpreting images of the city of Pittsburgh by Steven M. Rubin and Reddy.[176] The early work of Takeo Kanade examined the problems of interpreting three-dimensional blocks-world scenes, including shadows, highlights, and occlusions.[177] John R. Kender and Kanade used image regularities such as skewed symmetry to derive three-dimensional shape from textual cues.[178] Herman and Kanade studied the problem of recovering the three-dimensional shape of an object from a single view. Steven A. Shafer and Kanade studied the problem of recovering three-dimensional shapes using color, perspective, and shadows.[179] Shafer investigated the problem of removing highlights from color images. David McKeown studied the problem of terrain modeling from aerial images.[180]

In addition to its groundbreaking work in graphics, the University of Utah computer science group pursued projects in the area of sensory information processing for research in speech and vision, directed by Thomas G. Stockham. Researchers at Utah attacked the problem of image understanding by constructing images from raw data (synthesis), processing models for human vision, and using advanced mathematics in signal algebra. The speech work involved advanced signal processing, and processing and filtering based on models of human hearing. Some of the research on vision was related to basic problems in vision, but a considerable amount of this work related to applications for military systems. In the basic category, for example, they developed a method for improving image intelligibility, which became known as the Utah method; it applies the formal mathematics of signaling theory to the nonlinear mathematics of image formation. To eliminate background sounds, the researchers applied linear prediction techniques to the filtering of audio signals. Researchers at Utah also developed image-processing standards for bringing high-quality continuous tone images into and out of computers. Bridging the area between basic investigations and application, they developed a quantitative visual processing theory, that is, a model of what the human visual processing system does to images. They developed a high-speed convolution technique for use in high-speed digital processors in radar installations, as well as a sophisticated compression and expansion system for use in high-quality noise rejection systems to increase the quality of signal recognition on noisy channels. And they found a way to remove accidental blurs from scientific intelligence photographs, a technique called "blind" deconvolution.[181]

IPTO also supported the Image Understanding Program at the University of Southern California, which was directed by Ramakant Nevatia. Nevatia received his Ph.D. degree in electrical engineering from Stanford in October 1974 under the direction of Thomas Binford. For his thesis research, he developed a system that took depth maps of small objects — a doll, a toy horse, a glove, and a hammer — by means of a laser triangulation system. He continued his work in computer vision when he joined the staff at USC. There, he further developed his ideas of stereo vision and suggested the concept of "motion stereo," that is, stereo vision in a case in which the observer is moving with respect to the environment, the natural state for a moving robot or a moving aircraft. In the later 1970s, Nevatia worked with K. R. Babu to develop the Nevatia-Babu edge detector, which detected roads by locating parallel lines of opposite contrast, called "antiparallel lines." In 1982 Nevatia published *Machine Perception*, a review of the entire field of computer vision.[182]

Carrying out the research and development in the Image Understanding and Picture Processing subprograms and weaving the results into systems took more than a decade. By the end of the 1980s, picture processing had contributed to the achievement of DOD objectives for fast, reliable interpretation and distribution of intelligence images. Image-understanding techniques from AI systems were incorporated into military intelligence systems. The results of vision research found their way into a number of automated DOD systems, such as systems for map making and the repair of small objects.[183]

We can now return to our story of robotics developments and connect it with the development of vision systems. Like AI research in problem solving and speech understanding, research in robotics involved a set of complex, often difficult problems. As we have seen, researchers divided the problems into studies of interpretation and studies of motion. Interpretation involved pattern recognition, scene analysis, reconstruction of images from incomplete data, and determination of how to use the information acquired in this way for action. Studies of motion required better understanding of the sense data that a robot needed about its context; the relationship between sense data and courses of action; and the inverse problem of the performance of tasks.

As a result of this new work and the accomplishments in the mini-robot program, MIT researchers in the 1970s thought that "computer vision and manipulation will soon sweep through the manufacturing and maintenance industries."[184] In order to move from the blocks world to the real world, MIT went beyond the mini-robot program, created the Micro Automation Laboratory, and began work in electronics assembly. MIT researchers believed strongly that this technology should be transferred to industry.[185] They demonstrated the automatic assembly of a radial bearing, the orienting of integrated circuit chips in preparation for lead bonding, and the inspection of solder joints.[186]

With the establishment of the CMU Robotics Institute in 1979, the focus of image understanding research at CMU shifted to problems of robot navigation. This research led to a number of interesting solutions to problems in computer vision such as image registration, depth image processing, and stereo vision for problems involving road following, path planning, and obstacle avoidance, all of which were aspects of later IPTO attempts to build automated systems with these abilities as part of the Strategic Computing program.

At MIT, several areas ripe for being reoriented toward applications were also pursued during the late 1970s and early 1980s. For example, Gene Frouder's system for visual recognition used the continuously improved knowledge of a natural scene, while John Hollerbach's work dealt with the description of curved objects such as vases and pottery. David Marr and his students worked on translating images into symbolic descriptions without using "high level knowledge."[187]

Integration of the results of university AI research into industrial products was quite successful. Laboratories, both in universities and corporations, began to transfer the technology to the factory floor. And IPTO sought to incorporate the new designs into systems for use in military circumstances. To design and build demonstration systems, a greater number of industrial firms became involved in the IPTO AI program. In its Intelligent Systems Program, IPTO continued to support the basic studies needed to advance the state of the art in vision and robot systems. By the 1980s, IPTO also supported a number of companies to encourage the development of applications systems that could be incorporated into defense applications. The list included

— Westinghouse Electric Corporation
— Lockheed Palo Alto Research Laboratory
— Ford Aerospace and Communication Corporation
— Martin Marietta Corporation
— Technology Service Corporation
— ESL Incorporated
— Synectics Corporation
— Object Recognition Systems, Inc.
— General Electric Company
— Xerox Corporation.[188]

At a conference in 1980, three separate companies offered examples of systems that were "up-and-running." Representatives of other companies offered comments on systems that were all described as theoretical advances or as moving toward the testbed stage.

The first of the "up-and-running" applications belonged to Westinghouse, which described its new image-understanding application in automatic target acquisition. After more than a decade of algorithm and breadboard development, Westinghouse researchers had completed two breadboard systems for automatic recognition of tactical targets for the purpose of flight demonstration. The military used one of these systems aboard an observation helicopter in 1980.[189]

The second application was from Object Recognition Systems, which worked on real-time pattern recognition systems and based its work on rapid advancement in cost-effective sensors and microcomputers. This advancement made on-line implementation of pattern-recognition systems practical for a variety of industrial applications, one of which was real-time pattern recognition in the sorting of packaged/cartoned goods at high speed for automated warehousing and return goods cataloging.[190]

The third example came from the Xerox Palo Alto Research Center. Xerox PARC had developed a system designed for real-time optical inspection and analysis of complex or repetitive patterns. The primary application of the system was the inspection of etched and plated patterns. The system could inspect a twenty- by twenty-four-inch printed wiring board in approximately four minutes, utilizing an image resolution of one mil (one thousandth of an inch). Several other applications for this system included the inspection of the inner layers of multilayer printed wiring boards, the inspection of plated re-petitive structures, and optical character recognition.[191]

By the 1980s, computer vision research came under three main headings: machine, or low-level, vision; computational vision; and vision architecture. In low-level vision, the most significant development was "mathematical mor-phology," a method for simplifying image data on the basis of shape. It pre-served essential shape characteristics while eliminating irrelevancies. Mathe-matical morphology has a variety of applications in industrial vision systems. In computational vision, the most significant development was regularization theory, a technique for "making mathematically ill-posed surface-inference problems well posed." Regularization theory has been used to reconstruct in-tensity maps from limited samples, to analyze stereo images, and to compute optical flow from dynamic images.[192] Work on vision architectures looked at various forms of newly developing parallel computer architectures for data processing and their impact on vision work.

SCORPIUS combined a number of vision-related technologies developed over the years with support from IPTO's image understanding program into another integrated system at the end of the 1980s. Hughes Research Laborato-ries and the Electro-Optical and Data Systems Group of Hughes designed and built SCORPIUS, an image-understanding system for photograph interpreta-tion, with support from IPTO and its successor ISTO. SCORPIUS relied on the ACRONYM vision system developed in IPTO-sponsored work at Stanford by Rodney Brooks. At the time (1986), ACRONYM was the most comprehensive and integrated vision system in existence. In addition, SCORPIUS included a graphics-based prediction method with a symbolic interpretation module that was the result of work with IPTO support at Brigham Young University and Virginia Polytechnic Institute. The system used the BBN Butterfly and the CMU Warp VLSI microprocessors, which had been developed with IPTO support. When operational, SCORPIUS provided the strategic intelligence community with the ability to automatically detect and classify objects of interest at known facilities (e.g., submarines at a naval facility). The DOD set up a testbed with an operational prototype of SCORPIUS in 1990.[193] While robotic systems still have a long way to go to equal human capabilities, the systems available in 1992 demonstrated that much has been learned about human seeing and motion and methods of reproducing them in machines.

Our brief history of AI demonstrates four important points about AI. First and foremost, AI was and is a very dynamic subfield in computing, and this dyna-mism has had a substantial impact on its development. AI's accomplishments came about because of the enthusiastic dedication of the research teams. They believed that they were on a "quest"—at least, some researchers consciously thought of it that way. In fact, their minds were so absorbed with the quest that they often represented their work with respect to its final objective, the "intelligent machine," and suggested that it could be accomplished soon, rather than focusing on the difficulties and the long time span needed for the ultimate accomplishment. This is not a phenomenon peculiar to AI; development of other large technological systems often has the same character. In today's engineering world, NASA's space station may be another example. Insofar as it has stimulated research in AI, the notion of a quest has opened research paths that have benefited the field greatly. When people outside the AI field test the objective by examining the intermediate results, they seem not to focus on the short-range results and instead point out that these results are insufficient to accomplish the ultimate objective. Of course, some of this arises from AI researchers' exaggerated promises about how soon that objective will be reached.

This distinction between the public perception of a quest and the internal development of a field of computer science and engineering is an important one for evaluating the interests of funders of the various research programs. The AI research questions posed in the early 1960s were general in character and intelligible in terms of the ultimate objective, and were framed in general computer science contexts, such as natural language processing and machine translation. This research not only was seen as contributing to the need for more computing capability and reliability but also was of interest to the DOD. As a result, support for AI research came from the DOD's IPTO as part of the office's effort to develop more capable, flexible, intelligent, and interactive computing systems, and did not simply constitute NSF-style support for computer science and engineering. In the beginning, IPTO did not see itself as stimulating the growth of an area, except insofar as the area was part of that group of research interests associated with the development of an improved way of computing. In the 1960s, the perspective of IPTO toward AI was different than that of other support agencies. But as AI questions became more defined through research and implementation, NSF and others—such as NIH and NASA—became more interested in the field. By the 1980s, only about half of the funding for AI research was coming from IPTO. By recognizing the integral relationship of AI research to questions of computing, IPTO greatly stimulated the growth of AI in important ways.

Second, research results in AI have had significant impacts on other areas of computer science and engineering. Software development technology is a very important example, and computer architectures that are especially appropriate for doing interactive program development such as Windows is another. Expert systems have been used in the general programming environment. Programs developed in AI for decreasing data access time by interleaving data in different areas of the computer memory are being used in general computer systems. Schemes for representation and semantics are being used in database design — object-oriented coding, object databases, and deductive databases. The many interactions between AI and graphics, which make them almost inseparable in many ways, have advanced developments in both areas. For example, both areas rely on edge detection, scene analysis, and segmentation.[194]

Third, researchers in the cognitive sciences — psychology, linguistics, anthropology, and neuroscience, have repeatedly found in AI useful techniques and models that apply to their research programs. Vision, natural language, semantics, and learning are important questions in psychology as well as in AI. Semantics, syntax, and mental models are some of the overlapping concerns of AI and linguistics. The representation of knowledge, and story telling, are of central concern in anthropology, and are important in AI for simulating human activity. Neuroscience, in attempting to discover the mechanisms whereby brains function, is analogous in its aims to the AI search for mechanisms with which to construct an artificial "brain."[195] The relation of AI researchers to researchers in these other fields varies. Sometimes the relation is close, as in the case of Simon and Newell in cognitive modeling (production systems) and David Marr in vision; sometimes the relation is only peripheral, such as the AI work on continuous speech and parsing and their use in linguists' work on natural language. In short, starting from a group of brief but deep questions about the nature of intelligent behavior, AI researchers have made significant progress in the development of systems with the capability to know, learn, and act. And they have done this in less than four decades. Along the way, their modeling and implementations have influenced other disciplinary areas as well.

Fourth, and most important to IPTO and DARPA: as each of the IPTO subprogram elements of the 1970s contributed to the IPTO AI research program's objective of providing more capable, flexible, intelligent, and interactive computer systems, the potential contribution from these research activities to the solution of DOD problems became clearer. Not long after the speech understanding study group's first report in 1971, Lukasik told Congress that computers were too limited because of "the necessity of communicating with them according to very rigid formats and the necessity of providing information in the form of precisely timed electrical impulses."[196] IPTO, he noted, believed that it would be very useful to be able to communicate with a computer that could "understand" speech. In his 1972 presentation to the Senate Appropriations Committee, Lukasik blended his description of the subprogram's research objectives with statements about those objectives' application to DOD problems. He emphasized that the research would give computers the ability to analyze simple visual scenes, verify the correctness of programs, prove theorems in mathematics, and discover new results in organic chemistry. He did not mention by name the results of particular projects, but these descriptors appear to refer to the early results from MIT's programs MACSYMA, SHRDLU, and SAINT; to the CMU problem solvers such as GPS; and to the Stanford work on DENDRAL.

In the 1980s, IPTO-sponsored testbeds began to demonstrate how the results of the office's programs could contribute to the overall command, control, and communications mission of the DOD. Packet radio, packet satellite, wide-area networks, image-processing systems such as ACRONYM and SCORPIUS, various expert systems, and sensing systems became candidates for incorporation into defense systems. An anxious world saw the power of some of these new applications during the Persian Gulf War in 1991. The IPTO program had come together, and computing R&D had become not only integral to DOD thinking, but essential to planning and execution of the DOD mission around the world.

CHAPTER SIX

Serving the Department of Defense and the Nation

The progression of world political events over the last fifty years, and the U.S. government's stimulation of and reactions to these events, changed significantly the defense context of the United States during the period, which included World War II and the Cold War. The changed defense context resulted in a fluctuating defense posture and repeated requests by the armed services for new, more elaborate defensive systems. Consistent with its mission, DARPA continued to focus on the technological frontier of weapons and defensive systems throughout this period, and played a role in the evolving computing capability of the armed forces. Within DARPA, IPTO continued to pursue its mission of R&D on the frontier of computing by maintaining a focus on computing systems for resource sharing; the development of new computers with greater capability, flexibility, and intelligence; and the integration of research and results. Exploring this focus in earlier chapters, we concentrated on the mission, the programs to fulfill it, and the results of those programs. We have related in the previous four chapters how in the 1960s IPTO-supported researchers contributed new systems and techniques which altered the DOD command and control mechanisms, and have described how in the 1970s these contributions began to appear in systems such as ARPANET, MILNET, and the various commercial time-sharing systems. Later programs continued to contribute in much the same way, as we will show below. Our primary purpose in this chapter is to evaluate the influence of IPTO on computing, including computing for the military. The computer, especially the time-shared, networked, smart computer of today, is now an integral part of military systems, as we saw vividly during the Persian Gulf War. The capabilities of these systems are either directly due to IPTO-supported programs, or are based on ground-breaking work supported by IPTO which was later exploited by commercial interests to develop the effective systems of today.

As IPTO was having this influence, the defense context adjusted to the ebb and flow of world events. In the period from about the mid-1970s to the present, the nature of IPTO's mission changed only subtly in response to new demands by the DOD. There were, however, changes in the way the programs to accomplish IPTO's mission were defined. Therefore, in this chapter, a secondary purpose is to present further examples of IPTO activities that illustrate how the office altered its programs to assist in the design and development of new computing elements to take into account continually changing emphases in defense systems.

We begin the first section of this chapter with a brief review of IPTO's focus on direct resource sharing on-line, by way of summarizing our earlier discussions and setting the stage for our full evaluation of IPTO's influence. But our main emphasis is on the IPTO program of the 1970s and 1980s. Discussions of a range of projects on computer architecture systems illustrate IPTO's even wider perspective, in the 1970s and 1980s, both on the development of components for computer systems and on the entire systems based on new, nonmainstream computing concepts. In the late 1970s and 1980s IPTO strengthened its ability to assist the DOD mission without fundamentally altering its approach to R&D funding, which was described in chapter 1. In the 1980s, as the DOD increased its R&D for future defense systems in a series of cross-department and agency activities such as the Strategic Defense Initiative, IPTO, too, without changing its management approach, reorganized its program and enlarged its activities to aid in this new DOD effort. The concept statement of the reorganized IPTO program was the Strategic Computing program, which not only expanded IPTO's purview in R&D for computing but also became a model for cross-office cooperation in DARPA.

The IPTO program process changed over time. In the 1970s and especially in the 1980s, IPTO, sometimes through the forceful initiative of its own directors and sometimes through the prodding of DARPA directors, instituted testbeds for the evaluation of prototype computer systems, enlarged its perspective on the kind of R&D that could contribute to the DOD's newly developing effort to improve its technology base, and in overall programming for military systems incorporated concerns about national technological capability generally. Little change in management and office procedures was needed to adapt to these new concerns and practices,[1] but changes in the way IPTO programs were defined anchored the office more solidly in the DOD. The elements of program formation which we described in chapter 1—the holding of meetings with the research community, the establishment of advisory groups, discussion with DOD leaders, and so on—remained aspects of the process. What changed was the relation among them. In the early years, the research community participated in strategy discussions about how to achieve IPTO's (i.e., the DOD's) overall objectives using computing, a practice initiated by Licklider and followed by his successors. Then the programs devised were "sold" to DARPA management. The definition of IPTO programs in the 1980s was guided more by discussions inside the DOD, or with other government groups such as the Federal Coordinating Council for Science, Engineering, and Technology of the Office of Science and Technology Policy, than by discussions outside the DOD with the research community. Discussions with the research community at that time tended to be more concerned with tactics to achieve governmentally selected goals for specific programs. In some ways, IPTO's 1980s approach to R&D support is similar to its networking program, and can be seen as the extension of this practice to virtually all of IPTO activity.

The direct impacts of IPTO support for R&D and the office's wider, more subtle influences on computing can be seen through an examination of military uses of specific IPTO program results and through its contributions to generic computing technology. Perhaps most important was the effect of IPTO support on the crafting of an infrastructure for computing R&D which had barely existed before IPTO. This infrastructure is most evident in the size and strength of computer education and training programs and the diaspora of graduates of these programs to all the R&D settings of the nation's computing enterprise. From all these contributions we conclude that IPTO's focus on the search for more capable, flexible, intelligent, and interactive computer systems led to radical changes in computing; and as we will show, these changes helped to redefine mainstream computing.

A Constant Mission in a Changing Context

Direct Resource Sharing On-Line

The SAGE air defense system of the 1950s pointed the way to a more interactive computing system. So it is little wonder that looking forward in the early 1960s, Eugene Fubini and Jack Ruina saw advances in computing as one avenue of improvement in command and control systems. DOD analyses indicated that more research in computing was needed before computer systems could be more effectively incorporated into defense systems. To run the computing program in DARPA, Fubini and Ruina needed a person who possessed two abilities: the ability to develop research programs, and the ability to look ahead to the integration of the results into defense systems: in short, someone to envision radical future technology and prepare for it. They needed a person who possessed experience in the academic world and in the systems development world of the DOD. J.C.R. Licklider had all the experience needed. He had been involved in defense systems development during World War II, had been an able member of the faculties of MIT and Harvard, had served as a consultant to the DOD, and had most recently worked for one of the Cambridge area's important systems integrators, Bolt Beranek and Newman. Best of all, Licklider showed Fubini and Ruina that he had a vision about what needed to be done and a plan for bringing the vision to reality.

Picking up on research in the Cambridge community, Licklider astutely chose time sharing to illustrate a way to increase the productivity of a group at one location. Although time sharing was under development by several groups around the United States, the influx of a large amount of IPTO funds made it possible to implement the concept at a selected number of places around the country, and eventually to influence implementation outside the country as well. By establishing the IPTO program in this way, Licklider instituted as one of the IPTO program's basic elements the concept of resource sharing without intermediaries such as cards or tape — a concept that was to govern IPTO's activities throughout its history. Demonstrating how to share resources in new ways first at one site, and then across many sites, was an important step in learning how to share resources across many military sites linked by a command and control system.

The IPTO programs under Licklider and his successors stimulated resource sharing in several ways. First, the implementation of time sharing at one site made it possible for researchers to interact with the computer essentially in real time. Instead of transferring programs and files by paper or magnetic tape, researchers could obtain a needed file or program by accessing the files and programs of another person, thereby increasing their own productivity. Maintaining at each site a library of files from which a researcher could, as needed, transfer files and programs to his or her own terminal made research more productive. Second, IPTO set a goal of facilitating sharing among all IPTO-sponsored researchers, regardless of what computer they employed at their site. Robert Taylor initiated and Lawrence Roberts led the design and development of wide-area networking, specifically the ARPANET, which made resource sharing possible among all IPTO sites. And third, Robert Kahn and Vinton Cerf widened this sharing beyond ARPANET sites by making it possible to share across many networks.

Reflecting his own research interests and those of the Cambridge computing community, Licklider believed that another way to enhance the command and control mission was through the use of a "smart" machine that would serve as an assistant to a person. Licklider's concept of the smart machine has not been achieved yet, but the work that IPTO stimulated in the field of AI has brought us closer to an implementation of the concept. In AI, we see that the objective of an intelligent machine required the investigation of very many questions about how humans solve problems; how speech is understood; how sense data is used in motion; and how to conceptualize and design the component parts of an intelligent machine, and ultimately build a complete system that performs actions that can be said to be humanlike. An immensely tall order, but IPTO persevered in the search for such a machine, in spite of much criticism along the way. To date, researchers have introduced many "smart" capabilities into computer systems; and of course, many of these have found their way into command and control systems.

As we have pointed out throughout this book, the IPTO program took its direction largely from the interests of the leaders of the IPTO office. Licklider was already concerned with development of a smart machine when he arrived at IPTO — witness his own earlier research on human-machine interaction, speech recognition by machine, and time sharing. In the larger community, he had rubbed shoulders with scientists, engineers, defense system developers, and management people, who all were seeking ways to integrate the computer into larger, more complex systems — SAGE, artificial intelligence systems, production control systems, and so on. Many of these people saw the computer as a central element in each of these systems. It is no surprise, then, that Licklider carried to IPTO the idea that computers should be an integral element in any defense system. After leaving IPTO, he continued his activities in these areas, first at IBM, and then returning to MIT to direct Project MAC.

It is not surprising that the DOD would turn to the Cambridge community to find leaders such as Licklider for the defense R&D efforts. Since the 1920s, MIT project leaders and other researchers in the Cambridge area had been involved with various defense projects — in communications, aviation, computing, and air defense — the list goes on and on. MIT, for example, had utilized many of these projects as central elements in its education program. Many of the graduate students trained in these programs worked in environments such as the Lincoln and Instrumentation laboratories, where leaders placed an emphasis on the integration of basic research with applied research, and then on applications. Sutherland and Roberts both received their Ph.D. degrees from MIT in 1963 and did their dissertation research at Lincoln Laboratory. Both contributed in fundamental ways to graphics systems that were both basic and applied, and both recognized the importance of improving computer systems and integrating the results of research from separate projects as quickly as possible. After IPTO, Sutherland returned to research in graphics at Harvard and the University of Utah, later combining this with an interest in VLSI at the California Institute of Technology. Roberts, for his part, left IPTO to pursue the commercial exploitation of networking in several companies, some of which he founded.

Robert Taylor came from a different educational background, one that was unusual for IPTO personnel during the office's early years. Nevertheless, in his experience in human-machine systems and his knowledge of the importance of integrating research results for use in larger, complex systems, he resembled his co-workers from Cambridge, although his own experience had been gained in the nation's aerospace enterprise. Subsequently, Taylor became renowned for his leadership of the personal computer project Alto at Xerox's Palo Alto Research Center, in which the Ethernet local area network and a variety of other new concepts were developed.

Ivan Sutherland, and later Lawrence Roberts, who made some of their most important contributions to computing while at Lincoln Laboratory before joining DARPA, promoted programs that also enhanced the capability of computing systems. They believed that people used command and control systems more effectively when data was presented in images, especially moving images. Sutherland stimulated work on input and display components for graphics, on simulators using computer graphics, and on programming for better images. After he became IPTO director, Roberts pushed this work further and added to it programs in image understanding and picture processing to improve the analysis of the incoming data and the resulting images. By this time, many graphics projects were closely related to problems in AI and were supported by both programs. Here again, it was Sutherland's and Roberts's experience in the Lincoln environment that led them to insist on research that could be integrated into systems.

These heads of IPTO and their program managers served as "gatekeepers" (in the sense in which Nathan Rosenberg and Edward Steinmueller use the term) for the new technology.[2] Rosenberg and Steinmueller argue that gatekeepers develop an understanding of their organization's needs and perhaps have ideas for technological solutions that are feasible. The gatekeepers provide the links between researchers, designers, and users. IPTO directors and program managers provided the link between the R&D community and the military through use of the testbed. All the directors of IPTO served in succession as the DOD's gatekeepers for the integration of new computing research results into defense systems. This is not to say that they were the only gatekeepers for the DOD; many other people served a similar purpose in other DOD agencies, in systems integration groups such as the MITRE and Rand Corporations, and in industry. What we do assert is that IPTO personnel served a special role with respect to R&D in computing for command and control systems.

As impressive and significant as these developments were, they were only the beginning of IPTO's contributions. To see IPTO's history in its full light, we need to examine some other system developments that IPTO sponsored—developments that opened R&D in parallel computing, and attended to neglected areas in VLSI design and construction—and the assembling of these elements into a new, larger vision for IPTO, the DOD, and the high-technology standing of the nation.

Developments in Computer Architecture

IPTO entered the computing architecture area almost as a digression. In December 1964, after Wesley Clark left Lincoln Laboratory and took a position at Washington University in Saint Louis, he received a contract from Sutherland to investigate the possibility of constructing a computer using standard modules. The concept of constructing computers using predesigned and standardized modules, analogous to many military contracts for the design of field equipment, involved the design of a set of self-timed interconnectable modules—registers, adders, and the like—which could be used as electrically and logically distinct building blocks. These registers, adders, and memories could be used to construct complex computers. The idea was to allow an experimenter to construct a computer with an array of modules appropriate to the problem under consideration. The potential military application of this concept is apparent, but such a system was never completed.

In 1965, Sutherland went further. He issued a contract in the area of parallel processing, for the design and construction of the ILLIAC IV. According to Sutherland, the ILLIAC IV project was "the biggest thing I started."[3] After urging by Sutherland, researchers led by Daniel L. Slotnick at the University of Illinois, which had a long history of computer system development, proposed an experimental machine of sixty-four processors with common control logic coupled together to work simultaneously on a single problem.[4] The University of Illinois group would design the computer and then engage one of the major computer manufacturers to build it. According to Sutherland, IPTO at that time had two interests in this project: the development of a parallel architecture, and the stimulation of industry to produce new, more effective components. To evaluate the proposal with respect to these objectives, Sutherland sought advice from Sidney Fernbach, Allen Newell, and three or four others.[5] The group met with Charles Herzfeld, the recently appointed director

Fig. 7. A portion of the ILLIAC IV in about 1972 — control unit on the left, processor unit and extender on the right. Photograph courtesy of Charles Babbage Institute, University of Minnesota.

of DARPA, and concluded that the project was worthy of DARPA support. As we noted in chapter 1, IPTO spent more than $29 million on the project between 1966 and 1971. The selection of contractors for the project began as Sutherland's tenure in IPTO came to an end. Robert Taylor, Sutherland's successor, took over the work. After a major competition involving proposal assessments by people from the University of Illinois and IPTO, the construction contract went to the Burroughs Corporation. Taylor awarded other, related contracts: for the design and implementation of a FORTRAN compiler for ILLIAC IV (to Applied Data Research Corporation); for additional hardware and software to link ILLIAC IV and a trillion-bit memory with the ARPANET (to the Computer Corporation of America); for an interactive "front end" processor (to BBN); and for algorithm development for efficient use of the ILLIAC IV on the network (to the University of Illinois). As one of his duties, Roberts, as program manager and later director, oversaw the ILLIAC IV project and participated in important project decisions. (A portion of the operational ILLIAC IV can be seen in fig. 7.)

DARPA and some of the armed services research organizations planned to employ ILLIAC IV for the investigation of problems in climate dynamics, anti-submarine warfare, and ballistic missile defense.[6] The software developments for ILLIAC IV were meant to make the ILLIAC IV system useful to potential DOD users. One of the questions about ILLIAC IV at this time concerned its efficacy in weather prediction. Lukasik asserted that the science of weather was understood and that meteorologists knew how weather prediction calculations should be done. According to him, the problem was a shortage of computer capability. The ILLIAC IV was supposed to help to solve this problem; IPTO's support was intended to show that it was possible to complete this type of calculation quickly enough for the prediction to be useful.[7]

Three other potential users for ILLIAC IV were the Army Ballistic Missile Defense Agency, which saw that computer as one option for solution of the multiple target problem; the Prime Argus Project,[8] which saw it as a tool with which to approach a climate modification problem; and IPTO itself, which was interested in further research in cryptanalysis, simulation of fluid flow, seismic array processing, economics and logistics using the parallel-processing capabilities of ILLIAC IV.[9] In fact, ILLIAC IV was used for all these areas and more in the late 1970s.[10]

Even though some people criticized the ILLIAC IV project, it was just the type of project IPTO was organized to pursue—R&D on the frontier. ILLIAC IV was the world's first large-scale-array computer and a new development in parallel-processing architecture. The criticism may have been inevitable because in this project IPTO attempted to expand the boundaries of computer science in many directions simultaneously. According to Lukasik, IPTO and DARPA "were pushing machine architecture, software for parallel machines . . . medium scale integrated circuits—and then the application of that to a variety of important defense problems" simultaneously with the ILLIAC IV project.[11] The push on all these fronts simultaneously led to several shortcomings in ILLIAC IV.

The main problem with ILLIAC IV was in the medium-scale integrated circuits that were called ECL (emitter coupled logic) circuits. The ECL circuits were designed to have twenty electronic logic elements called gates per chip. Texas Instruments received the contract to develop these chips. As development progressed, however, it became clear that the chips would not work because the electronic noise in the chips was too high. Texas Instruments asked IPTO for an additional year to develop the chips, but instead of granting the request, the researchers at the University of Illinois made the decision to scale down the circuits to only seven gates per chip. This change altered the nature of the entire project. The smaller chips caused short circuits between leads and external circuit pins. These short circuits were very subtle and only became apparent over a long time, and to detect them, entirely new testing procedures had to be developed. Moreover, the new chips proved to be very sensitive to changes in humidity, and it was difficult to maintain proper functioning. Ironically, the smaller ECL chips took up more room in the computer. Consequently, the computer had no room for the planned thin-film memory. As a result, the ILLIAC IV employed all-semiconductor memories.

It was impossible for humans to design the incredibly complicated circuit boards for ILLIAC IV. Eventually researchers designed the circuit boards with computer-assisted design systems, but the designs contained flaws that were only uncovered later during day-to-day operation of the computer. Literally thousands of manufacturing flaws were subsequently discovered in the printed circuit boards and connectors, but these flaws only came to light after IPTO had exhibited significant interest in modular design of computers. In the early 1970s, as more was learned about the shortcomings of the ILLIAC IV design and production process, Roberts, then the director of IPTO, and several program managers in IPTO focused another program on shortcomings in circuit design generally. The later design of component circuitry using VLSI schemes, some developed with IPTO support, overcame many of the problems encountered in the design processes of ILLIAC IV and other machines.

Because of these shortcomings, it took several years of experimentation in the 1970s before ILLIAC IV became operational on any regular basis. Many of the early problems with ILLIAC IV experienced by contractors using ARPANET connections to the machine were attributed to software difficulties. Nevertheless, by 1977 ILLIAC IV was available, and IPTO and NASA began to fund application studies. In fact, the DOD and other federal agencies found ILLIAC IV very useful with some of their problems. A group at the University of Illinois developed an editor system that NASA, the U.S. Department of Agriculture, and the U.S. Geological Survey used to analyze Landsat images. NASA developed computational tools to simulate fluid flows around many new airplane and space shuttle parts. A three-dimensional seismic code called TRES was developed to represent earthquake effects in three dimensions. The Naval Surface Weapons Center funded NASA's implementation of a program to form matrices of preprocessed satellite observation data, an essential step in determining accurate geodetic parameters for satellite tracking. These are only some of the problems explored on ILLIAC IV. [12]

In spite of the disagreements about the level of success in the design of ILLIAC IV, the introduction of the microprocessor by Intel in 1971 showed that integrated circuits (ICs) on a chip were the wave of the future for computer design. The semiconductor companies focused their attention on those integrated circuits that were in demand in the commercial market. High-end, very fast chips, and chips for use in specialized equipment such as command and control systems, were outside their interest. Here was an ideal problem for IPTO and DARPA.

IPTO began to attend to the development of VLSI architectures based on ICs in the mid-1970s. The IPTO program focused on VLSI circuits with a large number of gates and components with dimensions of less than one micron (one-millionth of a meter), in order to advance the state of the art of ICs. VLSI projects were to provide the infrastructure for the design and construction of several new parallel architectures. With IPTO support, several research laboratories organized research projects to design and build a new generation of computer architectures suitable for implementation in VLSI; VLSI components for peripheral equipment such as graphics display devices; and fast-turnaround facilities for production of research quantities of new IC designs. We describe some of these efforts below. These efforts resulted in many new designs and showed IPTO's ability to organize projects that integrated results effectively and quickly.

The VLSI component of the Strategic Computing program led to a number of significant developments, including various VLSI architectures such as reduced instruction set (RISC) computers at the University of California at Berkeley, and million-instructions-per-second (MIPS) computers at Stanford University. Other advances were silicon compilers, computer-aided design tools, and advanced workstations, such as the Sun Microsystems and Silicon Graphics products.[13] These will be described below in the section on the Strategic Computing program.

Over the two-and-a-half decades after 1965, IPTO and its successor ISTO concentrated on systems such as the ILLIAC IV in an attempt to dramatically improve the speed of computers and increase the capability of computer systems to solve the large problems of interest to the DOD. IPTO stimulated research on the incorporation of circuit designs that contained software to increase the flexibility of the machines, to increase the speed of certain functions (e.g., graphics computations), and to make the computer systems smarter, that is, able to handle the large computational load inherent in AI research. IPTO directors and program managers designed programs and awarded contracts for research using these systems as soon as they were available. In a sense, these uses were part of the larger concern in IPTO for the integration of results as soon as they were available.

We keep insisting that integration was an important feature of the IPTO approach. However, the programs of the 1960s—time sharing, networking, graphics, and AI—illustrate this less clearly than do the later programs of the 1970s and 1980s. Once time sharing was an integral part of computing, the network was being implemented, and AI problems were reaching better definition, IPTO began to shift its attention to other systems integration concerns. The office used two different kinds of integration in its approach to research and development: (1) the integration of results across projects and in testbeds, and (2) the funding of projects that facilitated integration.

The Integration of Results across Projects

IPTO directors and managers pursued an approach to program design and execution which integrated results across projects. As we pointed out earlier, IPTO encouraged researchers to share resources by assembling researchers in frequent meetings, and by making frequent site visits with visiting groups that included researchers from other IPTO-supported sites. The integrative approach can be seen in, for example, the time-sharing projects on the East and West Coasts, the hand-eye coordination projects in AI, the speech-understanding research in AI, the use of teams in the network working groups to develop the requirements for the packet-switched network, the design and funding of the intelligent systems program designed in the 1970s under Roberts, the various vision projects for picture processing and image understanding, and the use of new architectures. IPTO also, as illustrated many times in the preceding chapters, pursued larger integration by incorporating new computer systems into testbeds to illustrate how the new designs would function in a defense system. Those examples did not lend themselves well to illustration of the testbed concept, however, although several examples of testbeds were given. Here we need to add a discussion of testbeds, because they were one of the means by which IPTO increased its influence on the development of military systems. The various directors of IPTO effectively utilized testbeds as a result of their experience at organizations—such as Lincoln Laboratory and BBN—that functioned as integrators of new technology into defense systems.

As systems became more complex and sophisticated and more difficult to incorporate into command and control systems, it became necessary to introduce a more traditional form of defense system testing to convince the various DOD agencies that the new IPTO systems could be an effective adjunct to defense systems. This form was the testbed, in which an experimental system was constructed for use in a simulated defense setting. In IPTO's first ten or so years, it was not always necessary to use a testbed to demonstrate to the military the usefulness of a new development: because the military was a heavy user of computing equipment, military personnel could appreciate the new advances made by IPTO contractors. Since SAGE in the 1950s, military groups had seen the value of computing in command and control, and the incorporation of time sharing into computer systems was an evolutionary step in the real-time transaction systems of the SAGE type. Moreover, because commercial companies began to include time sharing as a feature of computer systems that the DOD purchased, IPTO did not need to go to great lengths to integrate time sharing into the DOD. Networking, however, was not received as favorably by the DOD.

DOD agencies, and especially the military units, already had a number of operational message-passing systems when ARPANET was implemented. Besides the usual technical and training concerns that dampened enthusiasm for the new network, there were a number of political impediments to its acceptance. For example, the new network would make it easier for subordinates to send messages to other agencies without the approval of commanding officers, possibly circumventing the military's chain of command. More significant, perhaps, was the inability of the ARPANET, in the form in which it was used by IPTO-sponsored researchers, to function in battle situations. The military used communications facilities in the field more compatible with the overall message-passing system. Thus, IPTO used a testbed to demonstrate to the military that the new network would enhance military message-passing activities rather than hurt them. Another early testbed, the Advanced Command and Control Architectural Testbed, illustrated that ARPANET could also be used for database management. IPTO demonstrated how this could be done by combining the ADEPT time-sharing system with the TDMS management system. In this case, SDC organized the testbed and demonstrated it not only to DOD groups but to industry groups as well.

Testbeds became more significant in IPTO's program when the people at IPTO's helm understood that they were useful in obtaining military acceptance of new systems. David Russell's experience in the military, especially in Vietnam, helped him push for the integration of packet radio and packet satellite systems into military field activities. Perhaps the best-known testbeds of the 1970s were the packet radio and packet satellite projects.

The use of testbeds became more prevalent in IPTO after 1970. This reflected IPTO's response to demands by the DOD and Congress that the results of IPTO research projects be more relevant to the defense mission. In the late 1970s, George Heilmeier restated the mission of DARPA to reflect a more integrated approach to technology development for DOD, making greater use of the test-bed concept as a testing ground for what works and what fails in a defense setting. As a result, by the 1980s testbeds were the prevailing element in IPTO programs. A typical IPTO approach to testbeds in the 1970s was the VLSI program. The results of research on AI systems, with the later addition of speech capability, contributed to the use of testbeds in the Strategic Computing program of the 1980s, in which three large-scale testbeds, involving many contractors, played an important part. (These testbeds demonstrated the Automated Land Vehicle, the Battle Management program, and the Pilot's Associate.) In SC, the origins of which are described below, we see not only the design around testbeds, but greater integration of all IPTO programs related to SC.

Systems for the Facilitation of Integration

In the mid-1970s, IPTO sponsored research in integrated circuits at the California Institute of Technology under the direction of Carver Mead. With this support, Mead was investigating simple parallel processors involving one-, two-, and three-dimensional interconnection patterns in microprocessors.[14] Ivan Sutherland left Utah to join Caltech in 1976 and participated in this research.[15] In addition, IPTO supported integrated circuit research projects at other institutions for the design of circuits used in graphics systems and AI activities. As it typically did when entering new research areas, IPTO funded the early research on fast integrated circuits out of available funds from other subprograms and arranged for an assessment of the state of the art to provide recommendations about needs, in order to develop a more structured program. Mead, Sutherland, and Thomas Everhart conducted the assessment through the Rand Corporation. Their 1976 report was very optimistic about the future.

> There is every reason to believe that the integrated circuit revolution has run only half its course; the change in complexity of four to five orders of magnitude that has taken place during the past 15 years appears to be only the first half of a potential eight-order-of-magnitude development. There seem to be no fundamental obstacles to 10^7 to 10^8 device integrated circuits.[16]

However, the authors were concerned that the demand for commercial IC designs caused the IC manufacturers to limit the focus of their R&D. The demand for more functions within smaller packages at reduced cost drove the commercial activities of these firms in such a way that continued R&D on innovative designs became a significant burden.[17] Manufacturers oriented new designs toward a mass commercial market in computer production, rather than toward new computer architectures and systems. In the view of Sutherland, Mead, and Everhart, the manufacturers were likely to make incremental advances slowly. These advances would be inadequate to the demand for faster and more powerful circuits for research and defense systems. The authors also worried that manufacturers in other countries might surpass the capabilities of U.S. manufacturers.

Within IPTO, Kahn also worried about this problem. He believed that industry would try for chips with smaller semiconductor layerings, trying methods such as electron beams, x rays, or focused ion-beams. But these developments would not be small enough for IPTO. Kahn wanted researchers to produce chips with semiconductor patterns less than one micron wide to permit the design of single chips with up to 25 million gates. The basic research in scaling devices to submicron dimensions would permit the development of systems with smaller size, weight, and power requirements, as needed for use on space missions or in weapons systems.[18] Kahn was fully cognizant of the problems associated with chips this dense. Nevertheless, he knew that the chips would have greater speeds and capacities and would meet the need for new architectures for IPTO-sponsored research projects.[19]

One source of new chip designs could be IPTO-supported university programs, but fabrication facilities were expensive. The small university demand for the production of experimental chips, and the large expense of the facilities, precluded universities from acquiring such facilities in the same way they acquired computers or electron microscopes. Thus, university researchers with new circuit designs approached the commercial firms to make the experimental chips for them. Researchers obtained access to a company fabrication facility through negotiations with company personnel and through personal friendships. But they might have to wait as much as a year until the company could fit their task into its own program.

After completing the overall design, the major steps in developing an integrated circuit were logic design and layout; making a template (a mask) to guide the laying down of the semiconductor material onto small, thin backgrounds (wafers) composed of many chips; and testing and debugging. In an informal survey by Stanford University, it was found that logic design and layout took between three and ten months; mask making and wafer fabrication, between five and twelve weeks; and testing and debugging, between one week and three months. When more complex chips were designed, these times lengthened. A major goal of the IPTO VLSI program was "to significantly reduce the time for each of the steps" listed above.[20] As fabrication facilities and new chip designs increased in complexity, outsiders' access to chip production facilities became more difficult and costly. DARPA and IPTO became concerned about this decreasing access and began to consider a very large scale integration (VLSI) program, a cooperative program between DARPA's Defense Sciences Office and IPTO. The university research community required faster access to chip fabrication in order to support progress in integrated circuit and computer developments. In turn, faster access required the development of a flexible IC-fabrication system with fast turnaround times. IPTO's plan was to facilitate the design and fabrication of ICs in order to enhance integration of R&D results in other projects.

IPTO funded research on new and simpler IC design schemes, research on design techniques for surmounting the one-micron barrier, and implementation of procedures to reduce turnaround times for the fabrication of new university IC designs. In a joint project, Mead's Caltech group and Lynn Conway from Xerox's Palo Alto Research Center investigated new schemes for vendor-independent design and fast turnaround. Conway wanted "to create, document, and debug a simple, complete, consistent method for digital system design."[21] In typical IPTO style, Mead and Conway's results at Caltech were transferred quickly to MIT and shortly thereafter to other universities, including CMU and Stanford. In 1980 Mead and Conway published a book entitled *Introduction to VLSI Systems,* which essentially told people how to do VLSI design in a way that did not require detailed knowledge of semiconductor physics and was vendor independent, so researchers would not have to sign proprietary agreements with a vendor in order to do VLSI design.[22]

In its research program to achieve widths below one micron and fast-turnaround schemes, IPTO focused on design methodologies, innovative architectures, and design tools to use VLSI technology effectively. The office organized demonstrations of multiple-chip VLSI projects in which eighty-two different designs were combined into two mask sets and then fabricated. The demonstrations used the ARPANET. Two of these projects acquired the names "MPC 78" and "MPC 79." We need only examine one of the projects to appreciate the flavor of the work. The MPC 79 demonstration was labor intensive and required special support from Xerox PARC, a company called Micro Mask, and Hewlett Packard, which fabricated the chips. The total time from completed design to fabricated chips was twenty-nine days, whereas the standard procedures described above took several months.[23] On the basis of this result, IPTO decided that it was now ready for a larger-scale project to demonstrate the feasibility both of the fast-turnaround concept and of the design methodology in which it was embedded.

When Kahn became director of IPTO, he turned his concern about sub-micron ICs into action, and increased the activity of the VLSI program. He issued a contract to Stanford University to conduct research at the submicron level and to design and test a fast-turnaround facility. The Stanford proposal had two major components. First, Stanford researchers pursued a basic effort on new circuits, devices, materials, and processes suited to ultra-large-scale integration. The goal was to achieve the lowest submicron sizes of gates, and equivalent gate densities well in excess of 1 million gates per chip.[24] Simulta-neously, the project explored the fundamental limits that were encountered as circuit feature sizes were scaled down to submicron dimensions. Second, the research group established a fast-turnaround testbed for its own use, with an initial capability to produce chips with feature sizes of five microns, to reach one micron within two years. Stanford devoted a portion of the research to the improvement and enhancement of the fast-turnaround facility itself, but this facility was available only to researchers at Stanford. Since Kahn and others in IPTO considered the fast-turnaround facility to be integral to the design methodology being developed in the VLSI program, the office hoped that this project would establish a much broader, and in some ways new, base of inte-grated circuit designers and "significantly aid in reducing the development time of custom integrated circuits for DOD application."[25]

The Stanford group's fast-turnaround facility was for its own research pur-poses. To provide a facility for other researchers, IPTO issued a contract to the Information Sciences Institute for the Metal Oxide Semiconductor Implemen-tation System (MOSIS). MOSIS became central to the development of a new range of hardware architectures and the education of students in this critical area of ICs throughout the IPTO-supported community. New hardware system architectures based on VLSI systems had previously been explored through paper studies and small breadboards, a time-honored practice in system de-sign. MOSIS permitted experimentation with complicated, risky, or speculative designs.[26] The MOSIS task at ISI was fourfold. First, ISI supported very general design rules, initially based on the Mead-Conway results in MPC 79, to make it possible to use any fabrication line available from various United States ven-dors. Second, ISI distributed these design rules to members of IPTO's VLSI research community to use for their designs. Third, MOSIS coordinated regu-larly scheduled fabrication runs to satisfy the community's chip-prototyping needs. Fourth, ISI arranged special small-volume production runs. Initially, ISI expected the customer base to be university researchers with a need for the fabrication of circuits designed in class projects or research. Once this process was effective, ISI hoped to attract other users as well.[27]

MOSIS operated as a central brokering facility. Designs arrived at MOSIS through the ARPANET. Those employing a given semiconductor material and chip size would all be scheduled by ISI for a particular time. Semiconductor companies were under contract to do the fabrication, and MOSIS selected the company with the best delivery schedule. The biggest time saver of this system was the elimination of the long negotiation stage with the company. Within four years MOSIS had used eleven vendors and processed 1,777 projects from eighty-eight organizations. ISI developed a manual that enabled users to order and receive a completed integrated circuit or PC board within three weeks.[28] Prior to MOSIS, all the chips on a wafer had to be the same design. Two major advantages of MOSIS were that up to sixty different designs could share a wafer, and that each design could be tested by the manufacturer before delivery without knowing the details about how the chip worked. Through this facilitating mechanism, IPTO researchers were now in a better position to concentrate on R&D for computing systems, since they could now design needed components, receive fabricated components quickly, and proceed to the defense system R&D that was the contractor's ultimate objective.

The Strategic Computing Program

The arrival of the Reagan administration, with its concern for the defense posture of the United States, led to a reevaluation of defense system needs, as noted above.[29] This reevaluation quickly focused on command and control systems, among others, and IPTO became involved in the discussions and planning. Discussions in 1981 between Cooper and Kahn led to the formulation of DARPA's Strategic Computing program.[30] Toward the end of the summer of 1982, and before the full achievements of the VLSI program were available, IPTO director Robert Kahn prepared a statement for a new defense program in computing, which would capitalize especially on the recent developments in VLSI and artificial intelligence. Kahn called attention to the then-recent Falklands and Middle East conflicts, stating that "the crucial importance of computers in the battlefield can no longer be disputed. In the future, battles will be won or lost on the ability of sophisticated microchips to outperform the opponent's microchips." He noted the semiconductor industry's belief that the U.S. lead in semiconductors had vanished because of progress by the Japanese. He went on to cite the AI emphasis in the Japanese Fifth Generation Project, and asserted that the AI work on which this Japanese project was based was the product of U.S. researchers. He asserted that "in the view of many experts, this field [AI and supercomputing] represents the last bastion of U.S. leadership in the computer field."[31] To avoid ceding this area to the Japanese, Kahn proposed a national plan to retain technological leadership in the United States through a transfer strategy in which ideas cultivated in universities could be developed in industry.

Kahn placed four key areas in this new program, the Strategic Computing program: microelectronics (based on the ongoing VLSI program), supercomputers, generic applications, and specific defense applications.[32] The key objective of SC was to take artificial intelligence technology and create an industrial base for it, to create multiprocessor technologies that could run AI programs a thousand times faster, and to create speech-understanding capability that would run on the new multiprocessor.[33] A grander IPTO vision for computing was emerging, and this time IPTO would take industry directly into its orbit. In Licklider's day, the IPTO vision was the aspiration to change computer science and defense systems. Basing their view on the achievements in IPTO programs, DARPA director Robert Cooper, and Kahn and his associates in IPTO, believed that they could effect a change in national computing capability. A decade of program transition ended with a new IPTO vision built on the model of the old vision (see table 7). By 1986, when the program was a reality, DARPA officials were claiming that SC "is, in its broadest sense, a program to develop an extensive U.S. base of computing and machine intelligence technologies to enhance our national security and increase our economic strength."[34]

As we noted, Kahn built SC on the need for faster, larger computer systems based on the use of parallelism in computation, and on advanced VLSI technology for use with AI programs that would become the basis for new intelligent defense systems—a convergence of all IPTO programs. Technology for networks and technology for designing complex VLSI systems were to be fundamental parts of the computing system with which these defense systems would be built and deployed. Problem-solving expert systems, advanced vision systems, and programs using speech and locomotion would provide the initial basis for new defense systems capabilities. In an effort that involved other DARPA offices, IPTO established three applications—grander testbeds—for the exploratory stage of this program. The testbeds were the Autonomous Land Vehicle, the Pilot's Associate, and the Battle Management program. By 1990, these systems had achieved various amounts of limited success.

Kahn proposed expanding IPTO's VLSI activity as an undergirding for the SC program. He expected to capitalize on the wealth of opportunities he saw in electronic and electro-optical materials, materials processing, and device concepts to provide entirely new capabilities for, and increase the performance and reliability of, computation-intensive military systems. To supplement the

Table 7 IPTO Program Areas and Specific Emphases, Fiscal Year 1985

Intelligent systems	Survivable networks
Image understanding	Suran (survivable networks)
Basic AI research	Multiple satellite network
Knowledge-base processing	Network security
Expert systems	VLSI fast turnaround
AI applications	Foundry development
Distributed problem solving	Foundry automation
Modernization	Integrated packet nets
Robotics technology	European cooperation
Support software	Wideband net
Computer resources	Network integration
ADDCOMPE (Army/DARPA Distributed Communications and Processing Equipment)	Distributed sensor nets
	Software systems
Automation technology	Distributed C^3 software
VLSI architecture and design	Programming environments
	Distributed computing
New generation software	
Visual programming	Strategic computing
AI-based software	Machine architecture/software
	Generic AI software
System and network concepts	Infrastructure
User interface technology	Applications
Advanced system concepts	
Adaptive networks	Other
Strategic C^3 experiment	

Source: Compiled from "Program Overview," in "FY86 President's Budget Review: Information Processing Techniques Office, Defense Advanced Research Projects Agency," Sept. 1984.

submicron work already under way, he proposed using research on maskless and selective-area ion, laser, and electron beam processing with dimensional control down to 100 angstroms, and three-dimensional microcircuit material systems that could eliminate interconnection limitations.

Kahn's second area was high-performance supercomputers. He designed the enhanced VLSI activity to lead the way by integrating architecturally new designs to decrease the distance between data processing elements, leading to faster speeds and providing more processing and memory space. Kahn wanted support for research in new system architectures to continue. He cited ongoing parallel-processing projects, such as BBN's work on the Butterfly machine and CMU's work on Cm*, as starting points for future developments. Such machines would require the use of automatic design aids, software tools for design, analysis and validation, and color graphic displays for user interfaces — all research areas in computing in which IPTO had a long history of interest and accomplishment.

The third area that Kahn proposed was also one already addressed in ongoing programs, although the programs had been small in scale up to this time. Here, Kahn was interested in generic technologies based on AI, notably speech recognition, image understanding, knowledge base development, natural language processing, and expert systems. He noted that "no effort has been made to develop technology industrially or even to integrate all the pieces at the research level,"[35] and he proposed to involve industry in a wide-scale effort to develop systems incorporating AI techniques. He listed examples of generic technologies in three categories, indicating that work in later categories would be loosely dependent on development in earlier ones.

Category I
Distributed Data Bases
Multimedia Message Handling
Natural Language Front Ends [preprocessing computer]
Display Management Systems
Common Information Services
Category II
Languages and development tools for Expert Systems
Speech Understanding
Text Comprehension
Knowledge Representation Systems
Natural Language Generation Systems
Category III
Image Understanding
Interpretation Analysis
Planning Aids
Knowledge Acquisition Systems
Reasoning and Explanation Capabilities
Information Presentation Systems[36]

We can see many subprograms of long standing in this list. Most of the research topics in these categories were present in the budget review document prepared in September 1984 for inclusion in the FY 1986 budget request to the Office of Management and Budget.[37]

Kahn argued in the fourth major area of his proposal that development in generic AI technologies would open the door to many new specific applications, many of them for the military: "Distributed data bases can be populated with specific knowledge about missile performance, tactics, engagement plans, etc. . . . The main challenge [in C² applications] will be to utilize AI technology to provide computer based capabilities that model human capabilities."[38] From this concern emerged the array of major applications of SC — the Pilot's Associate, the Autonomous Land Vehicle, and the Battle Management program — to integrate the new hardware and software systems and AI techniques. A detailed budget plan accompanied Kahn's proposal. The proposed DARPA budget for SC, subsequently shared by several offices, ranged from $15 million in FY 1983 to $225 million in FY 1988. The actual budgets came close to these numbers.

The congressionally funded program had four elements as well, which incorporated almost all of the features in Kahn's proposal. The four program elements were machine architecture and software, AI software, infrastructure, and applications. Emphasis was placed on artificial intelligence, especially speech recognition, vision, natural language processing, and expert systems; multiprocessor system architecture; and microelectronics, optical electronics, packaging, and interconnection technology.[39]

The DARPA plan for SC was to continue programs based on AI developments which would result in intelligent functional capabilities in machines, hardware and software system architectures, and advanced microelectronics. According to Kahn, "this [i.e., smart robot weapons that combine artificial intelligence and high-powered computing] is a very sexy area to the military, because you can imagine all kinds of neat, interesting things you could send off on their own little missions around the world or even in local combat."[40] The programs and their execution would continue to depend on the network to tie everything together. Through the SC program, IPTO provided computers for the development of new software systems, Internet access for electronic communication between program participants, and MOSIS for rapid prototyping of IC and printed circuit board designs.[41] DARPA finally saw the opportunity to collect on its long-term investment in AI through SC. SC was very ambitious and fraught with risk, but simultaneously, it provided "a startling opportunity" to integrate projects in pursuit of a set of ambitious goals.[42]

As we mentioned above, the SC program led to a number of significant developments using reduced instruction set computers. Based on work at the University of California at Berkeley and Stanford University, Sun Microsystems and Silicon Graphics companies developed advanced workstations using

RISC. The RISC concept emerged from work in the mid-1970s at the IBM Research Center on the 801 minicomputer project, an attempt to design a 12-million-instructions-per-second processor for a digital telephone exchange to handle 1 million calls per hour. The research group, led by George Radin, included John Cocke, who designed a thirty-two-bit RISC system. The name "RISC" was first used at the University of California at Berkeley, where David Patterson was its main proponent.[43] In 1982, Berkeley researchers requested funds from IPTO for VLSI research. They intended to use the IBM reduced instruction set concept to overcome the constraints of VLSI implementations of processors in a computer. Their goal was to support high-level programming languages (HLL) with maximum performance obtainable from a single-chip implementation. The funds would be used to improve on the RISC I, a very regular design using thirty-two-bit addresses, thirty-two-bit registers, and a thirty-two-bit bus. For RISC I, they created a compiler, optimizer, assembler, linker, and simulator for the programming language C, to evaluate their design. Relying on Berkeley CAD software, they implemented the RISC I architecture in nMos technology using forty-five hundred transistors, in only six months.[44] The proposal requested that continued work leading to RISC II be included in IPTO's VLSI funding to the university. With IPTO funding, researchers at Berkeley went on to implement operational prototypes, RISC I and RISC II, in the early 1980s, as well as a third Berkeley RISC design that was oriented to multiprocessing and symbolic programming.[45]

One of the first computers offered commercially using RISC, the 90X supermini by Pyramid Technology Corporation, was put on the market in 1983.[46] In 1984, MIPS, Inc., a privately held company, was formed to produce RISC chips under contract to the government, using design ideas developed by John Hennessy and his associates at Stanford. According to Gordon Bell, writing in *Datamation* in 1986, initial benchmarks of the MIPS chip indicated that data processing was five to ten times faster than on the non-RISC computers, the VAX-11/780 and the Motorola 68020, for the same clock speed. He went on to note that the MIPS company's chip was "simpler, considerably smaller, and significantly faster than any of today's microprocessors."[47] The early work at Stanford that led to these developments, which we described above, was part of the IPTO VLSI program and the Strategic Computing program.

In addition to these RISC machines, several other new experimental computing machines and design processes resulted from the convergence of IPTO activities in SC. The Connection Machine was a processor designed by Daniel Hillis during his Ph.D. research at MIT. A sixty-four-thousand-node processor version was built by the Thinking Machines Corporation (of Cambridge, Massachusetts) with support from IPTO. The four people involved with the building of the machine were Hillis (a founder of the company), Brewster Kahle, Craig Stanfill, and David Waltz. The first Connection Machine was completed and made available for DARPA use in February 1986. Interaction with the machine took place through a host computer. It was "best suited for problems that can be broken down into a large number of small elements with related parts, such as VLSI circuit design and low-level vision application." [48]

The 128-node-processor version of the Butterfly computer was built by researchers at BBN with funding from IPTO's Strategic Computing program. The Butterfly was based on switching technology for wideband communication. It used Motorola 68000 chips as the central processing units, AMD-2901 bit-slice processors for memory management, and a custom-designed VLSI switch circuit fabricated through MOSIS. [49] The large-scale Butterfly represented one of the first computers to emerge from the DARPA Multiprocessor Architecture Program in SC. In 1985, "candidate programs for use of the Butterfly [were] the Autonomous Land Vehicle Program, where it would be involved in route planning and other command and control functions; the Air-Land Battle Management Program, where it would provide a simulation engine; and the Packet Speech Program." [50]

The Warp array was completed in 1984 by researchers at CMU under funding from IPTO. The Warp prototype was developed using a custom-designed microprocessor fabricated via MOSIS as the basis of the processor elements. It was planned that Warp itself would use commercially available microprocessors interconnected by a custom-designed VLSI circuit. [51] CMU built its Warp machine by hardwiring processors together on a board. General Electric produced a printed-circuit board version of the Warp, with processors designed to operate at 10 million floating-point calculations per second, completed by the summer of 1987. At the same time, Intel completed the initial design of an integrated-circuit version named iWarp. [52] Stephen Squires, manager of the IPTO program that funded this work, noted that "a key element of the Warp program is the transition of the technology developed at CMU to two industrial partners, General Electric and Honeywell, each to provide a prototype machine to CMU in 1985." [53] Intel went its own way. As we noted in chapter 5, the Butterfly and Warp microprocessors were used in the SCORPIUS photograph interpretation system.

Many more examples could be given of IPTO programs that show the deliberate integration of results from research efforts in multiple areas. For example, the use of VLSI circuits for new graphic systems incorporated graphic techniques into microprocessors, making it possible to build smaller, more powerful graphic computer systems, or time-sharing systems that allowed more effective use of the ARPANET, or again, new parallel computer systems that, when coupled with AI research results, led to the ability to power and guide new systems, such as the Autonomous Land Vehicle developed for sc. But any such examples would demonstrate that IPTO's strategy was to formulate a highly directed program — managed by highly skilled researchers brought into IPTO for their experience in organizations devoted to the development of integrated systems — which could influence computing practice in fundamental and substantial ways.

IPTO's Influence

How IPTO Influenced the Military

Focusing on the needs of DOD command, control, and communications systems over the years, IPTO established a series of R&D programs to support the development of new and improved computer systems for use in C³ systems. Our account of the origins and evolution of many of these programs exhibits IPTO's fundamental objectives in these programs, as well as the intricate overlap of programs to achieve an integration of results for use in defense applications. IPTO's interaction with the R&D community was important for the generation of ideas, concepts, achievable projects, and results. In the technical areas of IPTO's interest, IPTO directors and program managers took the lead in this interaction with the community. In our examples from wide-area networking, AI, graphics, and very large scale integration, this leadership was determinative as well as facilitative. While we recognize that developments in these technical areas might have happened similarly in other settings, IPTO determined the specific character of the results, the contexts of the R&D, the timetables in which the results were accomplished, and the applications to which the results were put.

It is evident from congressional testimony that there has always been a clear understanding in the minds of the IPTO directors and program managers that their main purpose was to advance the mission of the DOD. Licklider's earliest program plans and ARPA Order requests contain general statements that justify the proposed contracts on the basis of the possible use of the results in military areas of information processing. The justifications of later IPTO directors became increasingly specific about possible contributions to the DOD mission. In the beginning, IPTO directors assumed that transfer of technology could occur in three ways. First, it could occur by means of the contacts that

IPTO maintained with the military services' basic R&D departments (the Office of Naval Research, the Air Force Office of Scientific Research, and the Army Research Office), which were responsible for category 6.1 research for the services. The heads of these offices, IPTO directors thought, could help to alert military personnel of new results useful to them. Second, significant technological developments would likely be transferred from the research laboratory to industry in due course, and thus be made available to the military. And third, IPTO introduced experimental testbeds in conjunction with the military services, to show the operation and usefulness of IPTO technology.

IPTO people visited military facilities so that their knowledge about applications would be current. The military development people—including, in IPTO's early years, people from organizations such as MITRE—were not consulted by IPTO personnel; thus, office directors made it a point to visit military sites regularly.[54] Such contacts with the services were easier to maintain as the number of IPTO staff members with military backgrounds increased. Gale Cleven, a military officer, joined Licklider in the office very soon after IPTO's establishment. In the early 1970s, John Perry and Bruce Dolan, air force officers, were program managers during Roberts's directorship. The army assigned David Russell to DARPA in 1969, and he joined IPTO from DARPA's Nuclear Monitoring Program in 1974. Russell became director of IPTO in September 1975. Many others followed, including Ronald Ohlander, Jack Dietzler, and Allen Sears, all of whom came from the navy, and Robert Simpson, who joined IPTO from the air force.

The process of transferring technology from IPTO to the armed services began slowly, owing to the nature of the office's projects. During the 1960s, DARPA occasionally reminded congressional committees about the significance of IPTO programs to the DOD. For example, in 1969 DARPA director Eberhardt Rechtin told the Senate Appropriations Committee,

> The material sciences and computer sciences efforts in ARPA were established to give a strong boost to two fields that were known to be important to Defense but lagging behind in being able to solve Defense application problems. We are engaged in coupling them to Defense needs by establishing more personal contact between university and government laboratories, more joint funding of programs between ARPA and the Services, intensive summer studies by researchers of specified research programs, and management of funds to exploit new research ideas rapidly.[55]

In the period after 1970, transfers of IPTO-developed technology to the military became more frequent—packet radio, ADEPT, and AALPS, among others.

The examples of such transfers discussed above indicate just a few of the interactions between IPTO and the military or other offices within DOD. We could have just as easily cited the transfer of the ARPANET to the Defense Communications Agency, the involvement of IPTO in the development of SIMNET (a military network for remote training of personnel using simulators), or the spread of distributed communications in the Military Message Experiment to the navy at CINCPAC headquarters. Or we could have explored the implementation of MULTICS at the Rome Air Development Center, the development of computer and network security in DOD systems generally, or the flight simulator graphics techniques developed by the University of Utah under IPTO contract and used in equipment manufactured by the Evans and Sutherland Company and purchased by the air force. The examples we described offer evidence of the IPTO concern for the use of research results in specific military settings. These applications were experimental in the sense that they were attempts to show how the results could be used in particular settings, but were not intended to be confined to those settings. The successful testbeds contributed to a generic understanding of the use of IPTO research results, consistent with the IPTO and DARPA missions of becoming involved in the frontier of high-technology development.

How IPTO Influenced Generic Computing Technology

While IPTO was affecting defense system development, its support of R&D produced results of generic quality, that is, results that could be used in any computing system for many purposes regardless of the nature of the organization. The technical leaps produced by IPTO and the research community that it supported over the twenty-five years covered by this study included the rapid development of time sharing, wide-area networking, connections across networks, interactive graphics, distributed computing, VLSI design systems, natural language communication systems for use with computers, and expert systems. The R&D programs produced broadly relevant results that found their way into both military and civilian systems. Indeed, the public reputation of IPTO rests on the generic technical accomplishments that resulted from IPTO funding.

These generic technologies were the result of IPTO's ability to pursue the development of more capable, flexible, intelligent, and interactive computer systems. After advancing the state of the art in time sharing and encouraging the use of the new time-sharing systems, IPTO turned to a consideration of how to increase the sharing of resources among different systems. The office displayed technological leadership in the development of packet-switching technology and in the development of packet-switching networks, especially the ARPANET and later the Internet; it also supported early graphics work and established a graphics center at the University of Utah. Artificial intelligence research, with IPTO funding, was for years one of the fastest-growing areas of computer science and engineering. The results from both of these areas became the basis for major industrial development in the 1980s. The graphics and the expert systems segments of the computer industry are direct results of IPTO R&D programs.

IPTO continuously expanded the frontiers of computer science and engineering. However, whereas the emphasis in the IPTO program had initially been on R&D on techniques and tools useful to computer scientists, or to anyone (whether military or commercial) trying to use the system interactively, the office later focused on making a greater attempt to apply what had already been learned to new defense designs. IPTO's attention to areas such as problem solving, speech understanding and recognition, robotics, and high-performance circuits paid dividends in the integration of components into systems. Submicron VLSI circuits were the basis for new graphics interfaces and for memory circuits in new parallel architectures. The process for new integrated circuit construction involved using the network to transmit design information, and scheduling production via the MOSIS fast-turnaround facility.

Starting in the 1960s, IPTO supported research in many technical areas and selected projects that would quicken the pace of research so as to produce a more interactive and flexible system. It selected projects in artificial intelligence to produce more intelligent and capable systems. The office selected network research as a means to promote the efficient sharing of resources, and community building. Through these and other research projects, the style of computing was changed markedly, and the new technology was integrated into the command, control, and communications arena of the military.

How IPTO Influenced the Computing R&D Infrastructure

The most general area of IPTO's influence was in the development of a personnel and institutional infrastructure for R&D in computing. The R&D infrastructure for computing, especially for command and control, was underdeveloped in 1962 when IPTO was founded. Only a few organizations outside the DOD — for example, the Rand Corporation, the Systems Development Corporation, MITRE, the Institute for Defense Analyses, and MIT's Lincoln Laboratory — were involved in command and control R&D. These organizations employed much of the computer science talent then available. In 1960, computing was taught in university mathematics or electrical engineering departments, or occasionally in separate programs, but not in separate departments. Research sometimes was done in organizations that had no direct connection to computer science. For example, at CMU early computer science was associated with the Graduate School for Industrial Management. For the computer science field to focus on IPTO problems, the field would have to be nurtured and a research infrastructure created.

IPTO deliberately set about to create a few university "centers of excellence" in computing. In setting up these centers, Licklider's view extended beyond the single institution. He had seen the synergistic effects of complementary research activities among the many organizations in the Cambridge, Massachusetts, area, and wanted to enhance the work being done in that region and replicate the effect in other sections of the country.

> I had this picture that Cambridge is a good reenforcing community. There were enough different places doing related things that if you just had a single group doing something, nobody knew whether to believe it or not. But if Harvard had something, and MIT had something, and there was something at the Lincoln Lab, and BBN had something, and maybe Octo Barnett had something at M[assachusetts] G[eneral] H[ospital], it reenforced. I would have liked to see that happen in Cambridge; I'd have liked to see that happen in Los Angeles or the San Francisco area; maybe one other place.[56]

In November 1963, Licklider prepared a memorandum for Eugene Fubini, assistant secretary of defense, in which he proposed the concept of centers of excellence in the information sciences, saying, "We are here particularly concerned with advancement and exploitation of digital information processing and communication." Licklider believed that there was "a great need to accelerate advances in the theoretical, the 'software,' and the over-all system aspects of information processing and communication, and there is a great need to incorporate those advances into the working knowledge of the coming generation." He proposed creating one center "to lead the effort to achieve balance in information technology, to harness the logical power of computers and make it truly available and useful to men." Another center would "lead the effort to achieve fundamental understanding, to develop the theoretical bases of information processing." He expected these roles to be filled by the various university programs at MIT, CMU, and to a lesser extent, at the University of California (at Berkeley and Los Angeles) and Stanford University. He pointed out to Fubini that the original awards to MIT and CMU had been designed "to create centers of excellence," and that these institutions were "selected with great care."[57]

MIT's Project MAC constituted a center of excellence for computer science and engineering research in almost all the areas of concern to IPTO: time-sharing systems, programming languages and programming systems, computer-aided design, computer-directed instruction, heuristic programming, information processing and communications, input/output systems (including graphic interfaces), and intelligent systems. Support for problems in these areas also went to the Lincoln Laboratory at MIT. At the time of the proposal by Licklider to Fubini, MIT was the only center dedicated to investigation of all the areas of computer science and engineering of interest to IPTO. This was certainly cause for concern on Licklider's part, because he wanted to benefit from the leverage that multiple research centers could provide.

In 1964, CMU submitted a renewal proposal requesting a substantially increased amount of money in order to become an "IPTO center of excellence." The stated aim of this center was to understand the nature of information processing, by which was meant both the systems that process and transform information, and the ways in which information-processing systems are used to control, integrate, and coordinate other systems. This also included the fields of control systems, information theory, computer engineering, programming, automata theory, documentation and information retrieval, linguistics, dynamic programming, modern logic, and statistical decision theory.[58] In short, everything that could be justified as information science would be included.[59] There was focus to CMU's proposal, which stated that CMU was "oriented toward" software and programming,[60] and categorized CMU researchers' interests as falling into several problem areas. The CMU center was to be fully integrated into the computing activity in its parent university (it was a year before the founding of the computer science department). In fact, the boundaries of computing research and the center were essentially identical—intentionally.[61]

By the end of 1967, it was clear that the support IPTO provided to universities was effectively promoting the development of the discipline of computer science and engineering. By this time, the IPTO centers of excellence program included eight university campuses: CMU, MIT, the University of California at Los Angeles, Stanford, the Universities of Michigan, Illinois, and Utah, and the University of California at Berkeley. The program had four objectives:

— to bring researchers from different disciplines together for the purpose of solving common problems

— to reap the advantages of applying research and problem-solving techniques from one field to another
— to take advantage of the economies of scale to increase the amount and value of research without a concomitant increase in cost (e.g., by taking advantage of central facilities and common resources)
— to increase the production of advanced-degree graduates with interdisciplinary training[62]

This development of the field of computer science in these universities was an intended consequence of the IPTO program, consistent with the office's desire to provide long-term support to ensure a stable environment for graduate education at a small number of universities.

But in developing the infrastructure, IPTO did not confine its attention to university centers. Rapidly, the office brought other organizations into its orbit. Even before IPTO was founded, the DOD had involved the Systems Development Corporation in research on command and control. Quickly, Licklider brought SDC activity in command and control into line with his vision, making SDC a center for time-sharing development on the West Coast using the MIT time-sharing model. The presence of Charles Rosen at the Stanford Research Institute, Edward Fredkin at his new company (Information International, Inc.), and Daniel Bobrow at Bolt Beranek and Newman attracted IPTO directors to those organizations, which were also drawn into the IPTO community. Research groups grew at these organizations and became an integral part of the infrastructure for research into IPTO problems.

Over time, IPTO directors enlarged the office's group of contractors as the growth of university computer science and engineering programs increased the capability of the research community. In the late 1970s and the 1980s, IPTO added to its group of funded schools the California Institute of Technology, Rutgers, and the Universities of Rochester, Maryland, Massachusetts, and Southern California. A range of established companies such as Hughes, Martin Marietta, General Electric, and Applied Data Systems became associated with the IPTO community. These organizations were involved in VLSI projects and the Strategic Computing program.

IPTO chose organizations to work on particular research problems or on aspects of research problems. We described how Project MAC, SDC, BBN, and the University of California (at Berkeley and Los Angeles) all worked on aspects of time sharing. We related how Taylor and Roberts chose BBN, along with the UCLA, SRI, and the Network Analysis Corporation, for the basic development of ARPANET. In addition to demonstrating that wide-area networking was feasible, the early ARPANET connected all the IPTO contractors with each other and with DARPA. This connectivity facilitated the exchange of information among projects and provided a strong identity for the group. IPTO selected the University of Utah as a center of graphics research, although graphics research was included in Project MAC and was done at Lincoln Laboratory. The Utah research results became the basis for much of the graphics development of the 1970s and 1980s at all institutions.

IPTO had representatives of the organizations it supported meet frequently either in meetings of principal investigators or in small group meetings of a topical nature. Through attendance at these meetings, IPTO staff members played a significant role in defining the R&D that researchers performed under IPTO support, and monitored the research carefully so as to coordinate the results. IPTO saw to the transfer of the results from university laboratories to industry, so as to facilitate transfer to military systems. Since most members of the IPTO staff came from the research community and were accomplished researchers themselves, they were involved in that community as equals. People flowed easily among these universities and industrial and research organizations and among the projects each organization worked on. Many more organizations eventually joined this select group, forming the infrastructure that aided IPTO in carrying out its mission. It is this combined group — comprising both IPTO and the research organizations — that constituted the infrastructure for research into DOD computing problems.

Through the funding of a few educational centers dedicated to research and the training of computer professionals, new researchers entered the field and sustained the growth of the infrastructure. Soon these new researchers were training the next generation of computer scientists and engineers and contributing markedly to new, faster, and more powerful computer systems and software.

IPTO provided ample funds for computer equipment as part of its support for university research, thereby making it possible for universities to do serious experimental research in computing. This support had a major impact on graduate education and on the quality of the students graduating from the universities supported by IPTO. Although the IPTO-supported computer facilities were not available for undergraduate education, they created an environment that in various ways increased the quality of undergraduate education in computer science.

As late as 1959, there were virtually no formal programs in the field; the vast majority of people came to computing with training in other areas, especially mathematics and engineering, and became computer scientists through on-the-job experience. By 1965, the Association for Computing Machinery reported that computer science had become a "distinct academic discipline," with doctorates being offered at more than fifteen universities, masters degrees at more than thirty, and bachelors degrees at about seventeen.[63] By 1967, computer science was taught in separate departments at CMU, Cornell, the University of Illinois, the University of Michigan, the University of North Carolina, Notre Dame, the University of Pennsylvania, Pennsylvania State University, Purdue, Stanford, the University of Texas, the University of Toronto, and the University of Wisconsin (the compiler of this list noted that it was not necessarily a complete one). Institutions where there was a computer science program without a separate department included Brown, the California Institute of Technology, Harvard, the University of Maryland, Princeton, the University of California, the University of Chicago, MIT, New York University, and the University of Virginia.[64] A significant amount, in some cases all, of the research support for computing at CMU, Illinois, Stanford, California (Berkeley and Los Angeles), and MIT came from IPTO.

Two more direct indicators of IPTO's influence on education can be understood through analysis of the degrees of the tenured faculty members at the top forty U.S. computer science departments, and of the growth of computer science groups in various institutions. At the beginning of the 1990s, more than 26 percent of the tenured members of those select computer science departments had received their Ph.D. degrees from one of the three main IPTO-sponsored institutions: MIT, Stanford, and CMU. This shows a remarkable penetration of IPTO in the computer education process. An even greater percentage, 42 percent, of the tenured faculty members in the ten top-ranked departments had received their Ph.D. degrees from one of the three schools, and more than 53 percent of the nontenured faculty members.[65] Even though this is not a systematic survey, we can be virtually certain that the education these faculty members had received, the research they participated in, and the education they provided to their students, was affected by the IPTO programs at their degree-granting institutions. Not infrequently, these people were able to maintain an association with IPTO after arriving at the new institution. For example, David Evans went from Berkeley to the University of Utah, John McCarthy went from MIT to Stanford, Vinton Cerf went from UCLA to Stanford, and Henry Fuchs went from Utah to the University of North Carolina.

Many of the graduates of these IPTO-supported institutions went to work for industry. Those with research experience became agents in technology transfer from university laboratories to industrial development projects. For example, James H. Clark received a Ph.D. degree from Utah, joined the

Stanford faculty, and then became one of the founders of Silicon Graphics. William N. Joy followed a similar path, going from Berkeley to Sun Microsystems. After receiving degrees from Michigan and Berkeley, and serving as lead designer and programmer on the Berkeley UNIX project, he joined Sun Microsystems as vice-president of research and development. W. Daniel Hillis co-founded Thinking Machines after being a Ph.D. candidate at MIT. Also, John L. Hennessy served as chief scientist at MIPS Computer Systems, while on leave from the Stanford faculty. All four of these people had been participants in IPTO-sponsored projects before moving on to these posts. Not only was IPTO an agent in support of R&D at educational institutions and industrial organizations, but it was an indirect agent in the transmission of technological knowledge to organizations uninvolved in its R&D programs.

The second indicator of IPTO's influence on education was the growth of computer science and engineering, especially the growth of research laboratories in the institutions funded by IPTO, and the involvement of these laboratories in education. For example, at MIT during the 1960s, forty-six graduate students were supported by Project MAC. Fano reported to Taylor at IPTO that in the four years from 1963 to 1967, 110 graduate theses—including twenty-one Ph.D. dissertations—and twenty-five (computer science) faculty members in five departments had been supported by Project MAC. These figures included support for people working in AI. The effect of Project MAC went beyond these people. Fano also reported that Project MAC had supported "a much larger body of faculty representing 14 academic departments, [who] have at some time interacted significantly with Project MAC, and used Project MAC's computer facilities."[66] Also, eighty-seven students in these departments had received some support for their thesis work from Project MAC. In 1974 Project MAC metamorphosed into the Laboratory for Computer Science. MIT's AI Laboratory was for the most part already separate, but it became independent at this time. In the mid-1980s, these two laboratories together supported 215 graduate students per year and had some sixty faculty members, as well as a host of visitors. Thus these laboratories continued to show the same kind of influence on the educational programs of MIT that Project MAC had in the 1960s. Though we did not compile the statistics, similar numbers can be used to show the influence of the other IPTO centers of excellence.

Through its generous support for computing R&D at selected educational institutions, IPTO helped several generations of scientists and engineers who were interested in computing problems applicable to defense. These individuals diffused throughout the chain of production, from R&D centers in universities to development groups in industry. Thus, IPTO became one agency to fulfill the 1940s military objective of developing strong ties with civilian scientists and engineers so as to continue the development of new defense systems and thus keep the U.S. military in a strong position. IPTO accomplished this for computing by insisting on sharply focused centers of excellence and a carefully designed R&D program to generate results useful to achieving defense systems objectives, by integrating the R&D results quickly into systems, and by maintaining a close alliance with students, faculty, and R&D specialists in industry. Careful integration of results, attention to the needs of groups inside DOD, and continued evaluation of programs kept IPTO on the R&D frontier.

The New World of Computing in the 1990s

The computing environment from 1960 to 1990 changed in a variety of ways. A significant number of institutions developed time-sharing systems, and a number of companies implemented time sharing in commercial computers, which altered the pattern of computing by individuals. Artificial intelligence research advanced on many fronts, and researchers learned a great deal about the nature of problem solving, both by humans and by machines. AI research groups pursued projects in speech, vision, and "hand" manipulation, as well as the integration of these three fields in robotics. Packet switching for computer networks made feasible a working national — indeed, international — network. Graphics research made substantial progress, producing aesthetically pleasing, informative displays integrated with sophisticated analysis programs. Theory of computation and numerical analysis became major enterprises within computing. The microcomputer made single-user machines the norm for many computer users, and sophisticated workstations, commonplace now in many research laboratories and universities, brought a new level of computing power and display capabilities to the desktop. Software packages and applications enhanced the use of data and the ability to manipulate computers, meeting the desires of both programmers and end users.

The software area, hardly a reality in the 1950s, developed in significant ways. Many new programming languages appeared, designed for tasks including numerical scientific computation (e.g., FORTRAN, ALGOL), formula manipulation (e.g., FORMAC), on-line analysis and programming (e.g., JOSS, BASIC), business data processing (e.g., COBOL), list and string processing (e.g., LISP, SNOBOL), and multipurpose use (e.g., JOVIAL, PL/1).[67] Researchers pursued work on data structures and database management systems. Last, and perhaps most significant, there was the range of concerns that fell under the general heading of software: the design of software, operating systems, and compilers; and testing and documentation. Because of the rapidly rising costs of software production relative to other areas of computer systems, people spoke of a "software crisis," reflecting both the number of programmers needed and the complexity of design. But concern about software issues has not gone away, in spite of all the work in that area over the last two decades, and the problems remain with us today.

Today, the computer is no longer a reserved or precious resource that is to be used sparingly. Indeed, we find computers virtually in everyone's hands. Programmers of the 1990s do not worry about "wasting" computer time. Interactive, on-line computing is pervasive. The programming process in the 1990s incorporates sophisticated editing programs, displays, and debugging tools. The programmer enters his or her program directly into machine-readable media using a keyboard; few programmers today would dream of writing programs in longhand. Existing code can be displayed in another "window" on-screen as the programmer works; sections of code can be copied from one program into another with a few clicks of a mouse. Running a program takes no more than a simple command, and debugging is interactive. Intermediate results can be displayed, programs rerun, and changes incorporated in a few minutes rather than a few days. When a command is entered at a terminal the response is expected to be, and often is, instantaneous.

Today, computer users are interconnected throughout the United States and internationally; it is possible and relatively easy to get to and use remote systems. Those "on the net" can easily communicate across the country or across the world. Electronic mail is becoming widespread and is already a common method of communication among computer scientists and engineers—indeed, all scientists and engineers. Computer scientists are closely connected through networks. All major universities in the United States, as well as much of the computer industry and many other research institutions, are tied into the Internet, and virtually all students have access to it. Worldwide, the Internet links millions of people. Outside of universities, electronic bulletin boards have thousands of subscribers, and people with common interests debate and discuss all manner of topics through electronic networks. By the end of the 1980s, computer communications networks had become increasingly important to scientific advancement, economic strength, and national security. To make even more of the network in commercial settings, a presidential initiative in the 1990s established the High Performance Computing and Communications program to accelerate significantly the availability and utilization of the next generation of high-performance computers and networks.[68]

We understand that many settings played a role in these computing developments—academic, industrial, and governmental. While IPTO cannot claim responsibility for all of these developments, a comparison of the computing environment of the 1990s with that of the early 1960s shows that the later period had four striking features that were not present in the earlier period and that were the result of IPTO-supported programs. These features are the interactive nature of computer use and programming, the connectivity of computer systems and users, the visual capabilities of input/output systems, especially those found in workstations, and the increased sophistication of systems. IPTO programs and contractors developed many of these new areas, such as networking; occasionally, as in the case of graphics, they started them on their way and others furthered their development later under different auspices; a few times, as in the case of time sharing, IPTO-sponsored researchers joined in a development and stimulated rapid change. And AI virtually owes its progress to IPTO support and leadership. No single IPTO program addressed each of these features of computer systems exclusively. Indeed, improved capability was the objective of every designer, whether connected to IPTO or not, and all AI researchers sought more intelligent and interactive machines. One distinguishing aspect is that IPTO personnel designed programs to develop new computing systems and techniques for better integration of computer systems into defense systems. Another distinction was the design of support programs with the intention of integrating results across programs. Although never explicitly stated in this way, IPTO sought to integrate program results to make computer systems more capable, flexible, intelligent, and interactive for command, control and communications systems *all at the same time*, and herein lies the key to IPTO's success. To accomplish this, programs had both long- and short-term goals. As we saw, the short-term goals were better time-sharing systems; a network to facilitate communications and research among IPTO-supported researchers; and research into AI questions which led to examination of useful system applications and to more realistic, easier-to-use graphics systems. IPTO's long-term goals were better resource sharing and the integration of these results into larger and larger systems to improve flexibility, intelligence, and so on. Achievement of these goals helped the military respond to a changing defense context, and in turn contributed ideas to change the general computing environment. Thus, IPTO program results had a primary influence in changing the style of computing, bringing us the computing world of today. As a result, we see that 1962 was a critical turning point in computing R&D generally.

To justify this assertion, we need to step back and briefly examine the R&D

perspective of the computer industry in the late 1950s. Prior to the establishment of IPTO, R&D funding from whatever source went primarily into hardware R&D. As we pointed out in the Introduction, the U.S. government between 1943 and 1960 supported the design of new computers. The armed services wanted computers for aggressive weapons and defensive systems. The Atomic Energy Commission funded R&D at the national laboratories and in industry for the computers needed in highly specialized research programs, such as the nuclear weapons programs. Out of this work came machines with memorable names and major abilities: UNIVAC, Maniac, Stretch, LARC, and the like. Independent government agencies sought equipment to conduct large-scale routine analytical work ranging from the census to administration of the social security system. The R&D for all these machines merged easily with the R&D and prototyping of hardware that was of interest to the industry that built and sold hardware and programs to government, industry, and schools. This industry R&D was for the express purpose of developing computer systems for a standard market, one that sought systems capable of handling the typical concerns of businesses around the world and computers that could be upgraded periodically without loss of data or time.[69] In 1962 this market was so potentially enormous that the computer industry had little or no interest or incentive to deviate from serving it, a situation similar to that of the semiconductor manufacturers in the 1970s. To take up the concerns of academics or the military—to alter the existing orientation of R&D to produce different computer system designs of highly limited and specialized use, and therefore of little perceived value to the commercial and government sectors—would have been unprofitable and possibly suicidal for a company in this exciting new enterprise.

Yet some companies were not blind to the need to involve themselves with the academy. Witness IBM's computer gift program to universities and colleges to prepare a new generation for life in the "high tech" world of the 1960s and beyond. IBM also saw the value in placating research scientists by assisting their efforts to modify equipment to serve new, highly innovative R&D objectives (see the discussion of time sharing at MIT in chap. 2, above). But the changes that were made were limited to those compatible with the fundamental design of the computer system for sale on the open market. No one expected the computer industry to finance and support innovative R&D in computing on the scale of the perceived need. Nor was NSF or NIH in a position to do so.[70] Had the DOD waited for industry to produce the kind of interactive machines desired by military leaders for command and control, it would have had to wait at least until the demand for basic computing machines subsided.

Even then it might not have received the new designs; a new trend might have superseded the old one. IPTO intervened to assist the DOD, and in the process, created a new trend—interactive computing.

In the 1970s, after time sharing and networking were available and flexible software production had been demonstrated in various IPTO-sponsored research projects, the stage was set for further developments. Industry's development of expert systems, local area networks, applications software packages, flexible and sophisticated chip designs, graphics, and the use of AI in design and operation of computer systems—in other words, the world of computing in the 1990s—became possible because of the deliberate efforts of IPTO to produce more capable, flexible, intelligent, and interactive computer systems. IPTO, through the efforts of men such as Licklider, Sutherland, Taylor, Roberts, and Kahn, and their superiors in the DOD—Fubini, Ruina, Lukasik, Heilmeier, Cooper, DeLauer, and others too numerous to name—established a path of computing R&D that was parallel to that of industry and led to today's highly interactive and electronically communicating world.

While there are many challenges for computing in the 1990s, some of which (e.g., security issues and network management) have been introduced by the on-line and interconnected computing environment, there is no question that computing in the 1990s has overcome many of the limitations of the 1960s environment. This was in large part due to the emphasis IPTO placed on the interaction between humans and computers and on making the computing environment more responsive to human needs. IPTO's programs achieved these results because the office focused on resource sharing, system development, integration of results, and testbeds for implementation—a four-pronged approach to computing system development. In addition to influencing changes in the computing environment, IPTO also stimulated the creation of closer ties among members of the computer science and engineering community, affected military systems through the introduction of new computing techniques and systems, and had a substantial effect on the computing education community.

IPTO accomplished this with a lean structure of very competent office directors and program managers who believed primarily in results. These directors led the R&D community to develop a new style of computing more suitable for integrated defense systems. The IPTO program changed continually to adjust to new technological accomplishments and changing requirements. But amid all the change was a constant vision of the type of program that would serve the defense objectives. By choosing leaders for IPTO who understood the place of R&D in system integration, DARPA maintained a strong IPTO, which kept its eyes on the future needs of command, control, and communications, and in many cases provided research results capable of advancing both defense and military capability in computing in the two-and-a-half decades after 1962. The Licklider vision that started it all can still be found in the Kahn design for the Strategic Computing program. But by the time of sc, the context around IPTO had substantially altered. DARPA and DOD carried the IPTO vision into the new context, although IPTO ceased to exist. Its legacy lives on in the accomplishments of its programs and in the hearts of the members of the DARPA-IPTO community.

List of Interview Subjects

All of the interviews listed below were transcribed; and edited transcriptions of the interviews are held in the Charles Babbage Institute Archives. Consult CBI for access.

Listed below are the interview subjects, the date of interview, and, in parentheses, the name of the interviewer.

Saul Amarel, 5 October 1989 (Arthur L. Norberg)
Paul Baran, 5 March 1990 (Judy E. O'Neill)
Allan G. Blue, 12 June 1989 (William Aspray)
Bruce Buchanan, 11 and 12 June 1991 (Arthur L. Norberg)
Vinton G. Cerf, 24 April 1990 (Judy E. O'Neill)
Wesley Clark, 3 May 1990 (Judy E. O'Neill)
Robert S. Cooper, 3 September 1993 (Arthur L. Norberg)
Fernando J. Corbato, 18 April 1989, 13 November 1990 (Arthur L. Norberg)
Stephen Crocker, 24 October 1991 (Judy E. O'Neill)
William R. Crowther, 12 March 1990 (Judy E. O'Neill)
Charles A. Csuri, 23 October 1990 (Kerry J. Freedman)
Jack B. Dennis, 31 October 1989 (Judy E. O'Neill)
Michael L. Dertouzos, 20 April 1989 (Arthur L. Norberg)
Robert M. Fano, 20 April 1989 (Arthur L. Norberg)
Edward A. Feigenbaum, 3 March 1989 (William Aspray)
Howard Frank, 30 March 1990 (Judy E. O'Neill)
Frank E. Heart, 13 March 1990 (Judy E. O'Neill)
George H. Heilmeier, 27 March 1991 (Arthur L. Norberg)
Charles M. Herzfeld, 6 August 1990 (Arthur L. Norberg)
Robert E. Kahn, 22 March 1989, 24 April 1990 (William Aspray; Judy E. O'Neill)
Leonard Kleinrock, 3 April 1990 (Judy E. O'Neill)

J.C.R. Licklider, 28 October 1988 (William Aspray and Arthur L. Norberg)
Stephen J. Lukasik, 17 October 1991 (Judy E. O'Neill)
Alexander A. McKenzie, 13 March 1990 (Judy E. O'Neill)
John McCarthy, 2 March 1989 (William Aspray)
Marvin L. Minsky, 1 November 1989 (Arthur L. Norberg)
Allen Newell, 10, 11, 12 June 1991 (Arthur L. Norberg)
Nils J. Nilsson, 1 March 1989 (William Aspray)
Ronald B. Ohlander, 25 September 1989 (Arthur L. Norberg)
Severo M. Ornstein, 6 March 1990 (Judy E. O'Neill)
D. Raj Reddy, 12 June 1991 (Arthur L. Norberg)
Lawrence G. Roberts, 4 April 1989 (Arthur L. Norberg)
Douglas T. Ross, 1 November 1989 (Judy E. O'Neill)
Jack P. Ruina, 20 April 1989 (William Aspray)
David C. Russell, telephone interview, 26 May 1992 (Judy E. O'Neill)
Jules I. Schwartz, 7 April 1989 (Arthur L. Norberg)
Robert L. Simpson Jr., 14 March 1990 (Arthur L. Norberg)
Ivan E. Sutherland, 1 May 1989 (William Aspray)
Robert W. Taylor, 28 February 1989 (William Aspray)
Keith Uncapher, 10 July 1989 (Arthur L. Norberg)
David C. Walden, 6 February 1990 (Judy E. O'Neill)
Frank H. Westervelt, 15 May 1990 (Judy E. O'Neill)
Terry A. Winograd, 11 December 1991 (Arthur L. Norberg)
Patrick H. Winston, 18 April, 1 May 1990 (Arthur L. Norberg)
Charles A. Zraket, 3 May 1990 (Arthur L. Norberg)

Notes

Introduction

1. Harvey M. Sapolsky, *Science and the Navy: The History of the Office of Naval Research* (Princeton, N.J.: Princeton University Press, 1990); Paul Forman, "Behind Quantum Electronics: National Security as Basis for Physical Research in the United States, 1940–1960," *Historical Studies in the Physical Sciences* 18 (1987): 149–229; Stuart W. Leslie, *The Cold War and American Science: The Military-Industrial-Academic Complex at MIT and Stanford* (New York: Columbia University Press, 1993); David Noble, *Forces of Production: A Social History of Industrial Automation* (New York: Knopf, 1984); Richard G. Hewlett and Oscar E. Anderson Jr., *A History of the United States Atomic Energy Commission,* vol. 1, *The New World* (Washington, D.C.: Government Printing Office, 1962); Ann Markusen and Joel Yudken, *Dismantling the Cold War Economy* (New York: Basic Books, 1992); and Roger L. Geiger, *Research and Relevant Knowledge: American Research Universities since World War II* (Oxford: Oxford University Press, 1993).

2. Donald MacKenzie, *Inventing Accuracy: A Historical Sociology of Nuclear Missile Guidance* (Cambridge: MIT Press, 1990); Noble, *Forces of Production;* Leslie, *Cold War and Science.* For another view of the partnership, see also Michael Aaron Dennis, "'Our First Line of Defense': Two University Laboratories in the Postwar American State," *Isis* 85 (1994): 427–55.

3. Hewlett and Anderson, *History of the Atomic Energy Commission;* Sapolsky, *Science and the Navy;* Noble, *Forces of Production;* Richard R. Nelson, ed., *Government and Technical Progress: A Cross-Industry Analysis* (New York: Pergamon, 1982); Kenneth Flamm, *Creating the Computer: Government, Industry, and High Technology* (Washington, D.C.: Brookings Institution, 1988); Anthony DiFilippo, *From Industry to Arms: The Political Economy of High Technology* (Westport, Conn.: Greenwood Press, 1990).

4. On Clinton administration initiatives, see, e.g., "FCCSET Initiatives in the FY 1994 Budget," 8 Apr. 1993, transmitted to the U.S. Congress by John H. Gibbons, director of the Office of Science and Technology Policy, on behalf of the president; and Mike Mills, "Clinton Technology Policy Aims at Partnership with Industry," *Congressional Quarterly,* 27 Feb. 1993, 453–55.

5. Sapolsky, *Science and the Navy.*

6. Claude Baum, *The System Builders: The Story of SDC* (Santa Monica, Calif.: Sys-

tem Development, 1981); Bruce L. R. Smith, *The Rand Corporation: Case Study of a Non-Profit Advisory Corporation* (Cambridge: Harvard University Press, 1966); Nick A. Komons, *Science and the Air Force: A History of the Air Force Office of Scientific Research* (Arlington, Va.: Office of Aerospace Research, 1966); Leslie, *Cold War and Science;* Jerome B. Wiesner, "A Successful Experiment," *Naval Research Reviews* 19 (July 1966): 1–4, 11.

7. Flamm, *Creating the Computer;* Kent C. Redmond and Thomas M. Smith, *Project Whirlwind: The History of a Pioneer Computer* (Bedford, Mass.: Digital Press, 1980); Paul E. Ceruzzi, *Beyond the Limits: Flight Enters the Computer Age* (Cambridge: MIT Press, 1989); I. Bernard Cohen, "Howard H. Aiken and the Computer," in Stephen G. Nash, ed., *A History of Scientific Computing* (New York: ACM Press, and Reading, Mass.: Addison-Wesley, 1990), 41–52; Deirdre La Porte and George R. Stibitz, "Eloge: E. G. Andrews, 1898–1980," *Annals of the History of Computing* 4 (1982): 4–19 (contains two papers by Andrews on Bell machines).

8. Nancy Stern, *From Eniac to Univac: An Appraisal of the Eckert-Mauchly Computers* (Bedford, Mass.: Digital Press, 1981); Alice R. Burks and Arthur W. Burks, "The ENIAC: First General-Purpose Electronic Computer," *Annals of the History of Computing* 3 (1981): 310–89; Thomas P. Hughes, "ENIAC: Invention of a Computer," *Technikgeschichte* 42 (1975): 148–65; Charles J. Bashe, Lyle R. Johnson, John H. Palmer, and Emerson W. Pugh, *IBM's Early Computers* (Cambridge: MIT Press, 1986), 26–27.

9. Erwin Tomash and Arnold A. Cohen, "The Birth of an ERA: Engineering Research Associates, Inc., 1946–1955," *Annals of the History of Computing* 1 (1979): 83–97; Flamm, *Creating the Computer;* Arthur L. Norberg, *Computers and Commerce,* forthcoming.

10. James E. Tomayko, *Computers in Spaceflight: The NASA Experience,* vol. 18 of *Encyclopedia of Computer Science and Technology* (New York: Dekker, 1987).

11. For an exciting and comprehensive history of the government's response to *Sputnik,* see Robert A. Divine, *The Sputnik Challenge: Eisenhower's Response to the Soviet Satellite* (New York: Oxford University Press, 1993). See also James R. Killian Jr., *Sputnik, Scientists, and Eisenhower* (Cambridge: MIT Press, 1977); and Herbert F. York, *Making Weapons, Talking Peace: A Physicist's Odyssey from Hiroshima to Geneva* (New York: Basic Books, 1987).

12. This role for DARPA is implied in a number of public statements from 1958 and later about the role of the agency in weapons development (see, e.g., the answers to congressional questions by Secretary of Defense McElroy concerning the new agency in House Subcommittee on Defense Appropriations, *Department of Defense, Ballistic Missile Program,* (85) H1654-2, 85th Cong., 1st sess., 20–21 Nov. 1957, 21. Explicit statements about DARPA's role in preventing technological surprise appear after 1968. (See "Statement by Dr. Eberhardt Rechtin," in Senate Committee on Appropriations, *Department of Defense Appropriations for Fiscal Year 1970, Part 1,* (91) S2010-0-A, 91st Cong., 1st sess., 10, 12–13, 16 June, 15–17, 25 Sept. 1969, 425: "To insure DOD against technological surprise, ARPA must search out new fields and ideas, accelerate R&D where surprise could be critical, and bring developments to a stage at which sound decisions can be made on their further exploitation.")

13. Statements about DARPA's mission can be found in a number of places. See for example, "Statement of Dr. Stephen J. Lukasik, Director, DARPA," in Senate Subcommittee on Defense Appropriations, *Department of Defense Appropriations for Fiscal Year 1973, Part 1*, S181-4.44, 92d Cong., 2d sess., 14 Mar. 1972, 725–830.

14. Two studies on the history of DARPA are Richard J. Barber Associates, *The Advanced Research Projects Agency, 1958–1974* (Washington, D.C.: Barber Associates, 1975); and Richard H. Van Atta, Seymour J. Deitchman, and Sidney G. Reed, DARPA *Technical Accomplishments*, 3 vols., IDA Paper P-2192 (Washington, D.C.: Institute for Defense Analyses, 1991).

15. Forman, "Behind Quantum Electronics"; Stuart W. Leslie, "Playing the Education Game to Win: The Military and Interdisciplinary Research at Stanford," *Historical Studies in the Physical and Biological Sciences* 18, no. 1 (1987): 55–88.

16. Barber Associates, *Advanced Research Projects Agency*, pp. I-8 to I-9.

17. Anthony Debons, "Command and Control: Technology and Social Impact," *Advances in Computers* 11 (1971): 319–90.

18. Norman C. Dalkey, "Command and Control — A Glance at the Future," Rand Report P-2675 (Santa Monica, Calif.: Rand Corporation, 1962); H. D. Benington, "Military Information — Recently and Presently," in Edward Bennett, James Degan, and Joseph Spiegel, eds., *Military Information Systems: The Design of Computer-Aided Systems for Command* (New York: Praeger, 1964), 1–18.

19. Frederick B. Thompson, "Design Fundamentals of Military Information Systems," in Bennett, Degan, and Spiegel, eds., *Military Information Systems*, 46–87.

20. W. F. Bauer, D. L. Gerlough, and J. W. Granholm, "Advanced Computer Applications," *Proceedings of the IRE* 44 (Jan. 1961): 296–303.

21. F. H. Tyaack, "Progress Report: Survey of the Information Sciences," IDA-IM-198, 23 May 1960, Contract SD-50, RG 330-74-107, box 1, unlabeled folder, National Archives Branch Depository, Suitland, Md. (hereafter cited as "NABDS"). The records of DARPA/IPTO are RG 330. The survey was done "mostly" in March and April 1960.

22. *Message from the President of the United States Relative to Recommendations Relating to Our Defense Budget*, H. Doc. 123, 87th Cong., 1st sess., 28 Mar. 1961, 8.

23. Earlier studies of the command and control program in DARPA claim that its origins are shrouded. For example, reference is often made to a September 1966 memorandum by Robert Taylor, then director of IPTO, in which he stated that command and control research was assigned to DARPA in June 1961 by the DDR&E. This date is consistent with the August 1961 program plan that outlines the DOD's interest in applying computers to war gaming, command system studies, and information processing related to command and control. Robert W. Taylor, "Memorandum for the Director, ARPA: ODDR&E Coordination of Computer Oriented Research and Development," 30 Sept. 1966, RG 330-78-0013, box 1, unlabeled folder, NABDS.

24. Institute for Defense Analyses, "Computers in Command and Control," Technical Report no. 61-12, Nov. 1961.

25. In fact, the work statement and research plan in the DARPA contract are identical to the 1960 SDC proposal. SDC Command-Control Research, Proposal 184, Nov. 1960, p. 6, BR-(L)-31 SDC Command Control Research, Burroughs Corporation Records,

System Development Corporation Series, Technical Series, box 1, Charles Babbage Institute (hereafter, "CBI"), Minneapolis, Minn.

26. Stuart Cooney, "Command Control Research and Development," *SDC Magazine* 5 (Feb. 1962): 2, Burroughs Corporation Records, System Development Corporation Series, CBI.

27. Advanced Research Projects Agency, "Command Control Research," Program Plan no. 307, 1 Aug. 1961, RG 330-78-0013, box 1, folder: "Program Plans," NABDS.

28. Ibid., 5.

29. The ARPA office responsible for most of the computing research support has had several names during its short history. We are concerned with the period from 1962 (when the office was for a short time called the Command and Control Program) to the 1980s. The lineal continuation of IPTO was the Information Science and Technology Office (ISTO), organized in 1986. ISTO was reorganized in May 1991 into the Computing Systems Technology Office (CSTO) and the Software and Intelligent Systems Technology Office (SISTO).

30. David Allison made a similar assessment of the effect of individuals in his study of U.S. Navy R&D programs. According to Allison, navy research programs in the 1950s and 1960s were guided by individuals and reflected the "initiative, advocacy, and entrepreneurship" of these individuals. He went further and noted that in the 1980s "programs are still defined primarily by the particular individuals involved in them and not by oversight by higher authority." D. K. Allison, "U.S. Navy Research and Development since World War II," in Merritt Roe Smith, ed., *Military Enterprise and Technological Change: Perspectives on the American Experience* (Cambridge: MIT Press, 1985), 328.

31. The appearance of a missionary style of approach is not unusual at the beginning of a research field. Another example of this can be found in attitudes of the cyclotron engineers of the 1930s. For a discussion on cyclotron engineering, see John L. Heilbron and Robert W. Seidel, *Lawrence and His Laboratory: A History of the Lawrence Berkeley Laboratory* (Berkeley: University of California Press, 1990), vol. 1.

32. This point was emphasized in virtually all our interviews with DARPA directors. See Ruina interview, CBI; Herzfeld interview, CBI; Heilmeier interview, CBI; and Cooper interview, CBI.

33. Barber Associates, *Advanced Research Projects Agency*, p. I-12.

34. Lukasik interview, CBI; Uncapher interview, CBI.

35. Heilmeier interview, CBI.

36. The two basic funding categories in the DOD system for IPTO budgets over the years have been Research, called category 6.1, and Exploratory Development, called category 6.2. Category 6.2 can contain projects that have a research component.

37. Blue Ribbon Defense Panel, *Report to the President and the Secretary of Defense on the Department of Defense* (Washington, D.C.: Government Printing Office, 1970).

38. Senate Subcommittee, *Department of Defense Appropriations for 1973*, 747.

39. "Statement Submitted by Dr. George H. Heilmeier," in Senate Subcommittee on Defense Appropriations, *Department of Defense Appropriations for Fiscal Year 1978, Part 5: Research, Development, Test, and Evaluation*, 77 S181-27.1, 95th Cong., 1st sess., 4 Feb. 1977, 44–172; see especially 57–60.

40. Ibid.

41. It should be kept in mind that often these categories were for accounting purposes; it was difficult to distinguish basic research from the research needed for development in IPTO problem areas.

42. "Statement of Dr. Eberhardt Rechtin," in House Subcommittee on Defense Appropriations, *Department of Defense Appropriations for Fiscal Year 1971, Part 6, 70* H181-54.5, 91st Cong., 2d sess., 5 May 1970, 723–80; "Statement of Dr. Eberhardt Rechtin," in Senate Subcommittee on Defense Appropriations, *Department of Defense Appropriations for Fiscal Year 1971, Part 1, 70* S181-36.6, 91st Cong., 2d sess., 15 Apr. 1970, 713–66; "Statement by Stephen J. Lukasik," in House Subcommittee on Defense Appropriations, *Department of Defense Appropriations for Fiscal Year 1975, Part 4: Research, Development, Test, and Evaluation, 74* H181-72.1, 93d Cong., 2d sess., 29 Apr. 1974, 632–50; "Written Statement Submitted by Robert R. Fossum," in Senate Subcommittee on Defense Appropriations, *Department of Defense Appropriations, Fiscal Year 1980, Part 4: Procurement/Research, Development, Test, and Evaluation, 79* S181-49.1, 96th Cong., 1st sess., 21 Feb. 1979, 210–35.

43. Saul Amarel, "Issues and Themes in Information Science and Technology," 5 Feb. 1987. Provided by Saul Amarel.

44. Senate Subcommittee, *Department of Defense Appropriations for 1978,* 44–172; see especially 57–60.

Chapter One: Managing for Technological Innovation

1. Two recent examples are Burton I. Edelson and Robert L. Stern, "The Operations of DARPA and Its Utility as a Model for a Civilian ARPA," background paper for Carnegie Commission on Science, Technology, and Government (Baltimore: Johns Hopkins Foreign Policy Institute, 1989); and the Carnegie Commission's report *Technology and Economic Performance: Organizing the Executive Branch for a Stronger National Technology Base* (New York: Carnegie Commission on Science, Technology, and Government, 1991).

2. Ann Markusen and Joel Yudken, *Dismantling the Cold War Economy* (New York: Basic Books, 1992); Anthony DiFilippo, *From Industry to Arms: The Political Economy of High Technology* (Westport, Conn.: Greenwood Press, 1990).

3. A statement of the work proposed in this project is contained in Charles E. Hutchinson to Paul A. Dittman, 17 Oct. 1960, RG 330-78-0085, box 3, folder: "Air Force," National Archives Branch Depository, Suitland, Md. (NABDS). Dittman was one of Licklider's co-workers on the project.

4. J.C.R. Licklider to Charles E. Hutchinson, 30 Nov. 1960, RG 330-78-0085, box 3, folder: "Air Force," NABDS.

5. Licklider interview, CBI.

6. J.C.R. Licklider, "Man-Computer Symbiosis," *IRE Transactions on Human Factors in Engineering* HFE-1, no. 1 (Mar. 1960): 4–11.

7. Herzfeld interview, CBI. Herzfeld recalled that Licklider was at the DOD discuss-

ing computing problems at least by the time he himself arrived in September 1961. We know that Licklider was not there in July 1962. A routing slip attached to an air force memorandum ["Command and Control, Revision #1"] dated 18 July 1962 has a handwritten note "Hold for Lick." Note and memorandum in RG 330-78-0013, box 1, NABDS. Additionally, an SDC memo noted that Licklider had been involved in the SDC contract since May 1962. P. D. Greenberg to T. C. Rowan, "SDC History," 9 July 1964, memorandum no. M-14368, folder: "PDG Archives," Burroughs Corporation Records, System Development Corporation Series, CBI.

8. J.C.R. Licklider to Charles Hutchinson, 13 June 1962, RG 330-78-0085, box 3, un-labeled folder, NABDS.

9. Ibid.

10. Taylor interview, CBI.

11. Ibid.

12. Howard Rheingold, *Tools for Thought: The People and Ideas Behind the* Next *Computer Revolution* (New York: Simon & Schuster, 1985), 149.

13. Sutherland interview, CBI.

14. See chapter 3 for a description of Sutherland's work on Sketchpad.

15. Sutherland interview, CBI.

16. Ibid.

17. See especially Herzfeld interview, CBI.

18. Rheingold, *Tools for Thought*.

19. Taylor interview, CBI.

20. Ibid.

21. Sutherland interview, CBI.

22. Office of Science and Technology Policy, "The Federal High Performance Computing Program" (Washington, D.C., 1989). A number of people who were part of IPTO and its successor ISTO served on the subcommittee that generated this report.

23. Taylor interview, CBI.

24. Kahn interview, CBI.

25. Roberts interview, CBI; DARPA, "Graphic Control and Display of Computer Processes," Program Plan no. 439, 1 Mar. 1965, RG 330-78-0013, box 1, folder: "Program Plans," NABDS.

26. Roberts interview, CBI.

27. Ibid.

28. Ibid.

29. Seymour Papert, "One AI or Many?" *Daedalus* 117 (winter 1988): 1–14.

30. James L. Penick Jr. Carroll W. Pursell Jr., Morgan B. Sherwood, and Donald C. Swain, eds., *The Politics of American Science, 1939 to the Present* (Cambridge: MIT Press, 1972), 343.

31. Ibid., 343.

32. On the Mansfield Amendment, see ibid., 338–49.

33. Ibid., 345.

34. Roberts interview, CBI.

35. Blue interview, CBI.

36. Winston interview, CBI.

37. Kahn interview, CBI.

38. Quotations taken from Kahn interview, CBI.

39. Kahn interview, CBI.

40. Amarel interview, CBI.

41. Squires served for a time as director of CSTO, one of ISTO's successors.

42. Simpson retired from the Air Force as a lieutenant colonel, and Ohlander retired from the navy as a commander.

43. Roberts interview, CBI.

44. This emerges from a reading of a number of the interviews with program managers and office directors. See Cerf interview, CBI; Crocker interview, CBI; Kahn interview, CBI; Ohlander interview, CBI; Roberts interview, CBI; Simpson interview, CBI.

45. See the Licklider interview, CBI.

46. Ibid.; Taylor interview, CBI.

47. Mina Rees, "The Computing Program of the Office of Naval Research, 1946–1953," *Annals of the History of Computing* 4 (1982): 102–20.

48. Licklider to Donald Drukey and Thomas Rowan, 8 July 1963, RG 330-69-A-4998, box 3, folder: "350-1; Licklider/Fredkin correspondence, July to December 1963," NABDS; RG 330-69-A-4998, box 2, folder: "Information International, Inc.," NABDS.

49. Dertouzos interview, CBI.

50. Licklider interview, CBI.

51. ARPA Contractors Meeting Briefing Book, 8–10 Jan. 1973, Keith Uncapher Papers, CBI.

52. Taylor interview, CBI.

53. Dertouzos interview, CBI.

54. Feigenbaum interview, CBI.

55. Cerf interview, CBI.

56. Taylor interview, CBI.

57. Ornstein interview, CBI.

58. Ibid.; emphasis in original.

59. Frank interview, CBI.

60. Ivan Sutherland, "Outline of a Talk to Be Given on Tuesday, June 15th, at the Course in Computer Graphics, University of Michigan, Ann Arbor: Ten Unsolved Problems in Computer Graphics," RG 330-76-034, box 1, folder: "Graphics," NABDS; a slightly elaborated version can be found in idem, "Computer Graphics: Ten Unsolved Problems," *Datamation* 12 (May 1966): 22–27.

61. D. W. Davies to I. Sutherland, 19 Aug. and 15 Sept. 1965, RG 330-77-0046, box 3, folder: "NPL Symposium," NABDS.

62. Robert L. Simpson Jr., "DOD Applications of Artificial Intelligence: Successes and Prospects," *Society of Photo-Optical Instrumentation Engineers (SPIE): Applications of Artificial Intelligence VI* 937 (1988): 158–64. Also see "An Update on Strategic Computer Vision: Taking Image Understanding to the Next Plateau," *Proceedings of the Society of Photo-Optical Instrumentation Engineers (SPIE)* (Bellingham, Wash.: SPIE, 1989), 52–58.

63. Newell interview, CBI.

64. J.C.R. Licklider, "Easter Message," 2 Apr. 1975, sent by e-mail, University Archives, Hunt Library, CMU.

65. L. G. Roberts, "Resource Sharing Computer Networks," *IEEE International Convention Digest* (Mar. 1969): 326–27; Bolt Beranek and Newman, "The ARPANET Completion Report," draft, 9 Sept. 1977, pp. III-25 to III-26; Roberts interview, CBI.

66. Allen Newell et al., *Speech Understanding Systems: Final Report of a Study Group* (Amsterdam: North-Holland, 1973). The other members of the group—and Newell's co-authors—were Jeffrey Barnett (of SDC), James W. Forgie (of Lincoln), Cordell Green (of Stanford), Dennis Klatt (of MIT), J.C.R. Licklider (of MIT), John Munson (of SRI), D. Raj Reddy (of CMU), and William A. Woods (of BBN).

67. Roberts interview, CBI.

68. Newell et al., *Speech Understanding Systems*, v.

69. Newell interview, CBI.

70. Ruina interview, CBI.

71. Licklider interview, CBI.

72. Herzfeld interview, CBI.

73. Taylor interview, CBI.

74. Ibid.; Herzfeld interview, CBI.

75. This was occasionally done by other government organizations as well. For example, NSF began its university computer facilities program this way in the late 1950s.

76. Lukasik interview, CBI.

77. Uncapher interview, CBI.

78. Heilmeier interview, CBI.

79. Blue interview, CBI.

80. Ibid.

81. Kahn interview, CBI.

82. Robert N. Grosse and Arnold Proschan, "The Annual Cycle: Planning-Programming-Budgeting," in Stephen Enke, ed., *Defense Management* (Englewood Cliffs, N.J.: Prentice-Hall, 1967), 23–41.

83. Blue Ribbon Defense Panel, *Report to the President and the Secretary of Defense on the Department of Defense* (Washington, D.C.: Government Printing Office, 1970), 63–64.

84. "Statement by Eberhardt Rechtin," in Senate Committee on Appropriations, *Department of Defense Appropriations for Fiscal Year 1970, Part 1*, (91) S2010-0-A, 91st Cong., 1st sess., 10, 12–13, 16 June, 15–17, 25 Sept. 1969, 442.

85. Convenience or congressional mandate occasionally dictated that funds for similar activities be placed in categories 6.1 and 6.2. For example, artificial intelligence was supported by both categories until consolidated in 6.2. In such cases, changes in the budgets are not indicative of a trend, and an examination of contracts must occur to be sure of spending patterns (Kahn interview, CBI). As another example of the need for caution, consider a comment from a September 1982 budget review document. In the "Program Summary," there is the following statement:

The 6.1 IPTO program consists of basic research efforts primarily at universities in areas of Intelligent Systems, Robotics, VLSI, and System and Network Concepts. In addition, the Distributed Sensor Net program and C³ System Applications are 6.2 programs that also support basic research on distributed systems and software technology.

"Program Summary," in "FY 84 President's Budget Review: Information Processing Techniques Office," Sept. 1982, p. 1, IPTO Office Files, available in CBI).

86. Taylor interview, CBI. Charles Herzfeld confirmed that he followed this approach during the years he was director of DARPA. Herzfeld interview, CBI.

87. Roberts interview, CBI.

88. Kahn interview, CBI.

89. Some of these aspects of the IPTO management style were borrowed from other agencies such as ONR and NSF, see Nelson R. Kellogg, "Civilian Technology at the National Bureau of Standards," and Toby Appel, "From Macromolecules to Museums: NSF and Support of Biology in the Waterman Era," papers delivered at a Smithsonian Institution symposium "Science and the Federal Patron: Post–World War II Government Support of American Science," 15–16 Sept. 1989, National Museum of American History, Washington, D.C. When we analyze the overall approach of IPTO and DARPA, however, it seems unique in comparison.

90. Licklider interview, CBI.

91. Ibid.

92. Roberts interview, CBI.

93. Newell interview, CBI.

94. Winston interview, CBI; Newell interview, CBI.

95. Kahn interview, CBI.

96. Feigenbaum interview, CBI; Roberts interview, CBI; Blue interview, CBI.

97. Blue interview, CBI.

98. Herzfeld interview, CBI.

99. Uncapher interview, CBI.

100. Allen Newell and D. Raj Reddy, "Multi-Sensor Image Cognition System, MICS, Proposal," 16 Sept. 1974, Computer Science Department Records, University Archives, Hunt Library, CMU; J.C.R. Licklider, "Comments on CMU Proposal," 3–4 Feb. 1975, Computer Science Department Records, University Archives, Hunt Library, CMU.

101. Roberts interview, CBI; and Licklider, "Easter Message."

102. Roberts interview, CBI.

103. Blue interview, CBI.

104. Roberts interview, CBI.

105. Heilmeier interview, CBI.

106. Barry Boehm, director of the IPTO successor program ISTO [later SISTO], private comment to A. Norberg, Jan. 1992.

107. Blue interview, CBI.

108. Roberts interview, CBI.

109. See a report submitted to the House Subcommittee on Defense Appropria-

tions, in House Subcommittee on Defense Appropriations, *Department of Defense Appropriations for Fiscal Year 1970: Procurement Reprogramming Actions*, (91) H2466-0-A, 91st Cong., 1st sess., 19 Mar. 1970, 1044.

110. Senate Committee, *Department of Defense Appropriations for 1970*, 433.

111. House Subcommittee, *Department of Defense Appropriations for 1970*, 1055.

112. Ibid., 1047.

113. Ibid., 1050.

114. "Statement Submitted by Dr. George H. Heilmeier," in Senate Subcommittee on Defense Appropriations, *Department of Defense Appropriations for Fiscal Year 1978, Part 5: Research, Development, Test, and Evaluation, 77* S181-27.1, 95th Cong., 1st sess., 4 Feb. 1977, 44–172.

115. Cooper interview, CBI.

Chapter Two: Sharing Time

1. Robert R. Everett, "Whirlwind," in N. Metropolis, J. Howlett, and Gian-Carlo Rota, eds., *A History of Computing in the Twentieth Century* (New York: Academic Press, 1980), 365–84.

2. Kent C. Redmond and Thomas B. Smith, *Project Whirlwind: The History of a Pioneer Computer* (Bedford, Mass.: Digital Press, 1980), 33.

3. Emerson W. Pugh, *Memories That Shaped an Industry: Decisions Leading to IBM System/360* (Cambridge: MIT Press, 1984).

4. Ibid.

5. Harry R. Borowski, *A Hollow Threat: Strategic Air Power and Containment before Korea* (Westport, Conn.: Greenwood, 1982), chap. 9. We are indebted to Roger Launius for calling this reference to our attention.

6. George E. Valley, "How the SAGE Development Began," *Annals of the History of Computing* 7 (July 1985): 196–226.

7. Air Defense Systems Engineering Committee, "Air Defense System," 24 Oct. 1950, 9-10, as cited in C. Robert Wieser, "The Cape Cod System," *Annals of the History of Computing* 5 (Oct. 1983): 362–69.

8. "At RAND: Massed Brains to Meet Air Threat," *Business Week* 1383 (3 Mar. 1956): 86–92.

9. Wieser, "Cape Cod System."

10. John F. Jacobs, "SAGE Overview," *Annals of the History of Computing* 5 (Oct. 1983): 323–29.

11. Redmond and Smith, *Project Whirlwind*, 126, 174.

12. Valley, "How the SAGE Development Began."

13. For a description of the modification of the Whirlwind design to the AN/FSQ-7 see Charles J. Bashe, Lyle R. Johnson, John H. Palmer, and Emerson W. Pugh, *IBM's Early Computers* (Cambridge: MIT Press, 1986), 242–48.

14. James A. Fusca, "First SAGE Direction Center Operating," *Aviation Week* 69 (7 July 1958): 34–35. For descriptions of SAGE, see Emerson W. Pugh, Lyle R. Johnson,

and John H. Palmer, *IBM's 360 and Early 370 Systems* (Cambridge: MIT Press, 1991), 569. Also see Valley, "How the SAGE Development Began"; "The Controversial SAGE," *Business Week* 1365 (29 Oct. 1955): 30–31.

15. "Pushbutton Defense for Air War," *Life* 42 (11 Feb. 1957): 62–67.

16. D. F. Parkhill, *The Challenge of the Computer Utility* (Reading, Mass.: Addison-Wesley, 1966), 56–57.

17. Pugh, Johnson, and Palmer, *IBM's 360 and Early 370 Machines*, 599.

18. Wieser, "Cape Cod System," 362–69.

19. "Pushbutton Defense for Air War," 62–67.

20. The first demonstration of remote on-line computer use was at the 1940 meeting of the American Mathematical Society. Using a terminal connected over telephone lines to the Bell Laboratory's Complex Number Computer, the mathematicians at the meeting could enter problems in the addition, subtraction, multiplication, and division of complex numbers, with answers received in less than a minute. Bell Laboratories continued using the machine in this mode via three operator stations located in various areas of the laboratory. See S. Millman, ed., *A History of Engineering and Science in the Bell System: Communications Sciences, 1925–1980* (New York: AT&T Bell Laboratories, 1984), 359.

21. Sherman C. Blumenthal, "An Approach to On-Line Processing," *Datamation* 7 (June 1961): 23–24.

22. A comparison of memory size is instructive. Memories in machines delivered in the early 1950s contained from 1,000 words of primary storage (UNIVAC I, using acoustic delay lines) to 16,384 (ERA 1101, using a magnetic drum). Later models in the 1950s used core memory. For example, the IBM 704 contained 8,192 words of primary storage as core memory. (Data on memory taken from Martin H. Weik, *A Second Survey of Domestic Electronic Digital Computing Systems*, Report no. 1010 [Aberdeen, Md.: Ballistic Research Laboratories, 1957].)

23. Montgomery Phister, *Data Processing Technology and Economics*, 2d ed. (Santa Monica, Calif.: Digital Press and Santa Monica Publishing, 1979), 289.

24. John Diebold, "Experience, 1959, in Automatic Data Processing—A Review," *Computers and Automation* 9 (July 1960): 9, 10–12.

25. Benjamin Conway, "What Business Needs Most from Manufacturers of Electronic Data Processors," *Computers and Automation* (Nov. 1960): 6–8.

26. For brief histories of some of these early programs, see Grace Murray Hopper, "Keynote Address," in Richard L. Wexelblat, ed., *History of Programming Languages*, ACM Monograph Series: Proceedings of the History of Programming Languages Conference (New York: Academic Press, 1981), 7–20; John Backus, "The History of FORTRAN I, II, and III," in ibid., 25–66.

27. H.R.J. Grosch, "Standardization of Computer Intercommunication," in *Proceedings of the Eastern Joint Computer Conference* [1955] (New York: Institute of Radio Engineers, 1956), 87–89.

28. T. B. Steel Jr., "SHARE," in Anthony Ralston, ed., *Encyclopedia of Computer Science and Engineering*, 2d ed. (New York: Van Nostrand-Reinhold, 1983), 1319–20; M. M. Maynard, "USE," in ibid., 1552–53.

29. George Ryckman, discussant in FORTRAN Session, commenting on Backus, "History of FORTRAN," in Wexelblat, ed., *History of Programming Languages*, 68.

30. F. J Corbato, M. M. Daggett, and R. C. Daley, "An Experimental Time-Sharing System," in *Proceedings of the AFIPS Spring Joint Computer Conference*, vol. 21 (Palo Alto, Calif.: National Press, 1962), 335–44.

31. Christopher Strachey, "Time Sharing in Large Fast Computers," in *Proceedings of the International Conference on Information Processing* [1959] (Paris: UNESCO, 1960), 336–41.

32. D. J. Breheim, " 'Open Shop' Programming at Rocketdyne Speeds Research and Production," *Computers and Automation* 10 (July 1961): 8–9.

33. Ibid., 8–9.

34. Corbato interview, CBI.

35. Nathaniel Rochester, "The Computer and Its Peripheral Equipment," *Proceedings of the Eastern Joint Computer Conference* [1955] (New York: Institute of Radio Engineers, 1956), 64–69.

36. E. F. Codd, "Multiprogramming," *Advances in Computers* 3 (1962): 77–153. Also see his description of real-time multiprogramming, p. 87.

37. For example, see W. F. Bauer, "Computer Design from the Programmer's Viewpoint," *Proceedings of the Eastern Joint Computer Conference* [1958] (New York: American Institute of Electrical Engineers, 1958), 46–51.

38. Maurice V. Wilkes, "Introductory Speech," *Proceedings of the International Conference on Information Processing* [1959] (Paris: UNESCO, 1960), 331–33.

39. Strachey, "Time Sharing in Large Fast Computers," 336–41. For additional information on the context of Strachey's work, see Martin Campbell-Kelly's "Christopher Strachey, 1916–1975: A Biographical Note," *Annals of the History of Computing* 7 (1985): 19–42, esp. 29–30.

40. F. J. Corbato, M. M. Daggett, and R. C. Daley, "An Experimental Time-Sharing System," in *Proceedings of the AFIPS Spring Joint Computer Conference*, vol. 21 (Palo Alto, Calif.: National Press, 1962), 335–44.

41. John McCarthy, "Reminiscences on the History of Time-Sharing," *IEEE Annals of the History of Computing* 14 (1992): 19–24.

42. A flexowriter is an electronic typewriter that could send and receive binary coded signals corresponding to the operations of the typewriter.

43. MIT, Computation Center, "Progress Report Number 4 of the Research and Educational Activities in Machine Computation by the Cooperating Colleges of New England," Dec. 1958, Academic Computing Collection, CBI.

44. Dartmouth was one of the New England schools that shared one of the three shifts on the IBM computer located at MIT.

45. John McCarthy, "History of LISP," in Wexelblat, ed., *History of Programming Languages*, 173–91.

46. John McCarthy to P. M. Morse, "A Time Sharing Operator for Our Projected IBM 709," 1 Jan. 1959, Collection MC 75, box 25, folder: "Computation Center," Institute Archives, MIT.

47. Ibid. Note that the memo was written eighteen months before the proposed system was scheduled to arrive. The details of the system were of course unknown.

48. Herbert Teager and John McCarthy, "Time-Shared Program Testing," *Preprints of the Papers Presented at the Fourteenth National Meeting of the Association for Computing Machinery* (New York: ACM Headquarters, 1959), sec. 12, pp. 1–2.

49. MIT, Computation Center, "Progress Report Number 6 of the Research and Educational Activities in Machine Computation by the Cooperating Colleges of New England," Jan. 1960, Academic Computing Collection, CBI.

50. "MIT Computation—Present and Future," Jan. 1961, Collection MC 75, box 25, folder: "MIT Computation: Present and Future #1," Institute Archives, MIT.

51. Other members listed are Dean Arden (of the Research Laboratory of Electronics and Electrical Engineering), Michael Barnett (of the Solid State Physics Group), Phyllis Fox (of the Computation Center), Daniel Goldenberg (of the Instrumentation Laboratory), Ronald Howard (of the Electrical Engineering Department), Norman Phillips (of the Meteorology Department), Alexander Pugh (of the Industrial Dynamics Group), and Victor Yngve (of the Research Laboratory of Electronics and Modern Languages). "Report of the Long Range Computation Study Group," Apr. 1961, Collection AC 12, box 17, folder: "Project MAC Proposal, Regular 1963," Institute Archives, MIT.

52. "MIT Computation—Present and Future."

53. Wesley Clark, who was part of the working committee, did not agree with the recommendation and did not sign either of the reports. Wesley Clark, "The LINC Was Early and Small," in Adele Goldberg, ed., *History of Personal Workstations* (New York: ACM Press, 1988), 358.

54. "Report of the Long Range Computation Study Group," Apr. 1961.

55. Albert Hill, "Meeting on MIT Computation Problems, 23 May 1961, Present: Dr. Stratton, Dr. Floe, Dean Brown, Professors Elias, Fano, Morse and Hill," memorandum, 24 May 1961, Collection AC 12, box 17, folder: "Project MAC General 1961-8," Institute Archives, MIT. It appears that Peter Elias replaced Jerome Wiesner on the ad hoc faculty committee.

56. Fano, Morse, and Hill to Carl Floe, 30 Mar. 1961, Collection AC 12, box 17, folder: "Project MAC General 1961-8," Institute Archives, MIT.

57. Computer science and its development as a research enterprise as seen from an NSF perspective are under investigation at the IEEE Center for the History of Electrical Engineering by William Aspray and his colleagues.

58. "Proposal to the National Science Foundation for Continuation of Support of Research and Development of a Compatible Time-Sharing System, Suitable for University Use," 12 Mar. 1963, Collection MC 75, box 25, folder: "Computation Center," Institute Archives, MIT.

59. John McCarthy, F. J. Corbato, Edward Arthurs, Jack B. Dennis, and Douglas T. Ross, "Working Committee Version of Proposal for MIT Advanced Computer System," 20 Oct. 1961, Collection AC 125, box 73 (73-7), folder: "Working Committee Proposal Draft," Institute Archives, MIT.

60. New Computer Working Committee (E. Arthurs, F. J. Corbato, J. Dennis, J. McCarthy [Chairman], and D. Ross), "Review of Specifications for the New Computer," undated, Collection AC 134, box 13, folder: "Computation Center 1958-7/64," Institute Archives, MIT.

61. Philip M. Morse, *In at the Beginnings: A Physicist's Life* (Cambridge: MIT Press, 1977), 308.

62. Corbato, Daggett, and Daly, "Experimental Time-Sharing System."

63. McCarthy et al., "Working Committee Version."

64. Information International, "Research on the Time-Sharing of Computers," Proposal no. 62-13, submitted to ARPA, 21 Nov. 1962, app. 2, RG 330-69-A-4998, box 2, folder: "Envelope 349-1," NABDS.

65. Edward Fredkin, "The Time-Sharing of Computers," *Computers and Automation* 12 (Nov. 1963): 12–20.

66. McCarthy interview, CBI. For the connection between Lincoln Laboratory and Digital Equipment Corporation see C. Gordon Bell, J. Craig Mudge, and John E. McNamara, *Computer Engineering: A DEC View of Hardware System Design* (Bedford, Mass.: Digital Press, 1978).

67. J. McCarthy, S. Boilen, E. Fredkin, and J.C.R. Licklider, "A Time-Sharing Debugging System for a Small Computer," in *Proceedings of the AFIPS Spring Joint Computer Conference*, vol. 23 (Baltimore: Spartan Books, 1963), 51–57.

68. Licklider describes the system as being too small to be usefully time-shared (Licklider interview, CBI). McCarthy says it wasn't used because there were so few users of the systems that dedicated mode sign-up was acceptable to them (McCarthy interview, CBI).

69. The design of the PDP-1 was related to that of the TX-0. See Bell, Mudge, and McNamara, *Computer Engineering*; and Clark, "LINC Was Early and Small."

70. Dennis interview, CBI.

71. J. E. Yates, "A Time-Sharing System for the PDP-1 Computer," Report AFCRL-62-519 (MIT Report ESL-R-140), June 1962, MIT, Electronic Systems Laboratory, Cambridge, Mass.

72. Thomas E. Kurtz, "BASIC," in Wexelblat, ed., *History of Programming Languages*, 515–49. McCarthy was connected with the Computation Center's CTSS, BBN, MIT's Electrical Engineering Department, SDC (on the advisory committee), Dartmouth, and Stanford.

73. Lewis E. Lacher, "Time-Sharing: Low-Cost Link to Computer," *Computer Digest*, charter issue (Sept. 1965): 5–10.

74. Pamela McCorduck, *Machines Who Think* (San Francisco: Freeman, 1979). Rand's JOHNNIAC was a first-generation computer modeled after the Institute for Advanced Study machine design developed by John von Neumann, and named after von Neumann. The term *open shop* meant that those other than professional programmers could use the computer.

75. J. C. Shaw, "JOSS: A Designer's View of an Experimental On-Line System," in *Proceedings of the AFIPS Fall Joint Computer Conference*, vol. 26 (Baltimore: Spartan Books, 1964), 455–64.

76. The school was known at this time as the Carnegie Institute of Technology.

77. Perlis cross-examination, in trial transcript, *United States v. IBM*, 2039, Computer and Communications Association Antitrust Records, CBI.

78. Licklider, "Man-Computer Symbiosis."

79. McCarthy et al., "Time-Sharing Debugging System," 51–57.

80. Licklider interview, CBI.

81. Ibid.

82. DARPA, "Computer Augmentation of Computer Programming," Program Plan no. 674, 7 Jan. 1963, RG 330-78-0013, box 1, folder: "Program Plans," NABDS.

83. DARPA, "A Research and Development Program on Computer Systems," Program Plan no. 683, 17 Jan. 1963, RG 330-78-0013, box 1, folder: "Program Plans," NABDS.

84. System Development Corporation, "SDC Command-Control Research, Proposal 184," Nov. 1960, Burroughs Corporation Records, System Development Corporation Series, CBI.

85. Stuart Cooney, "Command Control Research and Development," SDC Magazine 5 (Feb. 1962), Burroughs Corporation Records, System Development Corporation Series, CBI.

86. P. Greenburg to T. Rowan, "SDC History," 9 July 1964, memorandum no. M-14368, folder: "PDG Archives," Burroughs Corporation Records, System Development Corporation Series, CBI.

87. System Development Corporation, "SDC Command-Control Research, Proposal 184."

88. Erwin F. Czichos to Dr. Urner Liddel, 24 Aug. 1961, attachment D, p. 3, L-12810, folder: "Proposal 251," Burroughs Corporation Records, System Development Corporation Series, CBI.

89. Licklider to Ruina, 5 Nov. 1962, RG 330-69-A-4998, box 3, folder: "Contracts, Budget and Planning, 1961-1962," NABDS.

90. Licklider to Edward Fredkin, 12 Dec. 1963, RG 330-69-A-4998, box 2, folder: "Information International, Inc.," NABDS.

91. DARPA, "Research on the Time-Sharing of Computers in Command and Control Research," Program Plan no. 672, 4 Jan. 1963, RG 330-78-0013, box 1, folder: "Program Plans," NABDS.

92. Licklider to Fredkin, 12 Dec. 1963, RG 330-69-A-4998, box 2, folder: "Information International, Inc.," NABDS.

93. Edward Fredkin to George Mandanis, 13 Nov. 1962, RG 330-69-A-4998, box 2, folder: "Information International, Inc.," NABDS.

94. For further information about JOVIAL, see Jules I. Schwartz, "The Development of JOVIAL," in Wexelblat, ed., History of Programming Languages, 369–401.

95. Schwartz interview, CBI.

96. Claude Baum, The System Builders: The Story of SDC (Santa Monica, Calif.: System Development, 1981), 91.

97. Jules I. Schwartz and Clark Weissman, "The SDC Time-Sharing System Revisited," Proceedings of the ACM National Meeting [1967] (Washington, D.C.: Thompson Book, 1967), 263–71.

98. Licklider, Memorandum for the Record, 6 Nov. 1963, RG 330-69-A-4998, box 3, folder: "350-1," NABDS.

99. DARPA, "Remote Stations and Programs for Computer Networks," Program Plan no. 95, 5 Apr. 1963, RG 330-78-9913, box 1, folder: "Program Plans," NABDS.

100. It is also occasionally seen as "Man and Computer" (for example in Morse, *In at the Beginnings*), and derisively as "More Assets for Cambridge."

101. Fano interview, CBI.

102. Robert M. Fano, "Rough Outline of a Proposal to ARPA for a Comprehensive Project on Information Processing at MIT," p. 1, Collection AC 134, box 30, folder: "Project MAC," Institute Archives, MIT.

103. Ibid.

104. Charles Townes to Carl Overhage, 4 Dec. 1962, Collection MC 75, box unspecified, folder: "MIT Committee on Information Processing," Institute Archives, MIT.

105. "Proposal for Continuation of a Research and Development Program on Computer Systems," 20 Feb. 1964, pp. 1–14, Collection AC 12, box 17, folder: "Project MAC Proposals, Regular 1964," Institute Archives, MIT.

106. Fano interview, CBI; Licklider interview, CBI.

107. Licklider interview, CBI.

108. Fano interview, CBI. See also Robert Fano, "Project MAC," in Jack Belzer, Albert G. Holzman, and Allen Kent, eds., *Encyclopedia of Computer Science and Technology* (New York: Dekker, 1979), 339–60.

109. Licklider interview, CBI.

110. Licklider's involvement in setting these conditions is confirmed by Robert Fano in his interview, CBI.

111. DARPA, "Research and Development Program on Computer Systems."

112. Ibid.

113. Ibid.

114. MIT, "Proposal for a Research and Development Program on Computer Systems," 14 Jan. 1963, AC 12, box 17, folder: "Project MAC Proposals, Regular 1963," Institute Archives, MIT.

115. Ibid.

116. "Statement by Robert L. Sproull," in House Subcommittee on Defense Appropriations, *Department of Defense Appropriations for Fiscal Year 1965, Part 5: Research, Development, Test, and Evaluation, Appropriation Language, Testimony of Members of Congress, Organizations, and Interested Individuals,* (88) H2062-0-B, 88th Cong., 2d sess., 6, 11–13, 16–19 Mar. 1964, 127.

117. Fano, "Project MAC," 339–60.

118. "Proposal for Continuation," 20 Feb. 1964, 3–5.

119. DARPA, "Research and Development Program on Computer Systems."

120. Karl L. Wildes and Nilo A. Lindgren, *A Century of Electrical Engineering and Computer Science at MIT, 1882–1982* (Cambridge: MIT Press, 1985), 267, 294–95.

121. Fano, "Rough Outline."

122. "Status Report as of October 7, 1963," memorandum MAC-A-69, Collection 82-45, box 1, folder: "Project MAC 2/5," Institute Archives, MIT.

123. "Proposal for Continuation of Project MAC Research," Oct. 1965, p. 12, Collection AC 12, box 17, folder: "Project MAC Proposals, Regular 1965," Institute Archives, MIT.

124. Dennis interview, CBI.

125. "Acquisition of Computer Installation as a Replacement for the Present IBM 7094 Installation," 23 Nov. 1964, Collection AC 12, box 17, folder: "Project MAC Proposals, Regular 1964," Institute Archives, MIT.

126. R. M. Fano to J. A. Stratton, 29 June 1964, Collection AC 12, box 17, folder: "Project MAC General, 1961–1968," Institute Archives, MIT.

127. Dennis interview, CBI.

128. Weil testimony, in trial transcript, *United States v. IBM*, p. 7123, Computer and Communications Association Antitrust Records, CBI.

129. R. M. Fano to J. A. Stratton, 19 June 1964, Collection AC 12, box 17, folder: "Project MAC General, 1961–1968," Institute Archives, MIT.

130. Pugh, Johnson, and Palmer, *IBM's 360 and Early 370 Systems*.

131. Newell interview, CBI. Green went to CMU from MIT, where he was engaged in AI research. He developed the program BASEBALL.

132. Licklider to Fubini, 26 Nov. 1963, RG 330-78-0013, box 1, unlabeled folder, NABDS.

133. DARPA, "Heuristic Programming and Theory of Computation," Program Plan no. 56, 20 Mar. 1963, RG 330-78-0013, box 1, folder: "Program Plans," NABDS; Licklider to "Members and Affiliates of the Intergalactic Computer Network," 25 Apr. 1963, RG 338-69-A-4998, box 3, folder: "350-1," NABDS.

134. Taylor interview, CBI; W. W. Lichtenberger and M. W. Pirtle, "A Facility for Experimentation in Man-Machine Interaction," in *Proceedings of the AFIPS Fall Joint Computer Conference*, vol. 27 (Washington, D.C.: Spartan Books, 1965), 589–98.

135. Sidney Fernbach and Harry Huskey, "Introduction," in Walter J. Karplus, ed., *On-Line Computing* (New York: McGraw-Hill, 1967), 6–7; DARPA, "Man-Machine Interaction," Program Plan no. 610, 9 Mar. 1966, RG 330-78-0013, box 1, folder: "Program Plans," NABDS; see also Lichtenberger and Pirtle, "Facility for Experimentation," 589.

136. SDS news release on Las Vegas meeting and announcement of SDS 940 in SDS folder, Product Literature Collection, CBI; on Taylor's interaction with Max Palevsky and attempts to have SDS produce a machine like the Berkeley system, see Douglas K. Smith and Robert C. Alexander, *Fumbling the Future: How Xerox Invented, Then Ignored, the First Personal Computer* (New York: Morrow, 1988), 61–62.

137. DARPA, "Dynamic Computer Models of Cognitive Processes," Program Plan no. 318, 25 May 1964, RG 330-78-0013, box 1, folder: "Program Plans," NABDS; and idem, "Research in Conversational Use of Computers," Program Plan no. 470, 25 Mar. 1965, RG 330-78-0013, box 1, folder: "Program Plans," NABDS.

138. Licklider to Bartels, 2 Oct. 1963, RG 330-73-A-2108, box 1, folder: "University of Michigan," NABDS.

139. Daniel G. Bobrow, Jerry D. Burchfiel, Daniel L. Murphy, and Raymond S. Tomlinson, "TENEX, A Paged Time Sharing System for the PDP-10," *Communications of the ACM* 15 (Mar. 1972): 135–43.

140. Ibid., 135.

141. F. J. Corbato, "System Requirements for Multiple Access, Time-Shared Computers," MAC-TR-3, May 1964, Academic Computing Collection, CBI.

142. Ibid.

143. Wright testimony, in trial transcript, *United States v. IBM*, 12944.

144. Angeline Pantages, "The ACM Conference," *Datamation* 11 (Oct. 1965): 53.

145. David E. Weisberg, "Computer Characteristics Quarterly—Recent Trends," *Datamation* 12 (Jan. 1966): 55–56.

146. For details about various pricing schemes, see D. F. Parkhill, *The Challenge of the Computer Utility* (Reading, Mass.: Addison-Wesley, 1966).

147. "Two Firms Offer Time-Sharing Service," *Datamation* 11 (Aug. 1965): 71.

148. Parkhill, *Challenge*, 78.

149. Weil testimony, in trial transcript, *United States v. IBM*, 7106-7.

150. Parkhill, *Challenge*, 70.

151. Jeremy Main, "Computer Time-Sharing—Everyman at the Console," *Fortune* 76 (Aug. 1967): 88–91, 187–88, 190.

152. Neil MacDonald, "A Time-Shared Computer System—The Disadvantages," *Computers and Automation* 14 (Sept. 1965): 21–22.

153. Louis Fein, "Time-Sharing Hysteria," letter to the editor, *Datamation* 11 (Nov. 1965): 14–15.

154. R. Patrick, quoted in "A Panel Discussion on Time-Sharing," *Datamation* 10 (Nov. 1964): 38–44.

155. Maurice V. Wilkes, "An Evaluation of the Compatible Time-Sharing System," memorandum MAC-M-103, app. H-2 to "Status Report as of October 7, 1963," MAC-A-69, Collection 82-45, box 1, folder: "Project MAC 2/5," Institute Archives, MIT.

156. No studies convincingly compared different methods of computer use in quantitative terms. One of the most sought-after benefits of time sharing was the improvement in the productivity of programmers. The anecdotal evidence from the Dartmouth system was encouraging: "The time-sharing man-machine interaction produces a situation where staff and students can implement their concepts by programming their mathematical models, debug programs, and obtain answers 10 to 100 times faster than a person operating in an efficient batch processing system" (Paul T. Shannon, Myron Tribus, and Stanley A. Gembicki, "Time-Sharing Computers in Design Education," *IEEE International Convention Record* 15, pt. 10 [1967]: 3–6). However, productivity claims were hard to support. In a 1967 review of the experimental results on programmer productivity in time-sharing and batch systems, Patrick argued that the measurement techniques lacked validity, the conclusions drawn were suspect, and the studies did not adequately cover the differences in cost of the methods compared (Robert Patrick, "Time-Sharing Tally Sheet," *Datamation* 13 [Nov. 1967]: 42–47). A 1969 study concluded that the differences between individual programmers outweighed the method of machine use. It found no particular preference for time sharing when compared to rapid batch turn-around among the students in the study (Jeanne Adams and Leonard Cohen, "Time-Sharing vs. Instant Batch Processing," *Computers and Automation* 18 [Mar. 1969]: 30–34).

157. Marvin Emerson, "The 'Small' Computer versus Time-Shared Systems," *Computers and Automation* 14 (Sept. 1965): 18–20.

158. Martin Solomon Jr., "Are Small Free-standing Computers *Really* Here to Stay?" *Datamation* 12 (July 1966): 66–67.

159. Fred Gruenberger, "Are Free-Standing Computers Here to Stay?" *Datamation* 12 (Apr. 1966): 67–68.

160. See Ornstein interview, CBI; Clark interview, CBI.

161. Ornstein interview, CBI; emphasis in original.

162. "Acquisition of Computer Installation as a Replacement for the Present IBM 7094 Installation," 23 Nov. 1964, p. 11, Collection AC 12, box 17, folder: "Project MAC Proposals Regular 1964," Institute Archives, MIT.

163. Fano to Brown, "The MULTICS-645 System as an MIT Computer Utility," 17 Mar. 1966, Collection AC 12, box 17, folder: "MAC General 1966-9," Institute Archives, MIT.

164. Licklider to Smullin, 19 Dec. 1968, Collection AC 12, box 17, folder: "Project MAC 3/5," Institute Archives, MIT.

165. Weil testimony, in trial transcript, *United States v. IBM*, 7234.

166. Licklider to Smullin, "My Introductory Remarks to the Multics Review Committee," 18 Dec. 1968, Collection 82-45, box 2, folder: "Project MAC 3/5," Institute Archives, MIT.

167. Licklider to Wiesner, "Planning for Multics," 27 Dec. 1968, Collection 82-45, box 2, folder: "Project MAC 3/5," Institute Archives, MIT.

168. F. J. Corbato, "Foreword," in Elliott I. Organick, *The Multics System: An Examination of Its Structure* (Cambridge: MIT Press, 1972), ix–xi.

169. Fernando J. Corbato, J. H. Saltzer, and C. T. Clingen, "Multics—The First Seven Years," in *Proceedings of the AFIPS Spring Joint Computer Conference*, vol. 40 (Montvale, N.J.: AFIPS Press, 1972), 571–82.

170. L. G. Roberts, "Extension of Contract NOOO14-70-A-0362-0001 with Project MAC," 14 Jan. 1972, 2095 MIT ONR—File 1 of 2, IPTO Office Files, CBI.

171. L. G. Roberts, "Extension of Contract with MIT," 30 Jan. 1973, 2095 MIT ONR—File 1 of 2, IPTO Office Files, CBI.

172. Weil testimony, in trial transcript, *United States v. IBM*, 7236.

173. Fano, "Project MAC."

174. By 1986 Honeywell had fifty-four customers for its MULTICS product, nearly 50 percent of whom were at military and government locations (*Honeywell Monthly* 4 [Oct. 1986]).

175. J.C.R. Licklider, "Position Paper on the Future of Project MAC," 6 Oct. 1970, Collection 85-25, box 10, folder: "Project MAC," Institute Archives, MIT.

176. Ibid.

177. Franklin M. Fisher, James W. McKie, and Richard B. Mancke, *IBM and the U.S. Data Processing Industry: An Economic History* (New York: Praeger, 1983), 166.

178. Maria Eloina Pelaez Valdez, "A Gift from Pandora's Box: The Software Crisis" (Ph.D. diss. University of Edinburgh, 1988).

179. Pugh, Johnson, and Palmer, *IBM's 360 and Early 370 Systems*.

180. For an example, see Brandt Allen, "Time-Sharing Takes Off," *Harvard Business Review* 47 (Mar.–Apr. 1969): 128–36.

181. Phyllis R. Kennedy, "The ADEPT-50 System: A General Description of the Time-Sharing Executive and the Programmer's Package," 4 Apr. 1968, p. 7, SDC Document TM-3899/100/00, Burroughs Corporation Records, System Development Corporation Series, CBI.

182. Ibid., 8.

183. Ibid., 9.

184. Ibid., 7.

185. Baum, *System Builders*, 118.

186. Ibid., 152–55.

187. Kennedy, "ADEPT-50 System," 7.

188. Baum, *System Builders*, 118.

189. Kennedy, "The ADEPT-50 System."

190. SDC, "Research Proposal into Computer Network Security," SDC Proposal 73-5459 (internal company document), undated, p. I-3, Burroughs Corporation Records, System Development Corporation Series, CBI.

191. Alfred H. Vorhaus and Robert D. Willis, "The Time-Shared Data Management System: A New Approach to Data Management," 13 Feb. 1967, p. 4, SDC Document SP-2747, Burroughs Corporation Records, System Development Corporation Series, CBI.

192. SDC, *Catalog of Products and Services*, 31 May 1968, pp. 2–5, Burroughs Corporation Records, System Development Corporation Series, CBI.

193. R. R. Linde, C. Weissman, and C. E. Fox, "The ADEPT-50 Time-Sharing System," in *Proceedings of the AFIPS Fall Joint Computer Conference*, vol. 35 (Montvale, N.J.: AFIPS Press, 1969), 39.

194. Baum, *System Builders*, 119.

195. R. M. Fano and F. J. Corbato, "Time-Sharing on Computers," *Scientific American* 215 (Sept. 1966): 129–40.

196. Ivan Sutherland, speech for presentation at ACM chapter meeting, 20 Jan. 1966, RG 330-78-0013, box 1, folder: "Program Plans," NABDS.

197. The Project MAC "CTSS Programmer's Guide" lists a MAIL command in July 1965. Although the command is not listed in the 1963 version of the manual, the July 1965 version appears to be an update to the command. Academic Computing Collection, CBI.

198. It could be argued that having multiple processes working on a user's behalf in modern workstations is time sharing, but this would require reverting to an earlier definition of the term. It should be noted, however, that the ability of computers to work on more than one process at a time was developed in part for time-sharing systems.

Chapter Three: Getting the Picture

1. To appreciate the overall character of graphics developments and the sweep of the field's present activity, see a readable essay by Robert Rivlin: *The Algorithmic Image: Graphic Visions of the Computer Age* (Redmond, Wash.: Microsoft Press, 1986).

2. Michael S. Mahoney, Norman H. Taylor, Douglas T. Ross, and Robert M. Fano, "Retrospectives I: The Early Years in Computer Graphics at MIT, Lincoln Laboratory, and Harvard," SIGGRAPH '89 Panel Proceedings, *Computer Graphics* 23 (1989): 19–38.

3. Cynthia Goodman, *Digital Visions: Computers and Art* (New York: Abrams, 1987).

4. Ibid.

5. Ibid.

6. John F. Jacobs, "SAGE Overview," *Annals of the History of Computing* 5 (Oct. 1983): 325.

7. Steven A. Coons, "An Outline of the Requirements for a Computer-Aided Design System," in *Proceedings of the AFIPS Spring Joint Computer Conference,* vol. 23 (Baltimore: Spartan Books, 1963), 299–304.

8. D. T. Ross and J. E. Rodriquez, "Theoretical Foundations for the Computer-Aided Design System," in *Proceedings of the AFIPS Spring Joint Computer Conference,* vol. 23 (Baltimore: Spartan Books, 1963), 305–22.

9. Karl L. Wildes and Nilo A. Lingren, *A Century of Electrical Engineering and Computer Science at MIT, 1882–1982* (Cambridge: MIT Press, 1985), 350.

10. To appreciate the computational load, consider the way in which a vector display system, the most common CRT display system of the 1960s, produces an image on the computer screen Vector display screens came with phosphors in white, green, yellow, blue, and red. The displays produced lines by lighting parts of the screen, in any sequence directed by the computer, to produce the colored image. The lines were divided into small units called pixels. (For example, a screen of 512 lines, where each line is divided into 512 pixels, contains more than 262,000 pixels requiring continuous attention.) Each pixel had to be programmed. Add to this the application elements that help to change the computer image in desirable ways, and the amount of calculating the computer needed to do, especially in view of how small computer memories were at that time, became enormous.

11. B. M. Gurley and C. E. Woodward, "Light Pen Links Computer to Operator," *Electronics* 32, no. 47 (1959): 85–87.

12. It should be noted that NSF and NIH support in the 1970s and 1980s led to significant graphics programs for the imaging of scientific and engineering models. It is our contention that these were possible as a result of the IPTO-supported research in the 1960s.

13. MIT, "Project MAC Status Report," 7 Oct. 1963, pp. 82–85, Collection AC 82-45, box 1, folder: "Project MAC 2/5," Institute Archives, MIT.

14. See for example, Douglas T. Ross, "The NATO Conferences from the Perspective of an Active Software Engineer," *Annals of the History of Computing* 11 (1989): 133–41.

15. Ross interview, CBI.

16. Robert Stotz, "Man-Machine Console Facilities for Computer-Aided Design," in *Proceedings of the AFIPS Spring Joint Computer Conference* 23:323–28.

17. Wildes and Lingren, *Electrical Engineering and Computer Science at MIT,* 351.

18. J.C.R. Licklider, "Graphic Input: A Survey of Techniques," in F. Gruenberger, ed., *Computer Graphics: Utility, Production, Art* (New York: Thompson Book, 1967), 46.

19. DARPA, "CRT-Aided Semi-Automated Mathematics," Program Plan no. 454,

15 Mar. 1965, RG 330-78-0013, box 1, folder: "Program Plans," National Archives Branch Depository, Suitland, Md. (NABDS).

20. ARPA, "Graphic Control and Display of Computer Processes," Program Plan no. 439, 1 Mar. 1965, RG 330-78-0013, box 1, folder: "Program Plans," NABDS.

21. Thomas O. Ellis, J. F. Heafner, and W. L. Sibley, "ARPA Semiannual Report No. 10," 22 June 1964, p. 10, Keith Uncapher Papers, box 1, folder: "Grail," CBI.

22. M. R. Davis and T. O. Ellis, "The Rand Tablet: A Man-Machine Graphical Communication Device," in *Proceedings of the AFIPS Fall Joint Computer Conference*, vol. 26 (Washington, D.C.: Spartan Books, 1964), 325–31.

23. Thomas O. Ellis, J. F. Heafner, and W. L. Sibley, "The Grail Project: An Experiment in Man-Machine Communications," RM-5999-ARPA (Santa Monica, Calif.: Rand Corporation, 1969).

24. G. F. Groner, "Real-Time Recognition of Handprinted Text," RM-5016-ARPA (Santa Monica, Calif.: Rand Corporation, 1966).

25. Thomas O. Ellis, "Some Promising Areas of Graphics Research," 15 July 1966, Uncapher Papers, box 1, folder: "Grail."

26. William R. Sutherland, James W. Forgie, and Marie V. Morello, "Graphics in Time-Sharing: A Summary of the TX-2 Experience," in *Proceedings of the AFIPS Spring Joint Computer Conference*, vol. 34 (Montvale, N.J.: AFIPS Press, 1969), 629–36.

27. Ivan E. Sutherland, "Computer Inputs and Outputs," *Scientific American* 215 (Sept. 1966): 86–96.

28. L. Gallenson, "A Graphic Tablet Display Console for Use under Time-Sharing," in *Proceedings of the AFIPS Fall Joint Computer Conference*, vol. 31 (Washington, D.C.: Thompson Book, 1967), 689–95.

29. T. O. Ellis and W. L. Sibley, "On the Problem of Directness in Computer Graphics," in D. Secrest and J. Nievergelt, eds., *Emerging Concepts in Computer Graphics* (New York: Benjamin, 1968).

30. Gallenson, "Graphic Tablet."

31. Ivan E. Sutherland, "Sketchpad: A Man-Machine Graphical Communication System," MIT Lincoln Laboratory Technical Report no. 296 (1963).

32. We are grateful to Mary L. Murphy of Lincoln Laboratory for arranging to show us several versions of original films on Sketchpad and films of the successive versions of Sketchpad.

33. Sutherland, "Sketchpad"; Licklider, "Graphic Input."

34. John Lewell, *Computer Graphics: A Survey of Current Techniques and Applications* (New York: Van Nostrand Reinhold, 1985), 12.

35. At about the same time that Sutherland was developing Sketchpad, two companies were independently at work on similar systems. Itek, a Lexington, Massachusetts, company, developed the Digigraphics system, an interactive system for lens design (M. David Prince, *Interactive Graphics for Computer-Aided Design* [Reading, Mass.: Addison-Wesley, 1971], 5). This system was later purchased by Control Data Corporation and became the basis of its interactive computer graphics line, which evolved into a variety of products, including the CDC Digigraphics 270 System (C. Machover,

"A Brief, Personal History of Computer Graphics," *Computer* 11 [Nov. 1978]: 39). On the acquisition of the Itek system by CDC, see the "Acquisitions Notebook," CDC Collection, CBI. CDC expected to use the system in machine tool control and weapons systems ("President's Newsletter," June 1963, CDC Collection, CBI). General Motors developed a system called Design Augmented by Computers (DAC-1). The hardware for DAC-1 was built by IBM, and its display was the prototype for the IBM 2250 console, which became part of the commercially available IBM System/360 computer series (Machover, "Personal History"; Prince, *Interactive Graphics*, 5). DAC-1 was kept secret until its unveiling a year after Sketchpad was demonstrated (Prince, *Interactive Graphics*, 5).

36. Licklider, "Graphic Input," 44.

37. K. B. Irani and A. W. Naylor, "Memo to: Professor B. Herzog. A Proposal for Research in Conversational Use of Computers for Systems Engineering Problems," 30 Sept. 1966, RG 330-73-A-2108, box 1, folder: "U. of Michigan" no. 2, NABDS.

38. Ivan E. Sutherland to Robert Taylor, IPTO director, 19 Oct. 1966, RG 330-73-A-2108, box 1, folder: "University of Michigan," NABDS.

39. Taylor interview, CBI.

40. ARPA, "Graphic Control."

41. Ibid.

42. Ibid.

43. Lawrence G. Roberts, comment, in Michael S. Mahoney, John T. Gilmore, Lawrence G. Roberts, and Robin Forrest, "Retrospectives II: The Early Years in Computer Graphics at MIT, Lincoln Lab, and Harvard," SIGGRAPH '89 Panel Proceedings, *Computer Graphics* 23 (1989): 39–73.

44. Lawrence G. Roberts, "Machine Perception of Three-Dimensional Solids," MIT Lincoln Laboratory Technical Report no. 315, 1963.

45. Ivan Sutherland, "Outline of a Talk to Be Given on Tuesday, June 15th, at the Course in Computer Graphics, University of Michigan, Ann Arbor: Ten Unsolved Problems in Computer Graphics," RG 330-76-034, box 1, folder: "Graphics," NABDS; reprinted with slight alterations as idem, "Computer Graphics: Ten Unsolved Problems," *Datamation* 12 (May 1966): 22–27.

46. Lawrence G. Roberts, "The Lincoln WAND," in *Proceedings of the AFIPS Fall Joint Computer Conference,* vol. 29 (Baltimore: Spartan Books, 1963), 223–27.

47. Douglas Englebart, "The Augmented Knowledge Workshop," in Adele Goldberg, ed., *A History of Personal Workstations* (New York: Addison-Wesley, 1988), 187–232.

48. Robert W. Taylor, "Accomplishments in Calendar Year 1967, Internal Memorandum for the Acting Deputy Director, ARPA," 5 Jan. 1968, pp. 2–3, RG 330-74-107, box 1, folder: "Internal Memoranda 1968 through 1970," NABDS; Several published articles and reports prepared at this time suggested that the field was at a turning point. Like Taylor's description, these papers were attempts to put into perspective what had been accomplished and to project what had yet to be done. See, e.g., Sutherland, "Computer Graphics"; Frank D. Skinner, "Computer Graphics: Where Are We?" *Datamation* 12

(May 1966): 28–31; and J.C.R. Licklider, "Computer Graphics as a Medium of Artistic Expression," in *Computers and Their Potential Application in Museums* (New York: Arno, 1968).

49. Ivan E. Sutherland, "A Head-Mounted Three Dimensional Display," in *Proceedings of the* AFIPS *Fall Joint Computer Conference,* vol. 33 (Washington, D.C.: Thompson Book, 1968), 757–58.

50. Ibid.

51. Thomas G. Stockham Jr. and Martin E. Newell, principal investigators, "Sensory Information Processing and Symbolic Computation Research Proposal, 1 October 1975 through 30 September 1977," RG 330-78-0012, box 3, folder: "Utah AO2477," NABDS.

52. Marshall M. Lee, *Winning the People: The First Forty Years of Tektronix* (Portland, Oreg.: Tektronix, 1986), 218.

53. Ibid., chap. 8.

54. J. D. Foley and Andreas Van Dam, *Fundamentals of Interactive Computer Graphics* (Reading, Mass.: Addison-Wesley, 1982).

55. Ibid.

56. Ivan E. Sutherland, "Computer Displays," *Scientific American* 222 (June 1970): 56–81.

57. Thomas G. Hagan and Robert H. Stotz, "The Future of Computer Graphics," in *Proceedings of the* AFIPS *Spring Joint Computer Conference,* vol. 40 (Montvale, N.J.: AFIPS Press, 1972), 447–52.

58. T. O. Ellis, project leader, "IPT Exploratory Research and Supporting Efforts," June 1971, Uncapher Papers, box 1, folder: "ARPA Progress Report June 1971."

59. Thomas O. Ellis and I. D. Greenwald, "Video Graphics System," in "ARPA-IPT Proposal 1968–1969 Program," Uncapher Papers, box 1, no folder.

60. Foley and Van Dam, *Fundamentals.*

61. See for example, Hagan and Stotz, "Future of Computer Graphics."

62. Andreas van Dam, "Computer Graphics," in Anthony Ralston and Edwin D. Reilly Jr., eds., *Encyclopedia of Computer Science and Engineering,* 2d ed. (New York: Van Nostrand Reinhold, 1983), 319–33.

63. Taylor interview, CBI.

64. Taylor, "Accomplishments in 1967," 1.

65. Sutherland interview, CBI.

66. Rivlin, *Algorithmic Image,* 27.

67. N. Addison Ball, "Trip Report, 1 April 1968," RG 330-78-0085, box 2, folder: "Networking, 1968–1972," NABDS.

68. Martin E. Newell, "Sensory Information Processing Research Proposal: Image Understanding in the Context of a Three Dimensional Geometric Model, 1 October 1977 through 30 September 1980," RG 330-78-0012, box 3, folder: "AO2477 Utah," NABDS.

69. Sutherland, "Computer Displays," 67.

70. Stockham and Newell, "Sensory Information Processing Research Proposal," 60.

71. Ronald B. Resch, "The Topological Design of Sculptural and Architectural Sys-

tems," in *Proceedings of the AFIPS National Computer Conference and Exposition,* vol. 42 (Montvale, N.J.: AFIPS Press, 1973), 643–50.

72. Sutherland interview, CBI.

73. Stockham and Newell, "Sensory Information Processing Research Proposal," 81.

74. Ibid.

75. Rivlin, *Algorithmic Image,* 41.

76. I. E. Sutherland, R. F. Sproull, and R. A. Schumaker, "A Characterization of Ten Hidden-Surface Algorithms," *Computing Surveys* 6 (1974): 3.

77. Ibid., 19.

78. Ibid.

79. Ibid.

80. Stockham and Newell, "Sensory Information Processing Research Proposal," 65–66.

81. Newell's work is cited by Turner Whitted, "Some Recent Advances in Computer Graphics," *Science* 215 (12 Feb. 1982): 767–74.

82. J. F. Blinn, "Simulation of Wrinkled Surfaces," *Computer Graphics* 12 (Aug. 1978): 286–92; and J. F. Blinn and M. E. Newell, "Texture and Reflection in Computer Generated Images," *Communications of the ACM* 19 (Oct. 1976): 542–47.

83. J. F. Blinn, "Computer Display of Curved Surfaces" (Ph.D. diss., University of Utah, 1978); and R. L. Cook and K. Torrance, "A Reflectance Model for Computer Graphics," SIGGRAPH '81 Proceedings, *Computer Graphics* 15 (Aug. 1981): 307–16.

84. D. Greenberg, A. Marcus, A. H. Schmidt, and V. Gorter, *The Computer Image: Applications of Computer Graphics* (Reading, Mass.: Addison-Wesley, 1982), 21.

85. B. Mandelbrot, *Fractals: Form, Chance, and Dimension* (San Francisco: Freeman, 1977).

86. Lawrence G. Roberts, "Memorandum for the Director, Program Management," 16 July 1973, RG 330-78-0012, box 2, folder: "Purdue," NABDS.

87. "Memorandum for the Director, Program Management," 30 Sept. 1974, RG 330-82-0215, box 1, folder: "Purdue," NABDS.

88. Stockham and Newell, "Sensory Information Processing Research Proposal," 63.

89. Sutherland, Sproull, and Schumaker, "Hidden-Surface Algorithms," 23.

90. Rivlin, *Algorithmic Image.*

91. Martin E. Newell and James F. Blinn, "The Progression of Realism in Computer Generated Images," in *Proceedings of the ACM National Conference* [1977] (New York: ACM Press, 1977), 444–48.

92. See, e.g., Franklin C. Crow, "A Three-Dimensional Surface Design System," in *Proceedings of the ACM National Conference* [1977], 440–43; Franklin C. Crow, "The Aliasing Problem in Computer-Generated Shaded Images," *Communications of the ACM* 20 (Nov. 1977): 799–805.

93. See, e.g., H. Fuchs, Z. M. Kedem, and S. P. Uselton, "Optimal Surface Reconstruction from Planar Contours," *Communications of the ACM* 20 (1977): 693–702; H. Fuchs, S. M. Pizar, L. C. Tsai, and S. H. Bloomberg, "Adding a True 3-D Display to a Raster Graphics System," *IEEE Computer Graphics and Applications* 2 (1982): 73–78.

94. S. H. Chasen, "Historical Highlights of Interactive Graphics," *Mechanical Engineering* 103 (1981): 32–41.

95. Licklider, "Graphic Input," 61.

96. Illustrated in Sutherland, "Computer Inputs and Outputs."

97. Hagan and Stotz, "Future of Computer Graphics," 447; emphasis in original.

98. Ibid., 450.

99. Goodman, *Digital Visions.*

100. Ibid.; Patrice D. Prince, "The Aesthetics of Exhibition: A Discussion of Recent American Computer Art Shows," *Leonardo,* suppl. (1988): 88–98.

101. Csuri interview, CBI.

102. Goodman, *Digital Visions.*

103. Herbert W. Franke, *Computer Graphics, Computer Art,* 2d ed. (Berlin: Springer-Verlag, 1971); Jasia Reichardt, *The Computer in Art* (London: Studio Vista, 1971); Lewell, *Computer Graphics.*

104. Rivlin, *Algorithmic Image.*

105. Ibid.; Foley and Van Dam, *Fundamentals.*

106. Lewell, *Computer Graphics;* Rivlin, *Algorithmic Image.*

107. See, e.g., Ira W. Cotton and Frank S. Greatorex Jr., "Data Structures and Techniques for Remote Computer Graphics," in *Proceedings of the AFIPS Fall Joint Computer Conference* 33:533–44; and Edgar H. Sibley, Robert W. Taylor, and David G. Gordon, "Graphical Systems Communication: An Associative Memory Approach," in ibid., 33: 545–55.

108. David C. Russell, "Memorandum for the Director, Program Management," 8 Apr. 1977, RG 330-78-0085, box 1, unlabeled folder, NABDS.

109. IPTO, "AO3796 Carnegie-Mellon University," memorandum regarding CMU, 1982, IPTO Office Files, CBI.

110. T.M.P. Lee, "Report on the 1971 Conference on Computer Vision," 29 Nov. 1971, RG 330-77-0046, box 2, unlabeled folder, NABDS.

111. Martin E. Newell, "Sensory Information Processing Research Proposal: Image Understanding in the Context of a Three Dimensional Geometric Model, 1 October 1977 through 30 September 1980," RG 330-78-0012, box 3, folder: "Utah AO2477," NABDS.

112. IPTO, "AO4844 UNC Losleben," memorandum regarding University of North Carolina, 1983, IPTO Office Files, CBI.

113. Kahn interview, CBI.

114. Larry Bergman, Henry Fuchs, Eric Grant, and Susan Spach, "Image Rendering by Adaptive Refinement," in David C. Evans, ed., *Conference Proceedings on Computer Graphics and Interactive Techniques: 1986 SIGGRAPH* (New York: Association for Computing Machinery, 1986), 29–33; and Jack Goldfeather, Jeff P. M. Hultquist, and Henry Fuchs, "Fast Constructive Solid Geometry Display in the Pixel-Powers Graphics System," in ibid., 107–16.

115. Lewell, *Computer Graphics.*

116. Rivlin, *Algorithmic Image;* Chasen, "Historical Highlights."

117. Franke, *Computer Graphics, Computer Art.*

Chapter Four: Improving Connections among Researchers

1. Licklider to "Members and Affiliates of the Intergalactic Computer Network," 25 Apr. 1963, RG 330-69-A-4998, box 3, folder: "350-1," National Archives Branch Depository, Suitland, Md. (NABDS).

2. William G. Gerhard, ed., *Proceedings of the Invitational Workshop on Networks of Computers (NOC-68)* (Fort George G. Meade, Md.: National Security Agency, 1969), III. A distinction is made between a computer network, as defined here, and a network of computers.

3. Fano interview, CBI.

4. C. Stephen Carr, Stephen D. Crocker, and Vinton G. Cerf, "HOST-HOST Communication Protocol in the ARPA Network," in *Proceedings of AFIPS Spring Joint Computer Conference,* vol. 36 (Montvale, N.J.: AFIPS Press, 1970), 589–97. Also see F. E. Heart, R. E. Kahn, S. M. Ornstein, W. R. Crowther, and D. C. Walden, "The Interface Message Processor for the ARPA Computer Network," in ibid., 551–67; Taylor interview, CBI.

5. DARPA, "Resource Sharing Computer Networks," Program Plan no. 723, 3 June 1968, RG 330-78-0013, box 1, folder: "Program Plans," NABDS.

6. Lawrence G. Roberts, "ARPA/Information Processing Techniques Computer Network Concept," in Gerhard, ed., *Invitational Workshop on Networks of Computers (NOC 68).* The paper is dated 11 May 1967.

7. UCLA, "Final Technical Report," Contract SD 184, 4 Mar. 1970, RG 330-71-A-1647, box 2, folder: "UCLA Reports," NABDS.

8. The Western Data Processing Center was established in November 1956 by IBM and the regents of the University of California to encourage and support research and education in business management. The center began with an IBM 650 computer, graduating a year later to an IBM 709. Machines and service were supplied by IBM, and costs were shared by the company and the university. The available machine time—at least one shift per day—was divided between UCLA and the participating institutions. In the academic year 1959–60, for example, faculty members from fifty-seven campuses west of the Rocky Mountains were affiliated with the center. See "Progress Report 2, Western Data Processing Center, UCLA," 1960, Academic Computing Collection, CBI.

9. DARPA, "Computer Network and Time-Sharing Research," Program Plan no. 93, 5 Apr. 1963, RG 330-78-0013, box 1, folder: "Program Plans," NABDS.

10. "Statement by Robert L. Sproull," in House Subcommittee on Defense Appropriations, *Department of Defense Appropriations for Fiscal Year 1965, Part 5: Research, Development, Test, and Evaluation, Appropriation Language, Testimony of Members of Congress, Organizations, and Interested Groups,* (88) H2062-0-B, 88th Cong., 2d sess., 6, 11–13, 16–19 Mar. 1964, 127.

11. "Statement by Robert L. Sproull," in House Subcommittee on Defense Appropriations, *Department of Defense Appropriations for Fiscal Year 1966,* (89) H2128-0-C, 89th Cong., 1st sess., 30, 31 Mar., 5, 7, 9, 13 Apr. 1965, 535.

12. Roberts, "ARPA/Information Processing Techniques Computer Network Concept."

13. DARPA, "Remote Stations and Programs for Computer Network," Program Plan no. 95, 5 Apr. 1963, RG 330-78-0013, box 1, folder: "Program Plans," NABDS.

14. The CCA contract was a subcontract under the Lincoln Laboratory contract ("ARPANET Completion Report Draft" [Cambridge, Mass.: Bolt Beranek and Newman, 1977]).

15. Thomas Marill, "A Cooperative Network of Time-Sharing Computers: Preliminary Study," Computer Corporation of America Technical Report no. 11, 1 June 1966, CBI.

16. Thomas Marill and Lawrence G. Roberts, "Toward a Cooperative Network of Time-Shared Computers," in *Proceedings of the AFIPS Fall Joint Computer Conference*, vol. 29 (Washington, D.C.: Spartan Books, 1966), 425–31.

17. For a description of paper tape systems see Donald Davies and Derek Barber, *Communication Networks for Computers* (London: Wiley and Sons, 1973).

18. Lawrence G. Roberts and Barry D. Wessler, "Computer Network Development to Achieve Resource Sharing," in *Proceedings of the AFIPS Spring Joint Computer Conference* 36:543–49.

19. Sidney H. Gordon, "Autodin II System Overview," in *National Telecommunications Conference (NTC '77) Conference Record*, vol. 3 (New York: Institute of Electrical and Electronics Engineers, 1977), pp. 37:1–1 to 37:1–2.

20. C. E. Houstis and B. J. Leon, "Priority Queuing for the AUTODIN Store-and-Forward Network," in *Proceedings of the 1977 IEEE Conference on Decision and Control* (New York: Institute of Electrical and Electronics Engineers, 1977), 1:826–30.

21. Paul Baran, "Some Perspectives on Networks — Past, Present and Future," *Information Processing 77: IFIP Conference Proceedings* (Amsterdam: North-Holland, 1977), 449–64.

22. "On Distributed Communications: Rand Memorandum Series" (Santa Monica, Calif.: Rand Corporation). The following titles comprise the series: (1) Paul Baran, *Introduction to Distributed Communications Networks* (RM-3420-PR); (2) Sharla P. Boehm and Paul Baran, *Digital Simulation of Hot-Potato Routing in a Broadband Distributed Communications Network* (RM-3103-PR); (3) J. W. Smith, *Determination of Path-Lengths in a Distributed Network* (RM-3578-PR); (4) Paul Baran, *Priority, Precedence, and Overload* (RM-3638-PR); (5) idem, *History, Alternative Approaches, and Comparisons* (RM-3097-PR); (6) idem, *Mini-Cost Microwave* (RM-3762-PR); (7) idem, *Tentative Engineering Specifications and Preliminary Design for a High-Data-Rate Distributed Network Switching Node* (RM-3763-PR); (8) idem, *The Multiplexing Station* (RM-3764-PR); (9) idem, *Security, Secrecy, and Tamper-Free Considerations* (RM-3765-PR); (10) idem, *Cost Estimate* (RM-3766-PR); (11) and idem, *Summary Overview* (RM-3767-PR). The two remaining reports, on potential weaknesses of the proposed system and on cryptography, were classified.

23. There are certain variations of centralized networks which are called decentralized, but this only means that there is more than one control point. It is still a hierarchical structure and is not the same as a distributed system.

24. Paul Baran, "On Distributed Communications Networks," Rand Paper P-2626 (Santa Monica, Calif.: Rand Corporation, 1962).

25. Frank Collbohm to deputy chief of staff, Research and Development, "Recommendation to the Air Staff Development of the Distributed Adaptive Message-Block Network," 30 Aug. 1965 (included as part of Baran interview, CBI).

26. See Baran speech, *Advanced Computer Technologies One,* Aug. 1989, Collection 115, CBI. Baran recommended against continuing with the network because he felt that those responsible for the implementation did not understand the digital techniques required.

27. Martin Campbell-Kelly, "Data Communications at the National Physical Laboratory (1965–1975)," *Annals of the History of Computing* 9, no. ¾ (1988): 221–47.

28. Sutherland and Roberts coordinated their visits to British sites. See, e.g., Roberts to Gill, 5 Oct. 1965, RG 330-77-0046, box 3, folder: "National Physical Laboratory Symposium," NABDS.

29. Lawrence G. Roberts, "The ARPANET and Computer Networks," in Adele Goldberg, ed., *History of Personal Workstations* (New York: ACM Press, 1988), 144.

30. Donald Watts Davies, interview by Martin Campbell-Kelly, 17 Mar. 1986, Teddington, England. Transcript generously supplied by the interviewer.

31. Campbell-Kelly, "Data Communications," 221–47.

32. Donald W. Davies, "Proposal for a Digital Communication Network" (Teddington, England: National Physical Laboratory, June 1966).

33. Davies, interview by Campbell-Kelly. In the interview Davies claims to have sent a copy of his "Proposal" to Roberts at this time.

34. Davies and Barber, *Communication Networks for Computers.*

35. George W. Brown, James G. Miller, and Thomas A. Keenan, eds., *EDUNET: Report of the Summer Study on Information Networks (University of Colorado, 1966)* (New York: Wiley, 1967).

36. Both Network/440 and TSS are discussed in Randall Rustin, ed., *Computer Networks: The Third Courant Computer Science Symposium* (Englewood Cliffs, N.J.: Prentice-Hall, 1972). Some of the work on the TSS network was funded by IPTO. See Ronald M. Rutledge, Albin L. Vareha, Lee C. Varlan, Allan H. Weiss, Salomon F. Seroussi, James W. Meyer, Joan F. Jaffe, and Mary Anne K. Angell, "An Interactive Network of Time Sharing Computers," in ACM *National Conference Proceedings* [1969] (New York: Association for Computing Machinery, 1969), 431–41.

37. *Control Data Corporation Annual Report* (1968), Control Data Corporation Records, CBI.

38. These large-scale systems included the available time-sharing systems, such as Project MAC at MIT, and the ILLIAC IV project begun in 1965 at the University of Illinois.

39. Roberts interview, CBI; Taylor interview, CBI.

40. L. G. Roberts, "Message Switching Network Proposal," attached to Gerald Estrin to Robert Taylor and Larry Roberts, 27 Apr. 1967, RG 330-78-0085, box 2, folder: "Networking 1968–1972," NABDS.

41. Taylor interview, CBI.

42. Lawrence Roberts, "Expanding AI Research and Founding ARPANET," in Thomas C. Bartee, ed., *Expert Systems and Artificial Intelligence: Applications and Man-*

agement (Indianapolis, Ind.: Sams, 1988), 229–35; Roberts interview, CBI; Taylor interview, CBI.

43. "ARPANET Completion Report Draft," pp. III-25, III-26.

44. Roberts, "Message Switching Network Proposal."

45. J.C.R. Licklider to Dan Bobrow and Bert Sutherland, 6 June 1967, RG 330-78-0085, box 2, folder: "Networking 1968–1972," NABDS.

46. Leonard Kleinrock, "Principles and Lessons in Packet Communications," *Proceedings of the IEEE* 66 (Nov. 1978): 1320–29.

47. Lawrence G. Roberts, "The Evolution of Packet Switching," *Proceedings of the IEEE* 66 (Nov. 1978): 1307–13.

48. Vinton G. Cerf, "Packet Communication Technology," in Franklin F. Kuo, ed., *Protocols and Techniques for Data Communication Networks* (Englewood Cliffs, N.J.: Prentice-Hall, 1981), 1–34.

49. Lawrence G. Roberts, "Multiple Computer Networks and Intercomputer Communication," in ACM *Symposium on Operating System Principles* (New York: Association for Computing Machinery, 1967). The paper was apparently prepared at least a few months earlier, as it refers to September 1967 in the future tense.

50. D. W. Davies, K. A. Bartlett, R. A. Scantlebury, P. T. Wilkinson, "A Digital Communication Network for Computers Giving Rapid Response at Remote Terminals," in ACM *Symposium on Operating System Principles.*

51. Davies interview, CBI.

52. Roberts, "ARPANET and Computer Networks."

53. Roberts speech, *Advanced Computer Technologies One,* Aug. 1989, p. 27, session 1, Collection 115, CBI.

54. Leonard Kleinrock to Lawrence Roberts, 3 Nov. 1967, RG 330-71-A-1647, box 1, folder: "ARPA Computer Network Working Group," NABDS. Baran's name appears on the list of contractors attending the 13–15 Mar. 1968 PI meeting, RG 330-77-0046, box 3, folder: "PI Meeting 1967/1976," NABDS.

55. Lawrence G. Roberts, "Memorandum for the Director," 21 June 1968, RG 330-77-0046, box 1, folder: "A01260 BBN IMPS," NABDS.

56. Elmer Shapiro, 17 Nov. 1967, RG 330-71-A-1647, box 1, folder: "ARPA Computer Network Working Group," NABDS.

57. E. Rechtin, "ARPA Order 1137," RG 330-74-107, box 1, folder: "ARPA Orders Chrono thru 1970," NABDS.

58. Several memos sent to the group are in RG 330-71-A-1647, box 1, folder: "ARPA Computer Network Working Group," NABDS.

59. L. G. Roberts, "Memorandum for the Director, Program Management," 21 June 1968, requested that an ARPA order be written for $563,000 in FY1968 for the design and construction of an interface message processor (RG 330-74-107, box 1, folder: "Internal Memoranda 1968 through 1970," NABDS). This money was for the proposal process, not the actual design and construction. ARPA Program Plan no. 723, describing the requirements for a network, was attached to this memorandum; it can be found in RG 330-77-0046, box 1, folder: "AO1260 BBN IMPS," NABDS.

60. "ARPANET Completion Report Draft," p. III-35.

61. Statement of Work Annex "A," "Specifications of Interface Message Processors for the ARPA Computer Network," July 1968, RG 330-77-0046, box 1, folder: "AO1260 BBN IMPS," NABDS.

62. Ibid., 17.

63. "ARPANET Completion Report Draft," p. III-35.

64. The existence of bids by these four companies is known from the following sources: Raytheon is listed in notes on the IPT Weekly Reports (in collection of materials from A. McKenzie, CBI); the Bunker-Ramo proposal is in the CBI archives; Jacobi Systems is mentioned by Cerf and Crocker in their CBI interviews; and BBN received the contract. IPTO files related to the proposals are classified.

65. Amendment 1 to ARPA Order 1260, RG 330-77-0046, box 1, folder: "AO1260 BBN IMPS," NABDS. In response, DARPA nearly doubled the initial amount ($563,000), increasing the funding to $1,112,000.

66. For SDC, see "ARPA Order 773 Quarterly Management Report—Period ending 5/17/66," 6 June 1966, RG 330-71-A-1647, box 1, folder: "AO773 SDC (first folder)," NABDS. For Clark, see Clark interview, CBI. Clark reports that Roberts asked him who he thought could do the job, to which he responded that the only one was Frank Heart at BBN. Roberts may have encouraged many potential bidders, in addition to BBN. It may not be significant that he encouraged Heart. See Heart interview, CBI; and Heart and Taylor speeches, *Advanced Computer Technologies One,* Aug. 1989, Collection 115, CBI.

67. Elmer Shapiro, 15 Feb. 1968, RG 330-71-A-1647, box 1, folder "ARPA Computer Network Working Group," NABDS.

68. J. K. Reynolds and J. B. Postel, "Request for Comments Reference Guide," Request for Comments no. 1000, Aug. 1987.

69. Jeff Rulifson, "Decode Encode Language," Request for Comments no. 5, 2 June 1969.

70. Cerf interview, CBI.

71. S. D. Crocker, "Documentation Conventions," Request for Comments no. 3, 9 Apr. 1969.

72. Lawrence Roberts and Davis Bobrow, "Initiation of a Contract with Raytheon Corporation to Study 'User System Interaction via the Network,'" memorandum for the director, Program Management, 18 Sept. 1969, RG 330-74-107, box 1, folder: "Internal Memoranda 1968 thru 1970," NABDS.

73. A. McKenzie, "The ARPA Network Control Center," *Fourth Data Communications Symposium: Network Structures in an Evolving Operational Environment* (New York: Institute of Electrical and Electronics Engineers, 1975), pp. 5-1 to 5-6.

74. S. D. Crocker, "Documentation Conventions," Request For Comments no. 24, 21 Nov. 1969.

75. NIC Staff, "NIC Newsletter," 16 Jan. 1969, RG 330-71-A-1647, box 1, folder: "ARPA Computer Network Working Group," NABDS.

76. See Heart interview, CBI; Kahn interview, CBI; Ornstein interview, CBI; Walden

interview, CBI; and Crowther interview, CBI. Others at BBN later joined in the IMP development effort; these included Truitt Thatch, Bill Bartell, Jim Geisman, Ben Barker, B. P. Cosell, and Marty Thrope.

77. For example, see Robert E. Kahn to ARPA Network Committee, memorandum, 11 Mar. 1968, RG 330-77-0046, box 1, folder: "AO1260 BBN IMPS," NABDS.

78. Heart, Crowther, and Ornstein had experience with real-time computing and communications through their work on the Whirlwind and SAGE computer systems at Lincoln Laboratory. Ornstein states in his interview that he did not actually work on the Whirlwind computer. He learned to program the Whirlwind from manuals and instruction from a co-worker who had used the Whirlwind. See Heart interview, CBI; Crowther interview, CBI; and Ornstein interview, CBI.

79. Although they knew of the CCA/SDC network experiment that was conducted at Lincoln Laboratory while they were there, they were not involved with it in any way. See Crowther interview, CBI; Heart interview, CBI; and Walden interview, CBI.

80. See Heart interview, CBI; and Kahn interview, CBI.

81. Dennis G. Perry, Steven H. Blumenthal, and Robert M. Hinden, "The ARPANET and the DARPA Internet," Library Hi Tech 6, no. 2 (1988): 51–62.

82. Howard Frank, Robert Kahn, and Leonard Kleinrock, "Computer Communication Network Design — Experience with Theory and Practice," in Proceedings of the AFIPS Spring Joint Computer Conference, vol. 40 (Montvale, N.J.: AFIPS Press, 1972), 255–70.

83. The project was known as the SNUPER computer. See Cerf interview, CBI; Crocker interview, CBI.

84. Leonard Kleinrock, Communication Nets: Stochastic Message Flow and Delay (New York: McGraw-Hill, 1964), vii.

85. Kleinrock interview, CBI.

86. Carr, Crocker, and Cerf, "HOST-HOST Communication."

87. "Statement by Charles Herzfeld," in House Subcommittee on Defense Appropriations, Department of Defense Appropriations for 1968, Part 3: Research, Development, Test, and Evaluation, (90) H2270-0-A, 90th Cong., 1st sess., 20 Mar. 1967, 136.

88. "Statement by Eberhardt Rechtin," in Senate Committee on Appropriations, Department of Defense Appropriations for Fiscal Year 1970, Part 1, (91) S2010-0-A, 91st Cong., 1st sess., 10, 12–13, 16 June, 15–17, 25 Sept. 1969, 433.

89. Lawrence Roberts, "Expansion of Network Analysis Corporation Contract," memorandum for the director, Program Management, 11 Feb. 1970, RG 330-74-107, box 1, folder: "Internal Memorandum 1968 through 1970," NABDS.

90. Frank interview, CBI.

91. Ibid. Also see Roberts, "Expansion of Network Analysis Corporation Contract."

92. Walden interview, CBI.

93. Heart interview, CBI.

94. F. E. Heart, "Interface Message Processors for the ARPA Computer Network: 1 July 1970–30 September 1970," in BBN Quarterly Report, 1 Oct. 1970.

95. See Uncapher interview, CBI.

96. The nodes included Digital Equipment Corporation's PDP-10s and PDP-11s,

IBM 360s, General Electric's 645 MULTICS system, Burroughs's 6500s, Xerox Data Systems's Sigma-7, and the ILLIAC IV. Even if two nodes had the same machines, they might use different software systems. For example, the TENEX, TOPS-10, and ITS systems all ran on the PDP-10. See Cerf, "Packet Communication Technology," 1–34.

97. Kahn interview, CBI.

98. "ARPANET Completion Report Draft," p. III-115.

99. Lawrence G. Roberts, "Network Rationale: A 5-Year Reevaluation," *Seventh Annual IEEE Computer Society International Conference (Compcon 7): Computing Networks from Minis through Maxis—Are They for Real?* (Long Beach, Calif.: IEEE Computer Society, 1973), (1973): 3–5.

100. Howard Frank, Robert E. Kahn, and Leonard Kleinrock, "Computer Communication Network Design—Experience with Theory and Practice," in *Proceedings of the AFIPS Spring Joint Computer Conference* 40:255–70.

101. DARPA, "Resource Sharing Computer Networks."

102. Roberts and Wessler, "Computer Network Development"; Heart et al., "Interface Message Processor"; Leonard Kleinrock, "Analytic and Simulation Methods on Computer Network Design" in *Proceedings of the AFIPS Spring Joint Computer Conference* 36:569–80; H. Frank, I. T. Frisch, and W. Chou, "Topological Considerations in the Design of the ARPA Computer Network," in ibid., 581–88; Carr, Crocker, and Cerf, "HOST-HOST Communication Protocol in the ARPA Network."

103. Kahn interview, CBI.

104. Ibid.

105. Stanley Winkler, ed., *First International Conference on Computer Communication: Computer Communications. Impacts and Implications* (New York: Association for Computing Machinery, 1972).

106. *Scenarios for Using the ARPANET at the International Conference on Computer Communication,* NIC 11863 (Washington, D.C.: Network Information Center, 1972).

107. See Kahn interview, CBI; Cerf interview, CBI.

108. McKenzie interview, CBI. At this time, providing such information to requesting organizations was one of McKenzie's responsibilities.

109. Lawrence Roberts, "Resource Sharing Computer Networks," RG 330-77-0046, box 1, folder: "AO1260 BBN IMPS," NABDS.

110. Roberts states that this was not a goal in the ARPANET (Roberts, "ARPA/Information Processing Techniques Network Concept"). On the presence of a MAIL command in the Project MAC CTSS Programmer's Guide, see chap. 2, n. 197, above.

111. See McKenzie interview, CBI; Crowther interview, CBI; Ornstein interview, CBI.

112. Franklin F. Kuo, "Public Policy Issues Concerning ARPANET," in *Fourth Data Communications Symposium: Network Structures in an Evolving Operational Environment* (New York: Institute of Electrical and Electronics Engineers, 1975), pp. 3–13 to 3–17.

113. Stuart L. Mathison and Philip M. Walker, "The Regulation of Value Added Carriers," in *Fourth Data Communications Symposium,* p. 3–1.

114. Telenet was also formed in response to the formation of a separate company (Packet Communications) by former BBN employees (Kahn interview, CBI).

115. Leonard Kleinrock, "On Communication and Networks," IEEE *Transactions on Computers* C-25 (Dec. 1976): 1326–35.

116. Norman Abramson, "The ALOHA System—Another Alternative for Computer Communications," in *Proceedings of the Spring Joint Computer Conference*, vol. 37 (Montvale, N.J.: AFIPS Press, 1970), 281–85.

117. Vinton G. Cerf, "Network Interconnection, 1 July 1975–20 September 1976," RG 330-84-0006, box 2, no folder, NABDS.

118. Norman Abramson, "The ALOHA System," in Norman Abramson and Franklin F. Kuo, eds., *Computer-Communication Networks* (Englewood Cliffs, N.J.: Prentice-Hall, 1973), 501–17.

119. Robert Kahn, "Amendment to ARPA Order 3214—BBN," memorandum for the director, Program Management, 21 Sept. 1976, folder: "AO3412 Bolt Beranek and Newman 1 of 3," IPTO Office Files, CBI.

120. Kahn interview, CBI.

121. Delbert D. Smith, *Communication via Satellite: A Vision in Retrospect* (Boston: Sijthoff), 151–55.

122. David Russell, "Systems Design of a World-Wide Seismological Communications Network," memorandum to the director, Program Management, 15 Sept. 1972, RG 330-80-0002, box 7, folder: "AO2298 Communication Satellite Corp.," NABDS.

123. COMSAT Laboratories Final Report, 2 May 1975, "Operation of a SIMP/SPACE Interface Unit (SSI) for Use in the ARPANET Satellite Experiment," RG 330-80-0002, box 7, folder: "AO2298 Communication Satellite Corp.," NABDS.

124. "Amendment to ARPA Order 3214—BBN," 21 Sept. 1976, lists what had been accomplished to date.

125. Cerf, "Network Interconnection."

126. The contractors included Stanford Research Institute, the Network Analysis Corporation (NAC), UCLA, BBN, Collins Radio, the University of Hawaii, and Texas Instruments. E. W. Stubbs, "AC33T," 23 Apr. 1976, folder 1: "AO2302, SRI Packet Radio System Development, #1" IPTO Office Files, CBI.

127. Kahn interview, CBI.

128. Ibid.

129. A. McKenzie, "Some Computer Interconnection Issues," in *Proceedings of the* AFIPS *National Computer Conference*, vol. 43 (Montvale, N.J.: AFIPS Press, 1974), 857–59.

130. Max Beere to Robert Kahn, 30 Nov. 1972, RG 330-78-0085, box 2, folder: "Networking 1968–1972," NABDS.

131. Vinton G. Cerf and Robert E. Kahn, "A Protocol for Packet Network Intercommunication," IEEE *Transactions on Communications* COM-22, no. 5 (May 1974): 637–47.

132. Cerf interview, CBI.

133. Ibid.

134. David Walden and Alexander McKenzie, "The ARPANET, the Defense Data Network, and the INTERNET," in Fritz E. Froehlich and Alan Kent, eds., *Encyclopedia of Telecommunications*, vol. 1 (New York: Marcel Dekker, 1991).

135. Peter Kirstein, "Proposal to the Advanced Research Agency," Nov. 1975, RG 330-

84-0006, box 2, folder: "AO2516 University College, London—Folder 2," NABDS. Kirstein reported that University College in London had "largely completed an implementation of the Transmission Control Program proposed for Hosts attached to several different networks" (4–5). See Cerf interview, CBI, for details of the International Network Working Group.

136. Walden and McKenzie, "ARPANET."

137. Ibid.

138. Douglas Comer, *Internetworking with TCP/IP* (Englewood Cliffs, N.J.: Prentice-Hall, 1988), 6.

139. Perry, Blumenthal, and Hinden, "ARPANET."

140. Thomas C. Harris, Peter V. Abene, Wayne W. Grindle, Darryl W. Harris, Dennis C. Morris, Glynn E. Parker, and Jeffrey Mayersohn, "Development of the MILNET," *IEEE EASCON* (Electronics and Aerospace Conference and Exposition) (New York: Institute of Electrical and Electronics Engineers, 1982), 77–80.

141. Walden and McKenzie, "ARPANET."

142. "Statement Submitted by Dr. George H. Heilmeier," in Senate Subcommittee on Defense Appropriations, *Department of Defense Appropriations for Fiscal Year 1978, Part 5: Research, Development, Test, and Evaluation, 77* S181-27.1, 95th Cong., 1st sess., 4 Feb. 1977, 115–16; emphasis in original.

143. "Plan for the Implementation of a National Software Works on the ARPANET," RG 330-80-0002, box 2, folder: "AO2832 National Software Works," NABDS.

144. IPTO, "Request for a New ARPA Order," memorandum for the director, Program Management, 7 May 1974, RG 330-80-0002, box 2, folder: "AO2832 National Software Works," NABDS.

145. Frederick Brooks, "Why Is the Software Late?" *Data Management* 9 (Aug. 1971): 18–21.

146. Barry W. Boehm, "Software and Its Impact: A Quantitative Assessment," *Datamation* 19 (1973): 48–59.

147. "GAO Hits Wimmix Hard: FY'72 Funding Prospects Fading Fast," *Datamation* 17 (1971): 41; "Wimmix," in the title, referred to WWMCCS.

148. Battelle (Columbus Laboratories), "Final Report on an Overview of Research and Development Programs and Objectives, Vol. III: The Information Processing Techniques Office," 26 Jan. 1973, p. 11, IPTO Office Files, CBI.

149. Vinton Cerf, private communication, Aug. 1992; Crocker interview, CBI.

150. Crocker interview, CBI.

151. Senate Subcommittee, *Department of Defense Appropriations for 1978*, 57–59.

152. Richard H. Van Atta, Seymour J. Deitchman, and Sidney G. Reed, *DARPA Technical Accomplishments*, IDA Paper P-2192 (Washington, D.C.: IDA, 1991), vol. 1, p. 23–1.

153. ARPA Order 3813, 7 Jan. 1981, file: "BBN Ohlander," IPTO Office Files, CBI.

154. Van Atta, Deitchman, and Reed, *DARPA Technical Accomplishments*, vol. 1, pp. 23–1, 23–2.

155. Ibid., p. 23–6.

156. Ibid., p. 23–5.

157. CINCPAC is the office of the commander-in-chief of the U.S. Navy's Pacific Fleet.

158. Robert S. Engelmore, "Request for New ARPA Order—ONR," 20 Apr. 1981, Naval Message Disambiguator RSE, IPTO Office Files, CBI.

159. Van Atta, Deitchman, and Reed, DARPA Technical Accomplishments, vol. 1, pp. 23–3, 23–4.

160. Ibid., p. 23–6.

161. Ibid., p. 23–4.

162. Richard des Jardin, "DARPA Program Summary: ADDCOMPE," in "FY 86 President's Budget Review: Information Processing Techniques Office," Sept. 1984, IPTO Office Files, CBI.

163. Robert L. Simpson Jr., "DOD Applications of Artificial Intelligence: Successes and Prospects," Society of Photo-Optical Instrumentation Engineers (SPIE): Applications of Artificial Intelligence VI 937 (1988): 158–64.

164. Ronald P. Uhlig, "Human Factors in Computer Message Systems," Datamation 23 (May 1977): 126.

165. Ibid.

166. Walden and McKenzie, "ARPANET."

167. Douglas K. Smith and Robert C. Alexander, Fumbling the Future: How Xerox Invented, Then Ignored, the First Personal Computer (New York: William Morrow, 1988), 65.

Chapter Five: The Search for Intelligent Systems

1. Margaret A. Boden, Artificial Intelligence and Natural Man (New York: Basic Books, 1977), 17.

2. Patrick H. Winston, Artificial Intelligence (Reading, Mass.: Addison-Wesley, 1977), 3.

3. Avron Barr and Edward A. Feigenbaum, eds., The Handbook of Artificial Intelligence, vol. 1 (Los Altos, Calif.: Kaufmann, 1981), 3.

4. Edward A. Feigenbaum and Julian Feldman, eds., Computers and Thought (New York: McGraw-Hill, 1963), 6.

5. Ibid.

6. "Statement of Dr. Stephen J. Lukasik, Director, Defense Advanced Research Projects Agency, Before the Subcommittee on Research and Development of the Senate Armed Services Committee in Connection with the ARPA Research and Development Program for FY1975," 17–18. This document was provided to us by IPTO.

7. John McCarthy, "History of Lisp," in Richard L. Wexelblat, ed., History of Programming Languages, ACM Monograph Series: The History of Programming Languages Conference (New York: Academic Press, 1981), 174.

8. Philip Klahr and Donald A. Waterman, "Artificial Intelligence: A Rand Perspective," AI Magazine 7 (summer 1986): 54–64.

9. A. Newell and H. A. Simon, "The Logic Theory Machine," IRE Transactions on Information Theory 2 (1956): 61–79; and A. Newell, J. C. Shaw, and H. A. Simon, "Chess

Playing Programs and the Problem of Complexity," IBM *Journal of Research and Development* 2 (1958): 320–35.

10. McCarthy, "History of Lisp."

11. "Representation of Knowledge," in Barr and Feigenbaum, eds., *Handbook of Artificial Intelligence* 1:143.

12. Ibid.

13. H. L. Gelernter and N. Rochester, "Intelligent Behavior in Problem-Solving Machines," IBM *Journal of Research and Development* 2 (Oct. 1958): 336–45.

14. Frank Rosenblatt, "Two Theorems of Statistical Separability in the Perceptron," *Proceedings of a Symposium on the Mechanization of Thought Processes* (London: Her Majesty's Stationery Office, 1959), 421–56. See also Marvin Minsky and Seymour Papert, *Perceptrons: An Introduction to Computational Geometry* (Cambridge: MIT Press, 1969).

15. Nils J. Nilsson, *Learning Machines: Foundations of Trainable Pattern-Classifying Machines* (New York: McGraw-Hill, 1965).

16. Allen Newell, "Intellectual Issues in the History of Artificial Intelligence," in Fritz Machlup and Una Mansfield, eds., *The Study of Information: Interdisciplinary Messages* (New York: Wiley, 1983), 199.

17. Allen Newell and George Ernst, "The Search for Generality," in *Information Processing 1965: Proceedings of the IFIP Congress 65* (Washington, D.C.: Spartan Books, 1965), 17–24.

18. Saul Amarel, "On the Automatic Formation of a Computer Program Which Represents a Theory," in Marshall C. Yovits, George T. Jacobi, and Gordon D. Goldstein, eds., *Self-Organizing Systems, 1962* (Washington, D.C.: Spartan Books, 1962), 107–75.

19. J.C.R. Licklider, "Artificial Intelligence, Military Intelligence, and Command and Control," in Edward Bennett, James Degan, and Joseph Spiegel, eds., *Military Information Systems: The Design of Computer-Aided Systems for Command* (New York: Frederick Praeger, 1964), 118–33.

20. DARPA program plans in the National Archives for several of these projects are: (for Stanford University, Carnegie Tech, and SDC) "Heuristic Programming and Theory of Computation," no. 56, 20 Mar. 1963; (for BBN) "Natural Communication with Computers," no. 363, 15 Oct. 1964; (for Michigan, Carnegie Tech, and the University of California, Berkeley) "Dynamic Computer Models of Cognitive Processes," no. 318, 25 May 1964; (for Michigan) "Research in Conversational Use of Computers," no. 470, 25 Mar. 1965; and (for Project MAC) "A Research and Development Program on Computer Systems," no. 683, 17 Jan. 1963. All are in the Records of DARPA/IPTO, RG 330-78-0013, box 1, folder: "Program Plans," National Archives Branch Depository, Suitland, Md. (NABDS).

21. MIT, "Proposal for a Research and Development Program for Computer Systems," 14 Jan. 1963, pp. 19, 21, Collection AC 12, box 17, folder: "Project MAC Proposals Regular 1963," Institute Archives, MIT.

22. Project MAC, "Proposal for Continuation of a Research and Development Program on Computer Systems," 20 Feb. 1964, pp. 4–22, Collection AC 12, box 17, folder: "Project MAC Proposals Reg. 1964," Institute Archives, MIT.

23. Winograd interview, CBI.

24. DARPA, "Automata and Extensors," Program Plan no. 547, 28 Feb. 1966, RG 330-78-0013, box 1, folder: "Program Plans," NABDS.

25. MIT, "Research on Intelligent Automata," Sept. 1965, pp. 1–2, Collection AC 12, box 17, folder: "Project MAC Intelligent Automata, 1965," Institute Archives, MIT.

26. Ibid., pp. 1–3, emphasis in the original.

27. Winograd interview, CBI.

28. Herbert A. Simon, "The Architecture of Complexity," *Proceedings of the American Philosophical Society* 26 (1962): 467–82.

29. H. A. Simon, "Experiments with a Heuristic Compiler," *Journal of the Association for Computing Machinery* 10 (1963): 493–506.

30. Julian Feldman, "Simulation of Behavior in the Binary Choice Experiment," in Feigenbaum and Feldman, eds., *Computers and Thought,* 329–46.

31. Allen Newell, "Learning, Generality, and Problem Solving," in *Information Processing 1962: Proceedings of the IFIP Congress 62* (Amsterdam: North-Holland, 1963), 407–12.

32. H. A. Simon and K. Kotovsky, "Human Acquisition of Concepts for Sequential Patterns," *Psychology Review* 70 (1963): 534–46.

33. H. A. Simon and E. Feigenbaum, "An Information Processing Theory of Some Effects of Similarity, Familiarization, and Meaningfulness in Verbal Learning," *Journal of Verbal Learning and Verbal Behavior* 3 (1964): 383–96.

34. Lester Earnest, ed., "Recent Research in Artificial Intelligence, Heuristic Programming, and Network Protocols," July 1974, memorandum AIM-252, in *Artificial Intelligence Memoranda: AI Laboratory, Computer Science Department, Stanford University, 1963–1982* (New York: Scientific Datalink, Comtex Scientific, 1983), microfilm. (Hereafter, this is cited as *Artificial Intelligence Memoranda: Stanford.*)

35. R. C. Schank, "Identification of Conceptualizations Underlying Natural Language," in R. C. Schank and K. Colby, eds., *Computer Models of Thought and Language* (San Francisco: Freeman, 1973), 187–247.

36. DARPA, "Very Large Memory Facility," Program Plan no. 452, 12 Mar. 1965, RG 330-78-0013, box 1, folder: "Program Plans," NABDS.

37. Raphael received his Ph.D. degree from MIT in 1962. After a year at Lincoln Laboratory and a year at the University of California at Berkeley, he moved to SRI in 1964. Following a number of years on the faculty of MIT, Winograd moved to Stanford in 1979. Green became a program manager in IPTO in 1970.

38. DARPA, "Natural Communication with Computers," Program Plan no. 363, 15 Oct. 1964, RG 330-78-0013, box 1, folder: "Program Plans," NABDS.

39. C. Gordon Bell and Peter Freeman, "C.ai—A Computer Architecture for AI Research," in *Proceedings of the AFIPS Fall Joint Computer Conference,* vol. 41 (Montvale, N.J.: AFIPS Press, 1972), 779–90.

40. Allen Newell and George Robertson, "C.mmp: A Project Report on Synergistic Research," in Anita K. Jones, ed., *Perspectives on Computer Science: From the Tenth Anniversary Symposium at the Computer Science Department, Carnegie Mellon University* (New York: Academic Press, 1977), 147–82.

41. Amarel interview, CBI.

42. "Statement of Dr. Stephen J. Lukasik, Director, DARPA," in Senate Subcommittee on Defense Appropriations, *Department of Defense Appropriations for Fiscal Year 1973, Part 1*, S181-44.4, 92d Cong., 2d sess., 14 Mar. 1972, 778.

43. Newell and Ernst, "Search for Generality."

44. Ibid.

45. Edward A. Feigenbaum, "Artificial Intelligence: Themes in the Second Decade," in *Information Processing 1968: Proceedings of the IFIP Congress 68* (Amsterdam: North-Holland, 1969), 1010.

46. Marvin Minsky and Seymour Papert, "Proposal to ARPA for Research on Artificial Intelligence at MIT, 1971–1972," pp. 7–8, Dean of Engineering Records, Institute Archives, MIT; emphasis in the original.

47. Feigenbaum, "Artificial Intelligence," 1014.

48. Marvin Minsky, ed., *Semantic Information Processing* (Cambridge: MIT Press, 1968), 5, 18.

49. Nils J. Nilsson, *Problem-Solving Methods in Artificial Intelligence* (New York: McGraw-Hill, 1971), 212.

50. Minsky, *Semantic Information Processing*, 4.

51. Paul Cohen, "Models of Cognition," in Paul Cohen and Edward Feigenbaum, eds., *The Handbook of Artificial Intelligence*, vol. 3 (Los Altos, Calif.: Kaufmann, 1982), 36.

52. J. McCarthy and P. J. Hayes, "Some Philosophical Problems from the Standpoint of Artificial Intelligence," *Machine Intelligence* 4 (1969): 463–502.

53. B. Chandrasekaran, "Artificial Intelligence — The Past Decade," *Advances in Computers* 13 (1975): 201.

54. Minsky and Papert, "Proposal, 1971–72," 7.

55. Carl Hewitt, "Description and Theoretical Analysis (Using Schemata) of Planner, A Language for Proving Theorems and Manipulating Models in a Robot," Apr. 1972, AI memorandum 251, MIT Project MAC; Terry Winograd, "Procedures as Representation for Data in a Computer Program for Understanding Natural Language," technical report AI TR-17, Artificial Intelligence Laboratory, MIT, 1971 (also published as *Understanding Natural Language* [New York: Academic Press, 1972]); Gerald J. Sussman, "A Computational Model of Skill Acquisition," technical report AI TR-297, Artificial Intelligence Laboratory, MIT, 1973 — all in *Artificial Intelligence Memoranda: AI Laboratory, Massachusetts Institute of Technology, 1958–1979* (New York: Scientific Datalink, Comtex Scientific, 1982), microfilm; E. C. Charniak, "Toward a Model of Children's Story Comprehension," technical report AI TR-266, Artificial Intelligence Laboratory, MIT, 1972, in ibid. (Hereafter these memoranda in the Comtex microfilm edition are cited as *Artificial Intelligence Memoranda: MIT.*)

56. Winograd interview, CBI.

57. Barr and Feigenbaum, eds., *Handbook of Artificial Intelligence* 1:295.

58. Boden, *Artificial Intelligence*, 127.

59. Ibid., 124.

60. Anton Nijholt, *Computers and Languages: Theory and Practice* (Amsterdam: North-Holland, 1988), 392.

61. Feigenbaum, "Artificial Intelligence," 1010.

62. See the "Overview" to chap. 3, "Representation of Knowledge," in Barr and Feigenbaum, eds., *Handbook of Artificial Intelligence* 1:144.

63. Saul Amarel, "On Representations of Problems of Reasoning about Actions," *Machine Intelligence* 3 (1968): 131–71.

64. This assessment is found in Feigenbaum, "Artificial Intelligence," 1018; see Saul Amarel, "On the Representation of Problems and Goal-Directed Procedures for Computers," in R. Banerji and M. D. Mesarovic, eds., *Theoretical Approaches to Non-Numerical Problem Solving* (Berlin: Springer-Verlag, 1970), 179–244.

65. Nils J. Nilsson, "Artificial Intelligence," in *Information Processing 1974: Proceedings of the IFIP Congress 74* (Amsterdam: North-Holland, 1974), 786.

66. Ibid.

67. Ibid., 785.

68. Edward Feigenbaum and Bruce Buchanan, "Foreword," in Randall Davis and Douglas B. Lenat, *Knowledge-Based Systems in Artificial Intelligence* (New York: McGraw-Hill, 1982), xvi.

69. Buchanan interview, CBI.

70. Feigenbaum, "Artificial Intelligence," 1016.

71. Ibid.

72. Avron Barr and Edward Feigenbaum, eds., *The Handbook of Artificial Intelligence*, vol. 2 (Los Altos, Calif.: Kaufmann, 1982), 143.

73. Ibid., 144.

74. Feigenbaum and Buchanan, "Foreword," xvi.

75. Bruce G. Buchanan and Richard O. Duda, "Principles of Rule-Based Expert Systems," *Advances in Computers* 22 (1983): 166.

76. For EMYCIN, see W. van Melle, "A Domain Independent System That Aids in Constructing Knowledge Based Consultation Programs" (Ph.D. diss. STAN-CS-80-820, Stanford University, 1980); for ROSIE, see J. Fain, F. Hayes-Roth, H. Sowizral, and D. Waterman, "Programming Examples in ROSIE," Tech. Rep. N-1647-ARPA (Santa Monica, Calif.: Rand Corporation, 1981); for KAS, R. Reboh, "Knowledge Engineering Techniques and Tools in the Prospector Environment," June 1981, AI-243, in *Artificial Intelligence Technical Reports: Artificial Intelligence Center, SRI International, 1968–1983* (New York: Scientific Datalink, Comtex Scientific, 1984), microfilm (hereafter this is cited as *Artificial Intelligence Technical Reports: SRI*; for EXPERT, S. Weiss and C. Kukikowski, "EXPERT: A System for Developing Consultation Models," *Proceedings of the Sixth International Joint Conference on Artificial Intelligence, 1979* (Palo Alto, Calif.: Morgan Kaufmann, 1979), 942–47; and for OPS, C. Forgy and J. McDermott, "OPS, A Domain-Independent Production System Language," *Proceedings of the Fifth International Joint Conference on Artificial Intelligence, 1977* (Palo Alto, Calif.: Morgan Kaufmann, 1977), 933–39.

77. Buchanan and Duda, "Principles," 166.

78. Edward Feigenbaum, Pamela McCorduck, and H. Penny Nii, *The Rise of the Expert Company* (New York: Times Books, 1988), 7–8.

79. Ibid., x.

80. Ibid., 169–73.

81. Ibid., 78–83.

82. Kenan Sahin and Keith Sawyer, "The Intelligent Banking System: Natural Language Processing for Financial Communications," in Herbert Schorr and Alain Rappaport, eds., *Innovative Applications of Artificial Intelligence* (Menlo Park, Calif.: AAAI Press; 1989), 43–50.

83. Simpson, "DOD Applications"; see also Simpson interview, CBI.

84. We are indebted to Robert L. Simpson Jr. for information on the expert systems used in the DOD.

85. George Johnson, *Machinery of the Mind: Inside the New Science of Artificial Intelligence* (New York: Random House, 1986), 125.

86. J. L. Flanagan, "Communication Acoustics," in S. Millman, ed., *A History of Engineering and Science in the Bell System: Communication Sciences, 1925–1980* (Indianapolis, Ind.: AT&T Bell Laboratories, 1984), 134.

87. J. D. Carroll, B. Julesz, M. V. Mathews, E. Z. Rothkopf, S. Sternberg, and M. Wish, "Behavioral Science," in Millman, ed., *History of Engineering and Science in the Bell System*, 461–62.

88. Wayne A. Lea, "Speech Recognition: Past, Present, and Future," in Wayne A. Lea, ed., *Trends in Speech Recognition* (Englewood Cliffs, N.J.: Prentice-Hall, 1980), 61.

89. Johnson, *Machinery*, 124.

90. P. Denes and M. V. Mathews, "Spoken Digit Recognition Using Time-Frequency Pattern Matching," *Journal of the Acoustical Society of America* 32 (Nov. 1960): 1450–55.

91. James W. Forgie and Carma D. Forgie, "Results Obtained from a Vowel Recognition Computer Program," *Journal of the Acoustical Society of America* 31 (Nov. 1959): 1480–89.

92. John F. Hemdal and George W. Hughes, "A Feature Based Computer Recognition Program for the Modeling of Vowel Perception," in Weiant Wathen-Dunn, ed., *Models for the Perception of Speech and Visual Form* (Cambridge: MIT Press, 1967), 440–53.

93. P. D. Sholtz and R. Bakis, "Spoken Digit Recognition using Vowel-Consonant Segmentation," *Journal of the Acoustical Society of America* 34 (Jan. 1962): 1–5.

94. James E. Dammann, "Application of Adaptive Threshold Elements to the Recognition of Acoustic-Phonetic States," *Journal of the Acoustical Society of America* 38 (1965): 213–23.

95. N. R. Dixon and C. C. Tappert, "Toward Objective Phonetic Transcription: An Online Interactive Technique for Machine-Processed Speech Data," *IEEE Transactions on Man-Machine Systems* MMS-11 (Dec. 1970): 202–10.

96. John McCarthy, "Project Technical Report," June 1969, AI-87, in *Artificial Intelligence Memoranda: Stanford*, 1; Bruce Buchanan, "Introduction to the COMTEX Microfiche Edition of the Stanford Artificial Intelligence Laboratory Memos," in ibid., 5.

97. DARPA, "Natural Communication with Computers."

98. Bruce Lowerre and D. Raj Reddy, "The Harpy Speech Understanding System," in Lea, ed., *Trends in Speech Recognition*, 342.

99. Lawrence Fagan, Paul Cohen, and Avron Barr, "Understanding Spoken Language," in Barr and Feigenbaum, eds., *Handbook of Artificial Intelligence* 1:332.

100. Reddy interview, CBI.

101. D. Raj Reddy, "An Approach to Computer Speech Recognition by Direct Analysis of the Speech Wave," 1966, p. 1, memorandum AI-43, in *Artificial Intelligence Memoranda: Stanford.*

102. S. R. Hyde, "Automatic Speech Recognition: A Critical Survey and Discussion of the Literature," in Edward E. David Jr. and Peter B. Denes, eds., *Human Communication: A Unified View* (New York: McGraw-Hill, 1972), 420–21.

103. R. Reddy, "Computer Recognition of Connected Speech," *Journal of the Acoustical Society of America* 42 (1967): 329–47.

104. Ibid.

105. Ibid.; and Hyde, "Automatic Speech Recognition."

106. Reddy, "Computer Recognition."

107. Pierre Vicens, "Aspects of Speech Recognition by Computer," Apr. 1969, p. 5, AI-85, in *Artificial Intelligence Memoranda: Stanford.*

108. Ibid.

109. Ibid.

110. McCarthy, "Project Technical Report."

111. B. Gold and L. Rabiner, "Parallel Processing Techniques for Estimating Pitch Periods of Speech in the Time Domain," *Journal of the Acoustical Society of America* 46 (1969): 442–48.

112. Daniel G. Bobrow and Dennis H. Klatt, "A Limited Speech Recognition System," in *Proceedings of the AFIPS Fall Joint Computer Conference,* vol. 33 (Washington, D.C.: Thompson Book, 1968), 305–18.

113. Newell interview, CBI.

114. Allen Newell, Jeffrey Barnett, James W. Forgie, Cordell Green, Dennis Klatt, J.C.R. Licklider, John Munson, D. Raj Reddy, and William A. Woods, *Speech Understanding Systems: Final Report of a Study Group* (Amsterdam: North-Holland, 1973).

115. Ibid., 2.

116. Dennis H. Klatt, "Review of the ARPA Speech Understanding Project," *Journal of the Acoustical Society of America* 62 (1977): 1345.

117. Ibid., 3.

118. Battelle (Columbus Laboratories), "Final Report on an Overview of Research and Development Programs and Objectives," vol. 3, "The Information Processing Techniques Office," 26 Jan. 1973, p. 8, IPTO Office Files, CBI.

119. Dennis H. Klatt, "Overview of the ARPA Speech Understanding Project," in Lea, ed., *Trends in Speech Recognition,* 249–71.

120. Nils J. Nilsson, "The SRI Artificial Intelligence Center: A Brief History," in *Artificial Intelligence Technical Reports: SRI;* Claude Baum, *The System Builders: The Story of SDC* (Santa Monica, Calif.: SDC, 1981), 248.

121. This work can be found in D. E. Walker, *Speech Understanding Research* (New York: North-Holland, 1976).

122. This technique was based on the Ph.D. work of Bruce Lowerre of CMU.

123. Bruce Lowerre and Raj Reddy, "The HARPY Speech Understanding System," in Lea, ed., *Trends in Speech Recognition,* 340–60.

124. Fagan, Cohen, and Barr, "Understanding Spoken Language," 349–50.

125. Ibid., 351.

126. Ibid., 331.

127. Dennis H. Klatt, "Review of the ARPA Speech Understanding Project," *Journal of the Acoustical Society of America* 62 (1977): 1354.

128. ARPA SUR Steering Committee, "Speech Understanding Systems: Report of a Steering Committee," *Artificial Intelligence* 9 (1978): 313–14.

129. Ibid., 314–15.

130. L. R. Bahl, J. K. Baker, P. S. Cohen, A. G. Cole, F. Jelinek, B. L. Lewis, and R. L. Mercer, "Automatic Recognition of Continuously Spoken Sentences from a Finite State Grammar," in *Proceedings of the 1978 IEEE International Conference on Acoustics, Speech, and Signal Processing* (New York: IEEE Press, 1978), 418–21.

131. R. Reddy and V. Zue, "Recognizing Continuous Speech Remains an Elusive Goal," 20 *IEEE Spectrum* (Nov. 1983): 84–87.

132. Craig I. Fields and Brian G. Kushner, "The DARPA Strategic Computing Initiative," in *Spring COMPCON '86: Thirty-First IEEE Computer Society International Conference* (Los Angeles, Calif.: IEEE Computer Society Press, 1986), 335.

133. Elizabeth Corcoran, "Strategic Computing: A Status Report," *IEEE Spectrum* 24 (Apr. 1987): 52–53.

134. Kai-Fu Lee, *Automatic Speech Recognition: The Development of the SPHINX System* (Boston: Kluwer Academic Publishers, 1989), 10–15.

135. Ibid., xiv.

136. Lawrence G. Roberts, "Pattern Recognition with Adaptive Networks," *IRE International Convention Record*, vol. 8 (New York: Institute of Radio Engineers, 1960), 66–70.

137. Leo Hodes, "Machine Processing of Line Drawings," Report 546-0028[u] Lincoln Laboratory, MIT (Mar. 1961).

138. O. G. Selfridge and U. Neisser, "Pattern Recognition by Machine," *Scientific American* 203 (Aug. 1960): 60–68.

139. L. G. Roberts, "Machine Perception of Three-Dimensional Solids," in James T. Tippett, David A. Berkowitz, Lewis C. Clapp, Charles J. Koester, and Alexander Vanderburgh Jr., eds., *Optical and Electro-Optical Information Processing* (Cambridge: MIT Press, 1965), 159.

140. Newell, "Intellectual Issues," 210–11.

141. Roberts, "Machine Perception," 185.

142. Project MAC, "Proposal for Research on Intelligent Automata," Collection AC 12, box 17, folder: "Project MAC Proposals Regular 1965," Institute Archives, MIT.

143. Ibid.

144. "Research on Intelligent Automata," Collection AC 12, box 17, folder: "Project MAC Intelligence Automata, 1966-7," Institute Archives, MIT.

145. Ibid., 4–5.

146. "Status Report II," pp. 15–16, Collection AC 12, box 17, folder: "Project MAC Intelligent Automata, 1966-7," Institute Archives, MIT.

147. Marvin Minsky and Seymour Papert, "Proposal to ARPA for Research on Arti-

ficial Intelligence at MIT, 1970–71," Dec. 1970, AI-185, in *Artificial Intelligence Memoranda: MIT.*

148. Ibid., 16.

149. Patrick H. Winston, "Learning Structural Descriptions from Examples," in Patrick H. Winston, ed., *The Psychology of Computer Vision* (New York: McGraw-Hill, 1975), 157–210.

150. Jerome A. Feldman, G. Feldman, G. Falk, G. Grape, J. Pearlman, I. Sobel, and J. Tenenbaum, "The Stanford Hand-Eye Project," in *Proceedings of the International Joint Conference on Artificial Intelligence—1969* (Ann Arbor, Mich.: University Microfilms International, 1969), 524.

151. Buchanan, "Introduction to the Stanford Artificial Intelligence Laboratory Memos."

152. Jerry W. Saveriano, "Pioneers of Robotics," in Richard C. Dorf, ed., *Concise International Encyclopedia of Robotics: Applications and Automation* (New York: Wiley, 1990), 667.

153. Buchanan, "Introduction to the Stanford Artificial Intelligence Laboratory Memos," 4.

154. John McCarthy, L. D. Earnest, D. R. Reddy, P. J. Vicens, "A Computer with Hands, Eyes, and Ears," in *Proceedings of the AFIPS Fall Joint Computer Conference* 33: 329–38.

155. Feldman, "Stanford Hand-Eye Project," 522.

156. Ibid.

157. Marvin Minsky, "Mini-Robot Proposal to ARPA," Jan. 1972, memorandum AI-251, in *Artificial Intelligence Memoranda: MIT.*

158. Ibid., 31.

159. Feldman, "Stanford Hand-Eye Project," 524.

160. Ibid.

161. Nils Nilsson, "A Mobile Automaton: An Application of Artificial Intelligence Techniques," in *International Joint Conference on Artificial Intelligence,* 509–20.

162. Nilsson, "SRI Artificial Intelligence Center," 3.

163. DARPA, "Automata and Extensors," Program Plan no. 547, 28 Feb. 1966, RG 330-78-0013, box 1, folder: "Program Plans," NABDS. This was a joint program with the Advanced Sensors Office under Paul Tamarkin.

164. Ibid.

165. Nilsson, "Mobile Automaton," 513.

166. William E. Burrows, *Deep Black: Space Espionage and National Security* (New York: Random House, 1986), 218–21.

167. Senate Subcommittee, *Department of Defense Appropriations for 1973,* 777.

168. Ibid.

169. Battelle, *Final Report,* 10.

170. Senate Subcommittee, *Department of Defense Appropriations for 1978,* 138–39.

171. Patrick H. Winston, "Proposal to the Advanced Research Projects Agency, 1 Jan. 1976 to 30 Dec. 1976," Aug. 1975, MIT AI memorandum no. 366, in *Artificial Intelligence Memoranda: MIT.*

172. Winston interview, CBI.

173. Lester Earnest, ed., "Final Report on Basic Research in Artificial Intelligence and Foundations of Programming," May 1980, p. 17, AIM-337, in *Artificial Intelligence Memoranda: Stanford.*

174. Ibid., 14.

175. R. B. Ohlander, "Analysis of Natural Scenes" (Ph.D. diss., Carnegie-Mellon University, 1975).

176. R. B. Ohlander, K. Price, and D. R. Reddy, "Picture Segmentation Using a Recursive Region Splitting Method," *Computer Graphics and Image Processing* 8 (1978): 313–33; S. Rubin, "The ARGOS Image Processing System" (Ph.D. diss., Carnegie-Mellon University, 1978).

177. T. Kanade, "Shape from Texture" (Ph.D. diss., Carnegie-Mellon University, 1980); and idem, "Region Segmentation: Signal vs. Semantics," *Computer Graphics and Image Processing* 13 (1980): 279–97.

178. J. R. Kender and T. Kanade, "Mapping Image Properties into Shape Constraints: Skewed Symmetry, Affine Transformable Patterns, and the Shape-from-Texture Paradigm," *American Association for Artificial Intelligence* 1 (1980): 4–6.

179. S. A. Shafer and T. Kanade, "The Theory of Straight Homogeneous Generalized Cylinders and Taxonomy of Generalized Cylinders," Technical Report CMU-CS-83-105, Carnegie-Mellon University, 1983.

180. D. M. McKeown and T. Kanade, "Database Support for Automated Photo Interpretation," in *Proceedings of the Image Understanding Workshop* (1981), 7–13.

181. University of Utah, "Sensory Information Processing and Symbolic Computation Research Program Proposal," 1 Oct. 1975–30 Sept. 1977, RG 330-78-002, box 3, folder: "Utah AO2477," NABDS.

182. Ramakant Nevatia, "Structured Descriptions of Complex Curved Objects for Recognition and Visual Memory," Oct. 1974, AIM-250, in *Artificial Intelligence Memoranda: Stanford;* idem, "Depth Measurement by Motion Stereo," *Computer Graphics and Image Processing* 5 (1976): 203–14; Ramakant Nevatia and K. R. Babu, "Linear Feature Extraction and Description," ibid., 13 (1980): 257–69; and Ramakant Nevatia, *Machine Perception* (Englewood Cliffs, N.J.: Prentice-Hall, 1982).

183. "Statement of Dr. Stephen J. Lukasik," in House Subcommittee on Defense Appropriations, *Department of Defense Appropriations for Fiscal Year 1975, Part 4: Research, Development, Test, and Evaluation, 74* H181-72.1, 93d Cong., 2d sess., 29 Apr. 1974, 632–50.

184. Ibid., 47.

185. Ibid., 49.

186. Ibid., 56–59.

187. Patrick H. Winston, "Proposal to the Advanced Research Projects Agency," May 1976, AI memorandum 366, in *Artificial Intelligence Memoranda: MIT,* 15–16.

188. James J. Pearson, ed., *The Society of Photo-Optical Instrumentation Engineers: Techniques and Applications of Image Understanding,* vol. 281 (Bellingham, Wash.: SPIE, 1981).

189. A. R. Helland, T. J. Willett, G. E. Tisdale, "Application of Image Understanding to Automatic Target Acquisition," in ibid., 281:26–31.

190. G. M. Fitton, "Real-Time Pattern Recognition—An Industrial Example," in Pearson, ed., *Society of Photo-Optical Instrumentation Engineers* 281:160–68.

191. W. M. Sterling, "Nonreference Optical Inspection of Complex and Repetitive Patterns," in Pearson, ed., *Society of Photo-Optical Instrumentation Engineers* 281:182–89.

192. Robert M. Haralick, Alan K. Mackworth, and Steven L. Tanimoto, "Computer Vision Update," in Avron Barr, Paul Cohen, and Edward Feigenbaum, eds., *The Handbook of Artificial Intelligence,* vol. 4 (Reading, Mass.: Addison-Wesley, 1989), 517–82.

193. David Y. Tseng and Julius F. Bogdanowicz, "Image Understanding Technology and Its Transition to Military Applications," in *Proceedings: Image Understanding Workshop* (Los Altos, Calif.: Morgan Kaufmann, 1987), 1:310–12; Robert L. Simpson Jr., "DOD Applications of Artificial Intelligence: Successes and Prospects," *Society of Photo-Optical Instrumentation Engineers (SPIE): Applications of Artificial Intelligence VI* 937 (1988): 158–64. The date of the testbed start and a copy of the article were provided by Simpson.

194. Ohlander interview, CBI; Amarel interview, CBI; and Schorr and Rappaport, eds., *Innovative Application of Artificial Intelligence.*

195. For brief histories of each of these areas and their use of AI, see Howard Gardner, *The Mind's New Science: A History of the Cognitive Revolution* (New York: Basic Books, 1985). Margaret A. Boden's work *Computer Models of Mind* (Cambridge: Cambridge University Press, 1988), discusses areas of overlap between AI and the behavioral sciences.

196. "Statement of Stephen J. Lukasik," in House Subcommittee on Defense Appropriations, *Department of Defense Appropriations for Fiscal Year 1972, Part 7,* H181-56, 92d Cong., 2d sess., 4 June 1971, 287.

Chapter Six: Serving the Department of Defense and the Nation

1. The sole exception was the procurement procedures imposed by Congress on the DOD in the mid-1980s, which did affect IPTO's ability to move contract requests through the system quickly.

2. Nathan Rosenberg and W. Edward Steinmueller, "The Economic Implications of the VLSI Revolution," in Nathan Rosenberg, *Inside the Black Box: Technology and Economics* (Cambridge: Cambridge University Press, 1982), 178–92; for their use of the term *gatekeeper,* see 189–90.

3. Sutherland interview, CBI.

4. Daniel L. Slotnick, "The Conception and Development of Parallel Processors— A Personal Memoir," *Annals of the History of Computing* 4 (1982): 25.

5. Sutherland interview, CBI.

6. Richard J. Barber and Associates, *The Advanced Research Projects Agency, 1958–1974,* (Washington, D.C.: Richard J. Barber Associates, 1975), p. ix-56; "Statement of

Dr. Stephen J. Lukasik, Director, DARPA," in Senate Subcommittee on Defense Appropriations, *Department of Defense Appropriations for Fiscal Year 1973, Part 1*, S181-44.4, 92d Cong., 2d sess., 14 Mar. 1973, 817.

7. House Subcommittee on Defense Appropriations, *Department of Defense Appropriations for Fiscal Year 1970, Part 5: Procurement Reprogramming Actions*, H181-53.18, 91st Cong., 2d sess., 19 Mar. 1970, 1053–54.

8. "Prime Argus" was formed within the Nuclear Monitoring Research Office and initially included "research on evasion (or detection countermeasures), nuclear test diagnostics, research related to nuclear weapons proliferation, and a new program in military geophysics." Later, the project was scaled back to what was only "essentially diagnostics related research." Barber Associates, *Advanced Research Projects Agency*, p. viii-39.

9. "Statement of Dr. Eberhardt Rechtin," in Senate Committee on Appropriations, *Department of Defense Appropriations for Fiscal Year 1970, Part 1*, (91) S2010-0-A, 91st Cong., 1st sess., 10, 12–13, 16 June, 15–17, 25 Sept. 1969, 432–33.

10. R. Michael Hord, *The ILLIAC IV: The First Supercomputer* (Rockville, Md.: Computer Science Press, 1982).

11. Lukasik interview, CBI. In a private conversation, Stephen Squires, later the director of the office that succeeded IPTO in supporting R&D in computer architectures, expressed the view that the project had tried to accomplish too much too soon, at a time when the enabling technologies were not ready. He reported discussing this issue with Licklider when Squires joined DARPA and began planning for DARPA's parallel computing program as part of SC.

12. All examples taken from material reprinted in Hord, *ILLIAC IV*.

13. Delbert D. Smith, *Communication via Satellite: A Vision in Retrospect* (Boston: Sijthoff, 1976), 151–55.

14. Ivan E. Sutherland and Carver A. Mead, "Microelectronics and Computer Science," *Scientific American* 237 (Sept. 1977): 228.

15. Sutherland moved to Caltech from the University of Utah and Evans and Sutherland, the company he had founded with David Evans.

16. Ivan E. Sutherland, Carver A. Mead, and Thomas E. Everhart, *Basic Limitations in Microcircuit Fabrication Technology*, Rand Report R-1956-ARPA (Santa Monica, Calif.: Rand Corporation, 1976), iv.

17. Science Applications International Corporation (SAIC), "The Transfer of ISTO-Developed Technologies into Current and Future Military Systems," draft report, 5 Nov. 1987, p. 52, under ARPA Order 5605, contract N00014-86-C-0700 ONR.

18. ARPA Order 4012, Stanford University, 11 Apr. 1980, IPTO Office Files, CBI.

19. "U.S. Semiconductor Lead Threatened, Officials Say," *Science* 200 (23 June 1978): 1364.

20. Ibid.

21. Quotation from Science Applications International, "Transfer of ISTO-Developed Technologies," 52.

22. Kahn interview, CBI; also Carver Mead and Lynn Conway, *Introduction to VLSI Systems* (Reading, Mass.: Addison-Wesley, 1980).

23. ARPA Order 4012.

24. Ibid.

25. Ibid.

26. D. Walden, "MOSIS Status, Examples, and Operation: An Assessment" (Cambridge, Mass.: Bolt Beranek and Newman, 1987), CBI.

27. "ARPA Order 4848, USC-ISI," IPTO Office Files, CBI.

28. Science Applications International, "Transfer of ISTO-Developed Technologies."

29. Alex Roland and Philip Shiman of Duke University are engaged in a study of DARPA's Strategic Computing program, so we will only indicate the general outlines of the program here. In many ways, the SC is a culmination of IPTO's twenty-five-year history. It set the stage for a second phase of DARPA's history in computing R&D, so it is appropriate to divide the studies in this way.

30. Cooper interview, CBI; Kahn interview, CBI.

31. Robert E. Kahn, "A Defense Program in Supercomputation from Microelectronics to Artificial Intelligence for the 1990s," in "FY 84 President's Budget Review: Information Processing Techniques Office," p. 1, Sept. 1982, IPTO Office Files, CBI.

32. Ibid.

33. Kahn interview, CBI.

34. Craig I. Fields and Brian G. Kushner, "The DARPA Strategic Computing Initiative," in *Spring Compcon 86* (San Francisco: IEEE Computer Society, 1986), 352.

35. Kahn, "Defense Program," 6.

36. Ibid., 7.

37. "FY86 President's Budget Review: Information Processing Techniques Office," Sept. 1984, IPTO Office Files, CBI.

38. Kahn, "Defense Program," 7.

39. "FY 86 President's Budget Review."

40. "The Race to Build a Supercomputer," *Newsweek* 102 (4 July 1983): 62.

41. Fields and Kushner, "DARPA Strategic Computing Initiative," 353.

42. Quotation attributed to Lynn Conway in Dwight B. Davis, "Assessing the Strategic Computing Initiative," *High Technology* 5 (Apr. 1985): 48.

43. John Cocke, "The Search for Performance in Scientific Processors," Turing Lecture, *Communications of the ACM* 31 (Mar. 1988): 249–53.

44. "Preliminary Research Proposal for VLSI Research," p. 1, Electronics Research Laboratory, University of California, Berkeley, 1982, IPTO Office Files, CBI.

45. C. Gordon Bell, "RISC: Back to the Future?" *Datamation* 32 (1 June 1986): 98.

46. Ibid.

47. Ibid.

48. Elizabeth Corcoran, "Strategic Computing: A Status Report," *IEEE Spectrum* 24 (Apr. 1987): 52–53; Willie Schatz, "DARPA Goes Parallel," *Datamation* 31 (1 Sept. 1985): 36, 38.

49. Schatz, "DARPA Goes Parallel," 36.

50. Paul B. Schenck, Donald Austin, Stephen L. Squires, John Lehman, David Mizell, and Kenneth Wallgren, "Parallel Processor Programs in the Federal Government," *Computer* 18 (June 1985): 48.

51. Ibid., 49.

52. Corcoran, "Strategic Computing," 51.

53. Schenck et al., "Parallel Processor Programs," 49.

54. Taylor interview, CBI; Uncapher interview, CBI; Licklider interview, CBI. The lack of contact with MITRE is confirmed in the Zraket interview, CBI.

55. Senate Committee, *Department of Defense Appropriations for 1970,* 426.

56. Licklider interview, CBI.

57. Licklider to Fubini, 26 Nov. 1963, RG 330-78-0013, box 1, unlabeled folder, NABDS.

58. Carnegie Institute of Technology, "Proposal for a Center for the Study of Information Processing," 27 Apr. 1964, obtained from Allen Newell's private papers.

59. Newell interview, CBI.

60. Carnegie Institute of Technology, "Proposal for a Center."

61. Newell interview, CBI.

62. Robert W. Taylor, "Memorandum for the Deputy Director, ARPA, Interdisciplinary Characteristics of IPT Centers of Excellence," RG 330-78-0013, box 1, folder: "Program Plans," NABDS.

63. Association for Computing Machinery, Curriculum Committee on Computer Science, "An Undergraduate Program in Computer Science: Preliminary Recommendations," *Communications of the ACM* 8 (Sept. 1965): 543–52.

64. George E. Forsythe, "A University's Educational Program in Computer Science," *Communications of the ACM* 10 (Jan. 1967): 3–11.

65. These statistics were compiled by William Aspray through an analysis of the bulletins of the forty top-ranked computer science departments.

66. R. M. Fano to R. Taylor, 17 Jan. 1968, Collection AC 12, box 17, folder: "Project MAC General 1966–9," Institute Archives, MIT.

67. Jean E. Sammet, "Programming Languages: History and Future," *Communications of the ACM* 15 (July 1972): 601–10.

68. Office of Science and Technology Policy, *Grand Challenges 1993: High Performance Computing and Communications,* Supplement to the President's Fiscal Year 1993 Budget (Washington, D.C.: Office of Science and Technology Policy, 1993).

69. James W. Cortada, *The Computer in the United States: From Laboratory to Market, 1930 to 1960* (Armonk, N.Y.: Sharpe, 1993).

70. William Aspray and Bernard O. Williams, "Computing in Science and Engineering Education: The Programs of the National Science Foundation," in *Electro/93,* vol. 2, *Communications Technology and General Interest* (Ventura, Calif.: Western Periodicals, 1993), 234–39; and idem, "Arming American Scientists: NSF and the Provision of Scientific Computing Facilities for Universities, 1950–1973," *Annals of the History of Computing* 16 (winter 1994): 60–74.

Index

Abelson, Robert, 202

Abramson, Norman, 180

Adams, Charles, 120

Adams, Duane, 40

Adobe Systems, 143

Advanced Command and Control Architectural Testbed (ACCAT), 147, 186, 270

Advanced Development Prototype System (ADEPT), 270

Advanced Research Projects Agency (ARPA). *See* DARPA

air defense: Cape Cod System, 71; U.S. system (1950s), 70–71

Akin, Omar, 250

Amarel, Saul, 39: artificial intelligence research, 39, 210, 218–19; background, 16, 39; IPTO director, 21; joins IPTO, 39; RCA Laboratories, 39, 204

Anderson, Robert, 134

architecture (computer): ILLIAC IV, 264–68; parallel processors, 279–81; RISC, 268, 280; VLSI, 268, 271–75; use of modules, 264

ARPANET, 64, 175, 210, 248, 261, 270, 273; Bolt Beranek and Newman, 170–72; contract specifications, 167–68; demonstration, 177–78; electronic mail, 178; expansion, 174–76; implementation, 171–73; initiation, 51; interface message processor (IMP), 164, 170–71; as IPTO management tool, 48; limitations, 192; National Software Works, 187–88; Network Information Center, 170; Network Measurement Center, 171; Network Working Group, 168–69; packet radio, 180–82; packet satellite, 181–82; packet switching, 165; planning, 48–49; Rand study, 166;

terminal IMP (TIP), 174; time sharing, 155

artificial intelligence: accomplishments, 201; applications, 249; "blocks world," 240–41, 243; Carnegie-Mellon University, 207, 214, 215, 249–50, 252; command and control, 204; commercialization, 252–54; expert systems, 207–8, 220–24 (*see also* expert systems); heuristics, 200; history, 200–212; image processing, 247–48, 250; impact, 254–57; knowledge engineering, 221–22; knowledge representation, 202; Massachusetts Institute of Technology, 214–15, 241–43; military use, 223–24; natural language processing, 227; pattern recognition, 240; perceptrons, 203; picture processing, 247–50; problem solving, 203, 209–10, 213–24; representation, 210; robotics, 208, 239–47, 251–54; speech understanding, 208–10, 224–39; Stanford Research Institute, 245–46; Stanford University, 207–8, 244–45, 249; Strategic Computing, 278; vision (computer), 239, 241–42, 254. *See also* artificial intelligence programs

artificial intelligence programs: ACRONYM, 249, 254; Advice Taker, 202; EPAM, 207; General Problem Solver, 202, 214; HACKER, 216; heterarchical (robotics), 242; hierarchical (robotics), 242; LEARNING, 242–43; Logic Theorist, 201–2; QA3, 216; SHRDLU, 202, 216–18, 241, 242; SIR, 215; STORIES, 216; STRIPS, 202, 216; STUDENT, 214

Association for Computing Machinery, Special Interest Group on Computer Graphics (SIGGRAPH), 149

Johns Hopkins Studies in the History of Technology (New Series)